Spatial Analysis in Geomorphology

Spatial Analysis in Geomorphology

**Edited for the
British Geomorphological Research Group**

by RICHARD J. CHORLEY

HARPER & ROW, PUBLISHERS
New York, Evanston, San Francisco, London

Contents

PART IV   Continuous Distributions

PART V   Space Partitioning

PART VI   Simulation

Prefatory Note

The year 1971 marked the coming-of-age of the application of modern statistical techniques to geomorphology. During the 1950's a whole range of 'linear' statistical methods were introduced, particularly those relating to regression and variance analysis. Towards the end of that decade multivariate methods began to be used in association with the advent of second-generation electronic computers. Aided by its scholarly links with geography, the 'revolution' in which was much concerned with the development of spatial analysis and spatial model building, geomorphology was technically ready in the early 1960's for the application of spatial analytical techniques, in which a clear association is maintained between quantitative data and the spatial co-ordinates which locate them. That we have had to wait almost a decade for spatial model building to begin a wide maturation in geomorphology is now a matter of history. The purpose of the present volume is to highlight this coherent area of scholarship under the general headings of spatial point systems, networks, continuous distributions, space partitioning and simulation. Seventeen authors from Britain and the United States have been brought together to produce the fourteen chapters of which the present book is composed. This is not a textbook. Its aim is to focus attention on the body of spatial techniques necessary to enable the building of dynamic spatial models of landforms which will undoubtedly form the keystone of much geomorphic work during the next decade.

PART I
General

1 Spatial analysis in geomorphology

RICHARD J. CHORLEY

Department of Geography, University of Cambridge

For many geomorphologists the year 1971 marks the coming-of-age of the application of modern statistical techniques to their science. The first decade following Strahler's (1950) pioneer use of simple statistical techniques in a genuine attack on the classic problem of slope development was characterised by the introduction into geomorphology of the whole range of 'linear' (as distinct from spatial) statistical techniques and especially of regression and variance analyses (Chorley 1966). Towards the end of the 1950's more complicated types of multiple regression methods began to be used, first by means of laborious manual calculation (Melton 1957, 1958A and 1958B) and soon afterwards by the use of the early second generation of electronic computers (Krumbein 1959). At the beginning of the 1960's, at least from the technological standpoint, geomorphology seemed to be well placed for the application of quantitative techniques to its traditional problems and, particularly, for the employment of spatial analysis in the study of landforms and their associated processes. The term spatial analysis is used in this context to include the assemblage of analytical techniques and models in which a clear association is maintained between quantitative data and the spatial co-ordinates which locate them.

Geomorphology was at that time heir to two quite distinct quantitative dynasties. The first was that concerned with the application of quantitative techniques to geology which had been pioneered by W. C. Krumbein since the middle 1930's and which had proliferated in the 1950's (Miller and Kahn 1962; Krumbein and Graybill 1965), particularly in terms of the spatial analysis of point data by trend-surface analysis (Chorley and Haggett 1965). The second scholarly family with which geomorphology had been traditionally associated, particularly in Europe, was human geography. By 1960 this discipline, building on the pioneer work of Hägerstrand (1953), Garrison (1955), Shaefer (1953) and others, was undergoing a rapid quantitative revolution which began to accelerate about 1958 at such a pace that at least one scholar believed it to have achieved its ends only five years later (Burton 1963). This revolution was, above all, characterised by the analysis of spatial relations between geographic phenomena and by the construction and testing of locational models of human activity (Haggett 1965). Added to these two inheritances was the increasing availability of ever more powerful electronic computers, and it appears in retrospect that in the early 1960's geomorphology potentially stood at the beginning of a decade which would be characterised by the rapid adoption of computer-based quantitative techniques, the development of the use of models (particularly the versatile general linear model), the increasing use of probabilistic models and, especially, by the adaptation and construction of spatial geomorphic models. Ten years later we are only just beginning to see the accelerated application of these lines of attack on geomorphic problems and one might legitimately ask what has retarded these developments.

Krumbein (1969) has recently analysed the application of computer-based techniques in geology and even his incomplete information on geomorphic publications shows that, except perhaps surprisingly in the field of trend-surface analysis, the timing of initial use of these techniques by geomorphologists has been about average for the seven major geological disciplines examined. Indeed multiple regression techniques were applied in geomorphology by Melton in 1957 (and later by Krumbein (1959), Wong (1963), Chorley (1964B) and Harrison and Krumbein (1964)), sequential multiple regression by Krumbein, Benson and Hempkins in 1964, polynomial trend-surface analysis by Chorley in 1964, harmonic and variance spectra analysis to stream channel patterns by Mackay in 1963 (followed by Speight 1965), spectral analysis to the microstructure of the bottom of a Texas lake by Horton, Hoffman and Hempkins in 1962, harmonic analysis to large-scale terrain by Piexoto, Saltzman and Telwes in 1964, factor analysis to terrain classification by Aschenbrenner *et al.* in 1963 and by Simons *et. al.* in 1964 (contemporaneously with the general applications in geology by Imbrie in 1963), Markov-type analysis to stream channels by Leopold and Langbein in 1962, and spatial simulation to stream networks by the latter authors and by Schenck in 1963. The initial infusion of spatial techniques into geomorphology from geography was more limited, but nevertheless present. Cassetti applied discriminant analysis in 1964, McConnell principal components analysis to slope measurements in 1964, and the same author employed quadrat counts to treat geomorphic point data in a number of mimeographed papers at about this date.

What emerges from this cursory examination is that before the middle of the 1960s most of the array of quantitative techniques available to the environmental and social sciences had been applied to geomorphic problems by individual innovators in the science. However, when one examines the general adoption of such techniques by geomorphologists one is forced to the opinion that, by contemporary standards of change, it has been very slow. Melton's pioneer use of correlation structures (1958B) was not substantially followed up until the work of Kennedy (1965) and of Towler (1969); the work of Chorley in 1969 represented at that time one of the few published applications of sequential multiple regression to geomorphology; space-filling models, which were innovations in geography in the 1930's and which formed much of the basis for the development of locational analysis, were not directly applied to terrain analysis until the work of McConnell in 1965 and, particularly, that of Woldenberg in a stream of publications after 1966 (Haggett and Chorley 1969, 314–18); and ideas of systems analysis, pioneered by Strahler in 1950, and more directly applied by Chorley in 1962 and by Leopold and Langbein in the same year, have yet to reach anything approaching their full potential (Chorley and Kennedy 1971). Listings of individual research in British geomorphology (British Geomorphological Research Group 1965, 1967 and 1969) show that in 1965 only a handful of the 304 projects seemed to lean at all heavily on statistical analysis and on the use of computers, whereas in 1967 22 of the 358 workers specifically referred to these techniques, and this figure had grown to 45 out of 335 by 1969.

Not only has the diffusion of quantitative techniques in geomorphology been slow but geomorphologists have not been particularly successful in applying spatial techniques to their science – despite their obvious place in model building of terrain and geomorphic processes, and despite the scholarly links between geomorphology and socio-economic geography. For example, a review article by Zakrzewska in 1967 lists many geomorphic articles with a spatial flavour but few which apply rigorous spatial techniques of analysis and King's (1969) treatment of spatial forms and their relation to geographic theory assembles little evidence of quantitative spatial analysis in geomorphology, in contrast with the flood of work in human geography. A recent survey by Tarrant (1971) of the availability since 1967 of computer programmes in the environmental sciences (including geography) mainly in the United States and United Kingdom has shown that, of the 87 programmes (including duplicates) seeming to have been applied directly or indirectly to geomorphical-type problems, almost a third of these had to do with the analysis of sedimentary parameters (15) and some form of trend-surface analysis (13). Of the total geomorphic programmes about half (44) can be said to have specific spatial concern, but two-thirds of them are concentrated on a narrow body of techniques previously pioneered in geophysics, geology and cartography – namely, trend-surface analysis (polynomial; spectral; Fourier; etc.) 13, regional taxonomy (factor analysis; cluster analysis; discriminant analysis etc.) 8, and automated cartography (interpolation of contours and elevations; contour drawing; etc.) 8. The remaining 15 were concerned with spatial simulation (hydrological networks; streams; spits; etc.) 5, analysis of azimuths (especially of slope data) 3, general terrain analysis 3, calculation of shape and volume (drainage basins; glaciers, etc.) 2, and point pattern analysis (sinkholes) 2. This impression of the late and patchy development of spatial techniques in geomorphology is, despite the incompleteness of Tarrant's listing, undoubtedly a correct one, and is borne out by the recent book on numerical analysis in geomorphology by Doornkamp and King (1971). Although these authors naturally devoted substantial space to reviewing the great amount of work analysing stream network characteristics during the last decade (Haggett and Chorley 1969), it is significant that of nine chapters on fluvial geomorphology fully four are partially concerned with slope profile analysis, and almost as much total space is devoted to the unpromising topic of 'morphological mapping' (Savigear 1965) as to factor, cluster, discriminant and multiple-discriminant analysis combined. This is not a criticism of the book which in this respect represents very faithfully the slow and sparse development of spatial applications to geomorphology.

It is necessary, therefore, to ask ourselves why so much of the promise of the early 1960s has had to wait almost a decade to begin maturation. Such a question naturally evokes a number of possible answers:

1. Many of the central geomorphic problems had not been cast traditionally in terms of spatial models. Regional taxonomy had never presented real problems. Three-dimensional morphometric problems relating to forms and processes had, because of their complexity, been reduced to a two-dimensional cross-sectional character (e.g. slope and hypsometric studies). Again, the time-bound preoccupations of many established geomorphologists did not encourage the ready adaptation of spatial statistical techniques to their needs, and isolated attempts (for example by Gregory and Brown 1966) to wed statistical methods to morphological mapping have proved rather unsatisfactory. Even the obvious application of trend-surface and variance analysis to the description and analysis of possible erosion surfaces has been slow and has evoked a distinctly timid response (Daniels, Nelson and Gamble

1970; Howarth 1967; King 1967; Rodda 1970; Tarrant 1970; Thornes and Jones 1969; Tinkler 1969).

2. For the reasons given above, many of the innovators and acceptors of quantitative techniques had insufficient mathematical background to exploit their use in geomorphology as rapidly as might have been expected, and it has been left largely to a later generation of research students to develop these mathematical skills to the point where they could be applied to complex geomorphic problems.

3. Economies of scale operated to the detriment of the spread of quantitative techniques in geomorphology, in comparison with other earth and social sciences. Geomorphology involved too few workers to generate quickly a coherent body of quantitative scholarship which would have provided the basis for substantial publications which, in turn, would have been rapidly influential across a wide academic spectrum. In short, one might well argue that there was little early opportunity for a work in geomorphology comparable to that by Haggett (1965) in human geography. Attempts in the early 1960's to give some sort of structural overviews of 'quantitative geomorphology' were thus on a much less impressive scale than in either geography or geology. It is also rather ironical that publication of two of the potentially most influential works (although at rather different levels) was long delayed by their involvement with symposia. The review of quantitative geomorphology written by Chorley in 1961 did not appear until 1966, and the proceedings of a symposium on quantitative physical geography organized by Garrison and Marble in 1960 had to await publication until 1967. Perhaps, however, one might prefer to regard these circumstances as effects, rather than causes, of the difficulties facing the general diffusion of quantitative work in the early part of the past decade.

4. In the United States and Sweden, where spatial analysis was most effectively applied in geography, the links between geomorphology and human geography were traditionally rather weak, at least when compared with those in Britain.

5. With one or two exceptions (e.g. Evans 1971; Sprunt 1970), work in automated cartography has centred mainly around the technical problems of data storage and retrieval, rather than on the use of such data in spatial model building. Even in this field of cartography, the applications of information theory (now almost a quarter of a century old) are only just beginning (Connelly 1968).

6. Spatial model building has proceeded most rapidly in those disciplines, like urban studies, where ideas of systems analysis have been

most readily adopted. Although recent work in geomorphology can be recognised to have significant systems overtones (Chorley and Kennedy 1971) much of the backwardness of geomorphic applications of spatial techniques can be attributed to a reluctance by geomorphologists to restate their problems in systems terms. This reluctance is partly explainable by the complex association in geomorphology of spatial systems having radically different relaxation times. Such systems are notoriously difficult to combine in model terms. When a systems approach is possible, however, it is clear that many interesting new lines of analysis present themselves (e.g. Thornes 1968).

With these reasons in mind, it is obvious why some spatial aspects of geomorphology have developed much more rapidly than others. Notably the study of the topology and geometry of stream-channel networks as directed graphs was a traditionally spatial one and it was possible to treat it mathematically in almost complete isolation from detailed considerations of fluvial processes. Thus, in contrast to other spatial work in geomorphology, there has been a flood of network analysis since about 1962 (largely reviewed in Haggett and Chorley 1969; see subsequently Coffman and Turner 1971; Chang and Toebes 1970; James and Krumbein 1969; Krumbein 1969, 1970; Krumbein and Dacey 1969; Krumbein and Shreve 1970; Surkan and Van Kan 1969; Thakur and Scheidegger 1970). The applications of trend-surface tech- niques, particularly to terrain, have also been somewhat more quickly diffused in geomorphology since the middle of the 1960's than other spatial techniques (Chorley and Haggett 1965; King 1967; Krumbein 1967; Miesch and Conner 1968). This work has concerned itself not only with the application of polynomial expansions but also with Fourier series (Preston and Harbaugh 1965; James 1966; Harbaugh and Pres- ton 1968; Nye 1969; Tobler 1969), spectral analysis (Preston 1966; Rayner 1967; Noskov 1969; Nye 1970) and more generalised harmonic analysis (Lee and Kaula 1967; Norcliffe 1969). Of particular interest have been studies comparing the applicability of trend-surface methods (Krumbein 1966A, 1966B; Bassett and Chorley 1971).

The application of other types of spatial analysis in geomorphology has been, by comparison, much more sparse and slow. Spatial simulation in geomorphology and related fields is now becoming significant but, until very recently, was limited in application and carried on by only a few workers. Apart from stream network simulation (reviewed in Haggett and Chorley 1969, 285-93; Harbaugh and Bonham-Carter 1970; Chorley and Kennedy 1971, 281-86), most of the effort has been concentrated on the simulation of spatial marine sedimentation (Harbaugh 1966; Harbaugh and Wahlstedt 1967; Bonham-Carter and Harbaugh 1968; Bonham-Carter and Sutherland 1968; Harbaugh and

Bonham-Carter 1970; McCullagh and King 1970). Although there has been much simulation of stream and sediment discharges in hydrology, it is only extremely recently that, for example, three-dimensional terrain simulation models are being constructed involving assumed time transformations (Moultrie 1970; Sprunt, in Chorley and Kennedy 1971, 285–8; Ahnert, 1971). The many remaining spatial techniques are also characterised by their very recent application to geomorphic problems, of which the following are representative examples: factor and cluster analysis (King 1968; Werritty 1969; Mather and Doornkamp 1970, Doornkamp and King 1971); point pattern analysis (Smalley and Unwin 1968); analysis of rough surfaces (Hobson 1967; Warntz 1968; Ronca and Green 1970); and Markov type analyses (Krumbein 1967; James and Krumbein 1969; Krumbein 1969B, Krumbein and Dacey 1969; Krumbein 1970).

It is clear that, despite delays in applying spatial techniques to established geomorphic problems and in formulating new problems appropriate to the analytical tools which have been available for many years, geomorphology is now entering a period of increasing experimentation in terms of spatial analysis. Although it is possible to point to some inadequacies in the existing armoury of techniques (for example, the lack of techniques for handling the correlation of spatial distributions (Rayner 1967, Chorley 1969), the very inadequate application of autocorrelation techniques (Agterberg 1970; Frolov and Yagodina 1970) and the marked lack of spatial model building in which linear and spatial processes are combined), it is clear that much of the coming work will be concerned with the application of existing techniques and models to significant geomorphic problems. It is also clear that some of the most exciting of this work will involve the construction of spatial simulation models in geomorphology in which a much better balance will be achieved than hitherto between deterministic and stochastic processes. For, when all is said and done, techniques are merely the artifacts of scholarship. We therefore look back at an academic landscape littered with quantitative geomorphic techniques with much the same eyes as an archaeologist who is trying to make some cultural assessment from pieces of broken pottery. The value of these tools lies solely in our ability to use them to solve existing problems or in their ability to prompt us to formulate and attack new problems.

*References

AGTERBERG, F. P. (1970) Autocorrelation functions in geology; In Merriam, D. F. (Ed.), *Geostatistics: A colloquium* (Plenum Press, New York), 113–41.

AHNERT, F. (1971) A general and comprehensive theoretical model of slope

* Including some not specifically referred to in the text.

profile development; *University of Maryland, Occasional Papers in Geography*, No. 1.

ASCHENBRENNER, B. C., BURGESS, W., DOYLE, F., IMBRIE, J. and MAXWELL, J. (1963) Applicability of certain multifactor computer programmes to the analysis, classification, and production of landforms; *ONR Task No. 387-030, Contract No. Nonr 4145(00), Prepared by the Autometric Corporation, N.Y.C.*

BASSETT, K. (1969) Alternative approaches to trend analysis: A discussion using one-dimensional series as examples; *Department of Geography, Bristol University, Seminar Paper Series A*, No. 17, 29p.

BASSETT, K. and CHORLEY, R. J. (1971) An experiment in terrain filtering; *Area* 3, 78-91.

BERRY, B. J. L. and MARBLE, D. F., 1968, *Spatial Analysis: A Reader in statistical geography* (Prentice-Hall, New York), 512p.

BONHAM-CARTER, G. and HARBAUGH, J. W. (1968) Simulation of geologic systems: An overview; In Merriam, D. F. and Cocke, N. C. (Eds.), Computer applications in the earth sciences: Colloquium on simulation, *Kansas Computer Contribution* No. 22, 3-10.

BONHAM-CARTER, G. and SUTHERLAND, A. J. (1968) Mathematical model and Fortran IV program for computer simulation of deltaic sedimentation; *Kansas Computer Contribution* No. 24, 56p.

BRITISH GEOMORPHOLOGICAL RESEARCH GROUP (1965, 1967 and 1969) *Current Research in Geomorphology* (Universities of Durham and East Anglia).

BURKE, M. J. (1969) The Forth Valley: An ice-moulded lowland; *Transactions of the Institute of British Geographers* 48, 51-9.

BURTON, I. (1963) The quantitative revolution and theoretical geography; *Canadian Geographer* 7, 151-62.

CASETTI, E. (1964) Multiple discriminant functions; *Office of Naval Research, Geography Branch, ONR Task No. 389-135, Technical Report No. 11*, 63p.

CHANG, T. P. and TOEBES, G. H. (1970) A statistical comparison of meander planforms in the Wabash Basin; *Water Resources Research* 6, 557-78.

CHORLEY, R. J. (1962) Geomorphology and general systems theory; *U.S. Geological Survey Professional Paper* 500-B, 10p.

CHORLEY, R. J. (1964A) An analysis of the areal distribution of soil size facies in the Lower Greensand rocks of east-central England by the use of trend surface analysis; *Geological Magazine* 101, 314-21.

CHORLEY, R. J. (1964B) Geomorphological evaluation of factors controlling shearing resistance of surface soils in sandstone; *Journal of Geophysical Research* 69, 1507-16.

CHORLEY, R. J. (1966) The application of statistical methods to geomorphology; In Dury, G. H. (Ed.), *Essays in Geomorphology* (Heinemann, London), 275-387.

CHORLEY, R. J. (1967) Application of computer techniques in geography and geology; *Proceedings of the Geological Society of London* 1642, 183-6.

CHORLEY, R. J. (1969) The elevation of the Lower Greensand ridge, South-East England; *Geological Magazine* 106, 231-48.

CHORLEY, R. J. and HAGGETT, P. (1965) Trend-surface mapping in geographical research; *Transactions of the Institute of British Geographers* 37, 47–67.

CHORLEY, R. J. and KENNEDY, B.A. (1971) *Physical Geography: A systems approach* (Prentice-Hall, London), 375p.

COFFMAN, D. M. and TURNER, A. K. (1971) Computer determination of the geometry and topology of stream networks; *Water Resources Research* 7(2), 419–23.

CONNELLY, D. S., 1968, *The Coding and Storage of Terrain Height Data; An introduction to numerical cartography* (unpublished M.Sc. Thesis, Cornell University), 140p.

DANIELS, R. B., NELSON, L. A. and GAMBLE, E. E. (1970) A method of characterizing nearly level surfaces; *Zeitschrift für Geomorphologie* 14, 175–85.

DOORNKAMP, J. C. and KING, C. A. M. (1971) *Numerical Analysis in Geomorphology* (Arnold, London), 372p.

EVANS, I. S. (1971) The implementation of an automated cartography system; In Cutbill, J. L. (Ed.), *Data Processing in Biology and Geology* (Academic Press, London), p. 39–55.

FROLOV, YU. S. and YAGODINA, L. L. (1970) Autocorrelation function and quantity character of relief (in Russian); *Vestrick Leningradskogo Universiteta, 18, Geologiva – Geografiva* 3, 115–28.

GARRISON, W. L. (1955) Studies of rural roads; *Proceedings of the Regional Science Association* 1, 7–14.

GARRISON, W. L. and MARBLE, D. F. (Eds.) (1967) Quantitative Geography: Part II Physical and Cartographic Topics; *Northwestern University Studies in Geography* No. 14, 324p.

GREGORY, K. J. and BROWN, E. H. (1966) Data processing and the study of landform; *Zeitschrift für Geomorphologie* 10, 237–63.

HÄGERSTRAND, T. (1953) *Innovationsföroppet ur Korologisk Synpunkt* (Lund).

HAGGETT, P. (1965) *Locational Analysis in Human Geography* (Arnold, London), 437p.

HAGGETT, P. and CHORLEY, R. J. (1969) *Network Analysis in Geography* (Arnold, London), 348p.

HARBAUGH, J. W. (1966) Mathematical simulation of marine sedimentation with IBM 7090/7094 computers; *Kansas Computer Contribution* No. 1, 52p.

HARBAUGH, J. W. and BONHAM-CARTER, G. (1970) *Computer Simulation in Geology* (Wiley–Interscience, New York), 575p.

HARBAUGH, J. W. and PRESTON, F. W. (1968) Fourier series analysis in geology; In Berry, B. J. L. and Marble, D. F. (Eds.), *Spatial Analysis: A reader in statistical geography* (Prentice-Hall, New Jersey), 218–38.

HARBAUGH, J. W. and WAHLSTEDT, W. J. (1967) Fortram IV program for mathematical simulation of marine sedimentation with IBM 7040 or 7094 computers; *Kansas Computer Contribution* No. 9, 40p.

HARRISON, W. (1970) Prediction of beach changes; *Progress in Geography* 2, 207–35.

HARRISON, W. and KRUMBEIN, W. C. (1964) Interactions of the beach–

ocean–atmosphere system at Virginia Beach, Va.; *U.S. Army Coastal Engineer Research Center, Technical Memorandum* No. 7.

HOBSON, R. D. (1967) Fortran IV programs to determine surface roughness in topography for the CDC3400 computer; *Kansas Computer Contribution* No. 14, 28p.

HORMANN, K. (1969) Geomorphologische Kartenanalyse mit Hilfe elektronischer Rechenanlagen; *Zeitschrift für Geomorphologie* 13, 75–98.

HORTON, C. W., HOFFMAN, A. A. J. and HEMPKINS, W. B. (1962) Mathematical analysis of the microstructure of an area of the bottom of Lake Travis; *Texas Journal of Science* 14, 131–42.

HOWARTH, R. J. (1967) Trend-surface fitting to random data – an experiment test; *American Journal of Science* 265, 619–25.

IMBRIE, J. (1963) Factor and vector analysis programs for analysing geologic data; *Office of Naval Research, Geography Branch, ONR Task No. 389–135, Contract Nonr 1228(26), Technical Report No. 6, Northwestern University*, 83p.

JAMES, W. R. (1966) The Fourier series model in map analysis; *Office of Naval Research, Geography Branch, ONR Task No. 388–078, Contract Nonr 1228(36), Technical Report No. 1, Department of Geology, Northwestern University*, 37p.

JAMES, W. R. and KRUMBEIN, W. C. (1969) Frequency distributions of stream link lengths; *Journal of Geology* 77, 544–65.

KENNEDY, B. A. (1965) *An Analysis of the Factors Influencing Slope Development on the Charmouthien Limestone of the Plateau de Bassigny, Haute-Marne, France* (unpublished B.A. Dissertation, Department of Geography, Cambridge University), 99p.

KENNEDY, B. A. (1969) *Studies of Erosional Valley-Side Asymmetry* (unpublished Ph.D. Thesis, Cambridge University), 289p.

KING, C. A. M. (1967) An introduction to trend surface analysis; *Bulletin of Quantitative Data in Geography, Nottingham University*, No. 12.

KING, C. A. M. (1968) Factor Analysis; *Bulletin of Quantitative Data in Geography, Nottingham University*.

KING, L. J. (1969) The analysis of spatial forms and its relation to geographic theory: Review Article; *Annals of the Association of American Geographers* 59, 572–95.

KOCH, G. S. and LINK, R. F. (1970) *Statistical Analysis of Geological Data* (Wiley, New York), 375p.

KRUMBEIN, W. C. (1959) The 'sorting out' of geological variables illustrated by regression analysis of factors controlling beach firmness; *Journal of Sedimentary Petrology* 29, 575–87.

KRUMBEIN, W. C. (1966A) A comparison of polynomial and Fourier models in map analysis; *ONR Office of Naval Research, Geography Branch, Task No. 388–078, Contract Nonr 1228(36), Technical Report No. 2, Department of Geology, Northwestern University*, 45p.

KRUMBEIN, W. C. (1966B) Classification of map surfaces based on the structure of polynomial and Fourier Coefficient matrices; In Merriam, D. F. (Ed.), Computer applications in the earth sciences: Colloquium on classification procedures, *Kansas Computer Contribution* No. 7, 12–18.

KRUMBEIN, W. C. (1967A) The general linear model in map preparation and analysis; *Kansas Computer Contribution* No. 12, 38–44.

KRUMBEIN, W. C. (1967B) Fortran IV computer programs for Markov chain experiments in geology; *Kansas Computer Contribution* No. 13, 38p.

KRUMBEIN, W. C. (1969) The computer in geological perspective; In Merriam, D. F. (Ed.), *Computer Applications in the Earth Sciences* (Plenum Press, New York), 251–75.

KRUMBEIN, W. C. (1970) Geological models in transition; In Merriam, D. F. (Ed.), *Geostatistics* (Plenum Press, New York), 143–61.

KRUMBEIN, W. C., BENSON, B. T. AND HEMPKINS, W. B. (1964) WHIRL-POOL, a computer program for 'sorting out' independent variables by sequential multiple linear regression; *Office of Naval Research, Geography Branch, ONR Task No. 389–135, Technical Report No. 14, Department of Geology, Northwestern University*, 49p.

KRUMBEIN, W. C. and DACEY, M. (1969) Markov chains and embedded Markov chains in geology; *Journal of the International Association of Mathematical Geology* 1(1).

KRUMBEIN, W. C. and GRAYBILL, F. A. (1965) *An Introduction to Statistical Models in Geology* (McGraw-Hill, New York), 475p.

KRUMBEIN, W. C. and SHREVE, R. L. (1970) Some statistical properties of dendritic channel networks; *Office of Naval Research, Geography Branch, ONR Task No. 389–150, Contract Nonr 1228(36), Technical Report No. 13, Department of Geology, Northwestern University*, 117p.

LAVALLE, P. D. (1965) *General Variation of Karst Topography in South-Central Kentucky* (unpublished Ph.D. Thesis, Department of Geography, University of Iowa).

LEE, W. H. K. and KAULA, W. M. (1967) A spherical harmonic analysis of the earth's topography; *Journal of Geophysical Research* 72, 753–8.

LEOPOLD, L. B. and LANGBEIN, W. B. (1962) The concept of entropy in landscape evolution; *U.S. Geological Survey Professional Paper* 500-A, 20p.

MACKAY, J. R. (1963) The Mackenzie Delta Area, N.W.T.; *Department of Mines and Technical Surveys, Canada, Geography Branch, Memoir 8*, 202p.

MATHER, P. M. and DOORNKAMP, J. C. (1970) Multivariate analysis in geography with particular reference to drainage basin morphometry; *Transactions of the Institute of British Geographers* 51, 163–87.

MCCONNELL, H. (no date), *Toward a More Incisive Assessment of Spatial Variation of Topographic Slope* (Mimeo, Department of Geography, University of Iowa), 26p.

MCCONNELL, H. (no date), *Randomness in Spatial Distributions of Terrain Summits* (Mimeo, Department of Earth Science, Northern Illinois University).

MCCONNELL, H. (no date), *Quadrat Counts of Terrain Summits Fit by Discrete Probability Distributions* (Mimeo, Department of Earth Science, Northern Illinois University).

MCCONNELL, H. (no date), *Evaluating Dispersal of Terrain Summits by Quadrat Methods* (Mimeo, Department of Earth Science, Northern Illinois University), 32p.

MCCONNELL, H. (1962) *On Spatial Variation of Average Slope: As related to difference in age and composition of materials* (Mimeo, Department of Geography, University of Iowa).

MCCONNELL, H. (1964) *Some Quantitative Aspects of Slope Inclination in Portions of the Glaciated Upper Mississippi Valley* (unpublished Ph.D. Thesis, Department of Geography, State University of Iowa).

MCCONNELL, H. (1965) A note on space filling by drainage systems; *58th Annual Meeting, Illinois State Academy of Science, DeKalb, Illinois,* 23p.

MCCONNELL, H. (1966) A statistical analysis of spatial variability of mean topographic slope on stream-dissected glacial materials; *Annals of the Association of American Geographers* 56, 712–28.

MCCONNELL, H., CHAPMAN, K. P. and KNOX, J. C. (1966) Topographic Slope in selected loess-mantled second order basins in Illinois: A multivariate analysis; *Department of Geography, University of Iowa, Discussion Paper No. 2*, 30p.

MCCULLAGH, M. J. and KING, C. A. M. (1970) Spitsym, a Fortran IV computer program for spit simulation; *Kansas Computer Contribution No. 50*, 20p.

MELTON, M. A. (1957) An analysis of the relations among elements of climate, surface properties and geomorphology; *Office of Naval Research, Geography Branch, ONR Task No. 389–042, Technical Report No. 11, Department of Geology, Columbia University*, 102p.

MELTON, M. A. (1958A) Geometric properties of mature drainage systems and their representation in an E_4 phase space; *Journal of Geology* 66, 35–54.

MELTON, M. A. (1958B) Correlation structure of morphometric properties of drainage systems and their controlling agents; *Journal of Geology* 66, 442–60.

MERRIAM, D. F. (Ed.) (1966) Computer applications in the earth sciences: Colloquium on classification procedures; *Kansas Computer Contribution No. 7*, 79p.

MERRIAM, D. F. and COCKE, N. C. (Eds.) (1967) Computer applications in the earth sciences: Colloquium on trend analysis; *Kansas Computer Contribution No. 12*, 62p.

MERRIAM, D. F. and COCKE, N. C. (Eds.) (1968) Computer applications in the earth sciences: Colloquium on simulation; *Kansas Computer Contribution No. 22*, 58p.

MIESCH, A. T. and CONNER, J. J. (1968) Stepwise regression and nonpolynomial models in trend analysis; *Kansas Computer Contribution No. 27*, 40p.

MILLER, R. L. and KAHN, J. S. (1962) *Statistical Analysis in the Geological Sciences* (Wiley, New York), 483p.

MOULTRIE, W. (1970) Systems, a computer simulation and drainage basins; *Bulletin of the Illinois Geographical Society* 12(2), 29–35.

NORCLIFFE, G. B. (1969) An alternative method for identifying the characteristic scale of periodic spatial series; *Area* 4, 21–8.

NOSKOV, V. F. (1969) A mathematical model of relief, and the connection of its parameters with geomorphological characteristics (in Russian); *Vestnik Moskovskogo Universiteta, Geografiva* 1, 105–110.

NYE, J. F. (1969) A calculation on the sliding of ice over a wavy surface using a Newtonian viscous approximation; *Proceedings of the Royal Society, Section A* 311, 445–67.

NYE, J. F. (1970) Glacier sliding without cavitation in a linear viscous approximation; *Proceedings of the Royal Society, Section A* 315, 381–403.

PIEXOTO, J. P., SALTZMAN, B. and TELWES, S. (1964) Harmonic analysis of the topography along parallels of the earth; *Journal of Geophysical Research* 69, 1501–5.

PRESTON, F. W. (1966) Two-dimensional power spectra for classification of landforms; In Merriam, D. F. (Ed.), Computer applications on the earth sciences: Colloquium on classification procedures; *Kansas Computer Contribution* No. 7, 64–9.

PRESTON, F. W. and HARBAUGH, J. W. (1965) BALGOL program and geologic applications for single and double Fourier series using IBM 7090/7094 computers; *Kansas Geological Survey, Special Distribution Publication* No. 24, 72p.

RAYNER, J. N. (1967) Correlation between surfaces by spectral methods; In Merriam, D. F. and Cocke, N. C. (Eds.), Computer applications in the earth sciences: Colloquium on trend analysis; *Kansas Computer Contribution* No. 12, 31–7.

RODDA, J. C. (1970) A trend-surface analysis trial for the planation surfaces of north Cardiganshire; *Transactions of the Institute of British Geographers* 50, 107–14.

RONCA, L. B. and GREEN, R. R. (1970) Statistical geomorphology of the lunar surface; *Bulletin of the Geological Society of America* 81, 337–52.

SALISBURY, N. E. (1962) Relief: slope relationships in glaciated terrain; *Annals of the Association of American Geographers* 52, 229–30.

SALISBURY, N. and LAVALLE, P. (1963) *Scale Variations in Morphometric Analysis* (Mimeo, Department of Geography, State University of Iowa).

SAVIGEAR, R. A. G. (1965) A technique of morphological mapping; *Annals of the Association of American Geographers* 35, 514–38.

SCHAEFER, F. K. (1953) Exceptionalism in geography: A methodological examination; *Annals of the Association of American Geographers* 43, 226–49.

SCHENCK, H. (1963) Simulation of the evolution of drainage-basin networks with a digital computer; *Journal of Geophysical Research* 68, 5739–45.

SIMONS, J. H., MAXWELL, J. C., IMBRIE, J. and MANSON, V. (1964) Expanded study of the applicability of multifactor computer programs to terrain analysis; *ONR Task No. 387–030, Contract Nonr 4523(00), Prepared by Autometric Facility, The Raytheon Co., Alexandria, Virginia.*

SMALLEY, I. J. and UNWIN, D. J. (1968) The formation and shape of drumlins and their distribution and orientation in drumlin fields; *Journal of Glaciology* 7, 377–90.

SPEIGHT, J. G. (1965) Meander spectra of the Angabunga River, Papua; *Journal of Hydrology* 3, 1–15.

SPRUNT, B. F. (1970) Geographics: A computer's eye view of terrain; *Area* 4, 54–9.

STRAHLER, A. N. (1950) Equilibrium theory of erosional slopes, approached by frequency distribution analysis; *American Journal of Science* 248, 673–96 and 800–14.

SURKAN, A. J. and VAN KAN, J. (1969) Constrained random walk meander generation; *Water Resources Research* 5, 1343–52.

TARRANT, J. R. (1970) Comments on the use of trend-surface analysis in the study of erosion surfaces; *Transactions of the Institute of British Geographers* 51, 221–2.

TARRANT, J. R. (1971) *Computers in the Environmental Sciences* (School of Environmental Sciences, University of East Anglia), 169p.

THAKUR, T. R. and SCHEIDEGGER, A. E. (1970) Chain model of river meanders; *Journal of Hydrology* 12, 25–47.

THOMPSON, W. F. (1964) *Determination of the Spatial Relationships of Locally Dominant Topographic Features* (Mimeo, Department of Geography, University of Iowa), 14p. plus figs.

THORNES, J. (1968) A queueing theory analogue for scree slope studies; *London School of Economics, Graduate School of Geography, Discussion Papers* No. 22, 16p.

THORNES, J. B. and JONES, D. K. C. (1969) Regional and local components in the physiography of the Sussex Weald; *Area* 2, 13–21.

TINKLER, K. J. (1969) Trend surfaces with 'low explanations': the assessment of their significance; *American Journal of Science* 267, 114–23.

TOBLER, W. R. (1969) Geographical filters and their inverses; *Geographical Analysis* 1, 234–53.

TOWLER, J. E. (1969) *A Comparative Analysis of Scree Systems Developed on the Skiddaw Slates and Borrowdale Volcanic Series of the English Lake District* (unpublished B.A. Dissertation, Department of Geography, Cambridge University), 84p.

TROEH, F. R. (1965) Landform equations fitted to contour maps: *American Journal of Science* 263, 616–27.

WARNTZ, W. (1968) A note on stream ordering and contour mapping; *Harvard Papers in Theoretical Geography* No. 18, 30p.

WERRITTY, A. (1969) On the form of drainage basins; *Department of Geography, Pennsylvania State University, Papers in Geography* No. 1, 33p.

WONG, S. T. (1963) A multivariate statistical model for predicting mean annual flood in New England; *Annals of the Association of American Geographers* 53, 298–311.

WOOD, W. F. (1955) The relationships of relief; *Annals of the Association of American Geographers* 45, 306–7.

ZAKRZEWSKA, B. (1967) Trends and methods in land form geography; *Annals of the Association of American Geographers* 57, 128–65.

2 General geomorphometry, derivatives of altitude, and descriptive statistics

IAN S. EVANS

Department of Geography, University of Durham

In the application of multivariate statistics to geographic problems, too little critical attention is paid to questions of operational definition and sampling. Yet these initial stages of the application of scientific method are the foundations on which any conclusion rests. The often implicit belief, that factor or principal components analysis of as many variables as can easily be measured, will solve problems of variable redundancy, runs counter to the first law of computer science: 'garbage in, garbage out'. Many of the operational definitions used in geomorphometry are extremely poor as measures of the intended concepts. Computers should not be subjected to such garbage.

I intend to use this opportunity to review critically various geomorphometric parameters, and to propose a simple system of geomorphometry which is both better integrated, and consistent with standard concepts of descriptive statistics. Its adoption would therefore lead to simplification of science and to improved connectivity, consistent with the general progress of 'reductionism'.

These proposals arose from work on the geomorphometry of glaciated mountains (Evans 1969A). If a glaciated land surface is distinctive, what are the essential ways in which it may be discriminated from other land surfaces? Can 'degree of glaciation' be expressed quantitatively? Study of cirque morphometry did not provide a complete answer to these questions. Measurement of summit magnitude and analysis of summit systems suggested one parameter (density of contour isolations) for degree of glaciation, but the glacial interpretation is not unique; further parameters are necessary if glaciated terrain is to be discriminated.

Hence attention was turned to the field of 'general geomorphometry', the measurement and analysis of those characteristics of landform which are applicable to any continuous rough surface. This is distinguished from 'specific geomorphometry', the measurement and analysis of specific landforms such as cirques, drumlins and stream channels, which can be separated from adjacent parts of the land surface according to clear criteria of delimitation. It is necessary to add the prefix 'geo' to the more commonly used 'morphometry', in order to distinguish the measurement of landform from that of zoologic forms and of sedimentary particles (Goldberg 1962; Tricart 1947).

Both branches of geomorphometry have their strong and weak points. General geomorphometry deals with surface altitude, gradient, distance and area, often in combination as in hypsometry, which relates area to altitude. In these fundamental concepts, it is quite objective; there are, however, derivative concepts such as convexity, relief, dissection and texture, which allow for more variability in operational definition. General geomorphometry as a whole provides a basis for the quantitative comparison even of qualitatively different landscapes, and it can adapt methods of surface analysis used outside geomorphology. Specific geomorphometry is more limited; it involves more arbitrary decisions, and leaves more room for subjectivity in the quantification of its concepts. On the other hand, its variables are more closely related to process, which is usually invoked in the operational definitions.

General geomorphometry provides an extremely useful basis for geomorphology, if only in suggesting the important spatial variations and structures that this science is to explain. In addition, some hypotheses concerning process and genesis, e.g. azimuthal and altitudinal variations, can best be tested in terms of relationships between general geomorphometric parameters. Yet little use is made of such techniques. I suggest that the current unhealthy state of general geomorphometry is due in part to the labour involved in producing useful quantities of data, and even more, to an adherence to nineteenth-century statistical concepts which fail to separate 'signals' from noise. The latter impediment can be removed; the former can be reduced greatly if geomorphometry takes advantage of the new opportunities for automation, in particular through data digitisation and computer processing.

For the discrimination of glaciated land surfaces from others, the parameters of general geomorphometry are more promising than those of specific geomorphometry. Such parameters may provide a measure of 'degree of glaciation'; the surfaces of glaciated areas are usually 'rougher' than those of others. However, the discussion which follows concerns the whole field of general geomorphometry, which is applicable to land surfaces of any origin and, significantly, of any combination of origins. Specific geomorphometry is less suited to land surfaces of mixed origin.

General geomorphometry has a long history among European (German, Polish, French) geographers (e.g. Frey 1965). Their rather descriptive approach, however, has not produced as much interest as the more recent analytical approach of American hydrologists and geologists (Strahler 1950). The American school has applied itself almost entirely to fluvial landforms and related slopes, using the drainage basin as its fundamental unit (Leopold, Wolman and Miller 1964, ch. 5; Chorley 1969). Most of its geomorphometry has been specific to channels or to basins, but its contributions to general geomorphometry in terms of gradient, relief, texture, dissection and hypsometry have also been of the first importance. A full synthesis of different landform parameters, however, has yet to be achieved, even for fluvial landscapes.

The present report attempts to propose a synthesis of general geomorphometric parameters, based on the simplest concepts of calculus and statistics. After a review of general geomorphometry as a whole, the new possibilities due to automation are discussed. Automation will be helpful in providing broad coverage by standardised methods, permitting comparisons of parameters for different areas. These general considerations provide a basis for evaluation of alternative operational definitions of the limited number of fundamental concepts in general geomorphometry, grouped (cf. Peltier 1955) under the headings relief, texture, slope, local convexity and regional convexity. This review reveals a number of new relationships between operational definitions, and permits some decisions to be made as to which are preferable. Where there are sound substantive or methodological reasons for preferring one operational definition to another, it is better to discard the latter rather than to throw both into a multivariate analysis.

The relationships between general geomorphometric parameters are then discussed, and it is concluded that the proposed approach should lead to clarification of the subject. Finally a small pilot study designed to test some of the relationships between preferred parameters and others is reported. It has not yet been possible to test the proposed new system in its entirety, but no major obstacles to its adoption are suggested by these preliminary results.

General geomorphometry

What is wrong with general geomorphometry today?

1. Though there are a limited number of concepts of significant characteristics of surface form, each researcher tends to develop his own operational definitions for the quantification of these, for example the

area over which relief is to be measured, or the way in which a characteristic 'high point' or 'low point' is to be selected. This prevents comparisons between areas studied by different authors. Any reduction in the arbitrariness of a definition is balanced by an increase in subjectivity.

2. Each parameter is abstracted independently and by manual methods. Wood and Snell (1960) defined six parameters of surface form, requiring four separate operations of map measurement. These parameters are: (i) 'grain size' (valley spacing) and (ii) relief (range in altitude over this spacing), both derived from the nick-point in a graph of relief against distance; (iii) average altitude, from nine random points; (iv) elevation/relief ratio, based on ii and iii – this is in fact equivalent to the hypsometric integral; (v) average gradient, from the number of contours crossing two randomly oriented traverses of length equal to the grain size; (vi) slope direction changes from uphill to downhill and vice versa, counted over longer traverses, to provide an estimate of channel density. On practical, conceptual and pedagogic grounds, it would be desirable to replace such an approach with a more integrated one.

3. Unstable statistics such as maximum altitude, minimum altitude and range are widely used, as for example in the hypsometric integral ((mean altitude-minimum altitude)/range of altitude). Such indices are influenced by unrepresentative extremes; their instability compounds the problems of operational definition. The extremes themselves may be unstable in relation to changes in the operational definition on which data capture is based. For example, Gerrard and Robinson (1971) have shown that maximum (and modal) gradients vary considerably according to the ground length over which gradient is measured, while mean gradient is relatively stable.

4. Geomorphometrists have been prone to develop 'indices' and coefficients, to measure their more complicated concepts. If two parameters seemed to relate to a particular concept, then they were multiplied together, even if highly correlated (in which case a squared value was in effect produced). Sometimes, indeed, researchers have produced results quite different from those intended.

Indices are the most primitive form of descriptive statistics; they involve *a priori* assumptions of simple relationships between variables. Recently they have been replaced by factors and principal components, which at first glance appear to be more objective and less arbitrary combinations of variables, since they are based on actual rather than hypothetical correlations. However, it seems unlikely that Principal Component Analysis or Factor Analysis will clarify the concepts involved in geomorphometry.

Components and factors include accidental contributions from irrelevant variables, and can be compared between studies only if exactly the same variables are subjected to exactly the same techniques: given the present fashion for novelty *per se*, such a coincidence is almost unknown. The relative magnitude of factors is principally a reflection of the choice of variables, yet it is uncommon to find this choice defended meaningfully, or for the equal weighting which is implicit in the standardisation of variables to be justified. Again, the number of factors to be extracted is critical in any solution involving rotation of axes; yet this decision is usually based on an arbitrary eigenvector value, and often there is no clear gap between the largest excluded and the smallest included factor. Most authors fail to test for normality of variables and linearity of relationships, and to make a careful study of the correlation matrix.

Though factor analysis may be of some use in the understanding of complex correlation matrices, more attention should be paid to choice and definition of variables. For example, Lewis (1969) included several variables which are simple functions of two other variables which also form part of his analysis. His operational definitions include extreme values (longest, steepest), and two involve the unsatisfactory measure 'basin perimeter', which is affected by irrelevant minor crenulations. Armstrong (1967) colourfully illustrated the need for prior hypotheses and theories, and for careful definition of variables: Miesch (1969) showed that factor techniques do not always fulfil what is claimed for them, while Matalas and Reiher (1967) provided a broader critique.

There seems to be some justification for regarding factors as indices without exact definitions. A different approach to the measurement of complex concepts is desirable, and can be found in relating them more simply to elementary concepts. Tricart (1965) noted the limitations of complex analyses such as those of Péguy (1942, 1948), and their dependence on very careful weighting of variables. He advocated study of the frequency distributions and the correlations of simple measures such as gradient, so that contact with reality would not be lost. It will be demonstrated that complex concepts can indeed be expressed in relation to frequency distributions of measures of simple concepts. Hence it seems better to define the simple concepts carefully, than to include many different parameters unselectively and to hope that factor analysis will be our salvation from the resulting confusion.

5. Several authors (e.g. de Smet 1951) show evidence of a 'Philosopher's Stone' syndrome, that is, they imply a belief that the essential nature of surface form can be summarised by a single parameter, usually an 'index of dissection'. Carr and Van Lopik (1962, 11) state that some measurements 'are more suitable than others in presenting a complete

picture of surface geometry'; they regard 'completeness' as a desirable characteristic of a geomorphic parameter. It seems unlikely that surface form is not multivariate; its dimensionality can be tested relatively objectively by multivariate statistics, bearing in mind the reservations noted above. The use of compound parameters mixing several concepts is likely to lead to confusion.

6. Finally, most parameters are defined in relation to areas and they therefore depend closely upon the choice of areal divisions. This arbitrary element would be reduced if point measures could be used.

Review of the literature of general geomorphometry shows that all of its parameters may be defined in terms of altitude. Relief is usually expressed as range in altitude, texture is the spacing of maxima or minima. Slope gradient is the first vertical derivative (rate of change) of altitude (z), and downslope convexity is the second vertical derivative (rate of change of gradient). By analogy, slope aspect is the first 'horizontal' derivative of the altitude surface (rate of change of x with y, for constant z), and cross-slope convexity is the second horizontal derivative (rate of change of aspect).

Building upon some recent work of Tobler (1969), I therefore suggest that altitude at a point, and its first two derivatives, provide the unifying concept around which general geomorphometry may be rebuilt. Tanner (1960) suggested use of the first four moments of altitude frequency distributions, but he did not relate these to existing geomorphometric parameters, or to slope and convexity.[1] Speight (1968) measured vertical and horizontal convexity, but allotted measurements to arbitrary classes and used separate manual procedures for extraction of these and other form properties.

Each derivative provides a map of point values, producing statistical distributions (five in all) which may be characterised, descriptively, by their first, second and third moments, expressed as mean, standard deviation and dimensionless skewness. The use of parametric moments in this way does not imply that any of these distributions is normal; simply that, given data on interval or ratio scales, it is best to avoid the loss of information involved in order-based statistics. Inclusion of

[1] Rowan, McCauley and Holm (1971) used the median, mean and standard deviation of apparent gradient to characterise the roughness of lunar terrain, with range in altitude as a 'second-best' fallback where gradients could not be measured. It should be noted that their distinction between 'absolute' and 'algebraic' frequency distributions is meaningful only in a restricted context where eastward or westward components of gradient along an east–west profile are the only gradient data available. Because of the use of gradient components, most modes are at zero; it is possible to convert these distributions to those of true gradient if it is assumed that contours on any slope are equally likely to intersect the line of traverse at any angle

skewness and kurtosis permits precise characterisation of any simple unimodal distribution. The resulting parametric description will, however, be less meaningful for distributions which are strongly multimodal. Aspect, the first horizontal derivative, ranges from 0° to 360°, and must be analysed by circular statistics.

Such a description is incomplete in that it does not specifically consider geographic distributions, i.e. relative locations. Some parameters do however provide related information, for example skewness of altitude reveals the extent to which the mode of area is placed high or low relative to the mean. Thus it provides a more stable replacement for the hypsometric integral (pp. 42–8).

Digitised contours and altitude matrices

The automation of geomorphometry involves using electronic computers to process digital data representing the earth's surface. The information storage device of the present non-automated system is the contour map. It is now possible to transform contour lines to machine-readable digital form as cartesian co-ordinates, with no more work than is involved in redrafting them. This is achieved by means of an analogue-to-digital converter, i.e. a digitising table; the d-mac Pencil Follower is the most suitable currently available (Evans 1970).

The source map is laid flat on the d-mac table, and the line to be digitised is traced with a free-moving 'pencil'. A magnetic coil centred on the point of the pencil sets up a field which is picked up by servo-magnets on a 'follower' under the table. This follower moves up and down parallel to the y-axis, on a gantry which moves along the x-axis, and its position is transmitted mechanically, by wires, to x and y shaft encoders. Rotation of the encoders digitises the position of the follower (which coincides with that of the pencil), and the resulting electronic signal is processed and punched in character form on magnetic tape or paper tape. Magnetic tape is preferable since its potential information density is much greater than that of paper tape. Four-digit x and y co-ordinates record the position of the pencil to the nearest 0·1 mm, at a spacing of 0·2 to 0·8 mm, depending on the speed of line tracing and on the setting of the machine's read-out speed. Absolute accuracy is about 0·3 mm.

Each contour must be labelled with its altitude, and the ends of contour segments must be marked. Such signals are typed in on a keyboard. To establish azimuth, it is also necessary to record the direction of north relative to the digitising table co-ordinates, e.g. by digitising the southern and northern ends of a north-pointing line. This is particularly important for the analysis of aspect. The recording of corner points is even more useful, if their positions are known in an absolute co-ordinate system; this permits the arbitrary table co-ordinates to be converted to absolute co-ordinates, so that adjacent sheets digitised separately may be joined together.

In future, the work involved in digitising a contour map may become unnecessary, since cartographic automation will probably begin at the stage of survey. Contours may be established by manipulating a 'floating dot' on an air-photo model in a direct-viewing stereoscopic plotting instrument ('stereoplotter'). The x and y 'lead screws' which transmit the motions of this dot to a plotting table can also operate shaft encoders, so that the position of the dot is digitised and rectangular co-ordinates are written on magnetic tape. The Ordnance Survey started such recording, on a trial basis, in 1970. In this way contours can be recorded in digital form *ab initio*, and the tape so produced can be used, after editing, to operate an automatic plotter and produce contours at any scale, on any projection and with any degree of generalisation. The edited tape would also be available for geomorphometric analysis on a computer.

From a digital contour record, a simple program can quickly calculate the length of each contour, and the area within closed contours. It is also possible to calculate the azimuth of contour trend, and to analyse azimuth in relation to length along the contour; this replaces the cumbersome measurement of angles by manipulation of a protractor. To establish slope or relief, however, it is necessary to search from one contour to another, which is a much more difficult procedure.

Here a fundamental limitation on the use of digitised contours emerges. Digitised contours are appropriate for answering the question 'where is the altitude z to be found?', rather than 'what is the altitude of this point (x, y)?' To answer the latter question, it is better to have data ordered in a scanning format, i.e. z values for changing x with constant y, for successive values of y. This implies having a square matrix of 'spot heights', i.e. altitudes at the intersections of a square grid. Because of the widespread importance of topography in military activities, the U.S. Army Map Service have produced numerous matrices of altitudes, which are usually referred to as 'numerical maps'. 'Figure field' is sometimes used in the same sense, but since both these terms logically have broader connotations, the more specific term 'altitude matrix' is preferred here. This is regarded as one type of geographic matrix, or 'geomatrix' (a matrix in which both rows and columns are in order according to their position in geographic space).

Accurate altitude matrices with fine meshes can be produced directly in a stereoplotter operated in traversing instead of contouring mode. In the absence of such facilities, altitude matrices are generated by visual interpolation from contour maps; this indirect process is necessarily less exact, and more time consuming since the matrix must be punched into machine-readable form manually. It is also possible to derive an altitude matrix from digitised contours by an interpolation program, but this is costly and somewhat inefficient.

The advantage of altitude matrices, relative to digitised contours, is that they make it easier to select all information concerning a restricted area, since the sequence of storage is known and regular. A point can be related to its neighbours in the grid without a prolonged search. This is particularly useful in the measurement of slope direction and local convexity. The derivation of volume, altimetric frequency distributions and related relief measures for subdivisions of the area digitised is also more direct. Their derivation from digitised contours involves area measurement, which is not difficult so long as the sequence of digitisation is planned carefully and sheet edges can be included to achieve closure of areas above contours; but for areas not digitised as units, area measurement may be very difficult and not fully automatic.

Absolute cartesian co-ordinates are usually used to record both altitude matrices (z) and digitised contours (x, y); this involves considerable redundancy of information (Connelly 1968). The relative efficiency of the two forms (altitude matrices and digitised contours) in information content per unit storage is not yet clear. The following questions may be posed:

(i) For an equal total data storage, which permits the more accurate estimate of altitude at any random point?

(ii) What are the relative computing costs of interpolating altitude in this way?

(iii) What is the relation between the altitude matrix grid mesh, and the mean digital contour spacing which achieves the same accuracy in depiction of terrain; and what are their relative data storage requirements?

(iv) What are the relative costs of conversion each way between numerical maps and digitised contours?

Boehm (1967) has compared the efficiency of the following forms of topographic information storage; (i) contour points sorted on x; (ii) contour tree ordering; (iii) uniform grid; (iv) uniform grid, incremental altitude; and (v) variable-mesh grid, incremental altitude. Of these, (iii) is a straightforward 'altitude matrix', while (ii) is close to a simple digitised contour set. Boehm excluded the use of polynomial approximations (to contours or to profiles), or of vectors (for contours), since he regarded these as less efficient than the above five methods. Incremental ('differential') methods are more economical in storage, but require much more time for retrieval from storage.

Boehm concluded (1967, 414) that 'The variable grid differential altitude method is superior on the whole, although the contour tree and

B

uniform grid differential altitude methods have advantages for particular problem areas.' His comparison between the variable and uniform grid methods, however, is that to achieve comparable precision in
specifying the altitude of a random point, the storage required for a
variable grid is inversely proportional to the square of *mean* distance
between contours, but that for a uniform grid is inversely proportional
to the square of *minimum* distance between contours (ibid., 410). In
other words, the uniform grid is required to have a mesh so fine that
there will never be two contours between grid points, while it is assumed
that the mesh of a variable grid can vary in perfect harmony with the
contour spacing so that it equals both minimum and maximum spacing,
and its average equals the mean spacing. The requirement seems un-
reasonable, and the assumption is extremely improbable given the
intricacy of many contour maps. If such spacings were used, the uniform
grid would be as accurate as the variable grid, at the point where the
mesh of the latter was finest; everywhere else, the uniform grid would
be more accurate. Minimum accuracy may be a realistic basis for comparison from the point of view of military science, but the pure scientist
is more interested in comparing the storage requirements of two
methods for equal average accuracy. Furthermore, contours are more
likely to run parallel where they are closely spaced, and hence interpolation will be more accurate there than elsewhere. Hence Boehm's
comparison between 3700 words with a variable grid, and over a
million words with a uniform one, is quite unrealistic.

The advantages of a variable grid in precision are unlikely to out-
weigh its disadvantages in data storage, retrieval and interpolation.
Though different grid meshes are suitable for different geomorphic
regions, it will usually be an unnecessary complication to vary the mesh
within a particular medium-scale map sheet, which serves as a unit
for digitisation. My conclusion from Boehm's paper is that for the same
number of points, contour methods require more storage, more time
for retrieval of relevant stored points, and more time for the interpolation of altitudes for a set of random or sequential points. Contour
methods may eliminate the need for an extra processing stage, for the
production of a grid mesh; but this stage is not involved if the altitude
matrix is established directly from the stereo model. Provisionally, then,
the grid method is to be preferred. A more definite conclusion awaits the
application of an empirical test, processing examples of various types
of terrain by efficient programs for each method. The assumptions involved in theoretical comparisons are not likely to be justifiable.

The problem is complex because the handling of large quantities of
data depends upon read-in and read-out techniques, and upon computer
configurations. Data Library storage is likely to be on magnetic tape,
and efficient techniques are needed to transfer appropriately selected

subsets of library data to core, probably via disc storage if available. Such techniques would be required to make sub-matrices of a large altitude matrix readily available, and to economise on search times. Retrieval would be difficult for a sequence of widely scattered random sub-matrices, but relatively easy for the scan of successive sub-matrices which is required in producing a matrix (map) of the overall distribution of slope. This suggests that altitude matrices should be stored not in long parallel scans, but in nested cells each of which consists of several short scans. Similarly, it would appear that digitised contours should be stored in fairly small cells, which might need to be processed 2 or 4 at a time to provide information on areas marginal to cells.

Altitude matrices can be portrayed graphically either by shading, by block diagrams, or by contours. It is probably easier to derive contours from altitude matrices than vice versa; almost all programs for contouring from point data first create a rectangular matrix, and then interpolate contours through this. A matrix is in a form ready for statistical manipulation, e.g. by smoothing, contrast enhancement or spectral analysis; it provides better control for trend surface analysis; and it can be related more easily to other distributions (especially other matrices). Hence it seems that an altitude matrix is likely to be in greater demand than the corresponding set of digitised contours, and if altitude data is to be stored in a machine-readable numerical library or data bank in one form only, that form should if possible be a matrix. Data initially in contour form should however be stored in contour as well as matrix form, since conversion to matrix form is likely to involve some loss of information. The structure of the library or data bank will then depend on the available forms of input, as well as upon the desirable forms of output.

The availability of machine-readable altitude data in either form has profound implications for geomorphometry. The availability of altitude matrices is particularly suggestive, since it facilitates the rationalisation of general geomorphometry, with derivatives of altitude as the unifying concept. The following discussion anticipates increasing importance of altitude matrices and of processing by computer.

Vertical dimension; relief

'Relief' is a concept intended to describe the vertical extent of landscape features, without reference to absolute altitude or to slope. There is no generally accepted definition of 'relief'; range in altitude is most commonly used (Smith 1935), but there are various ways of defining the area within which range is to be measured. Dury (1951) related relief to depth of dissection, the difference in altitude between a 'stream line surface' and a 'summit surface': but even if it is accepted that the

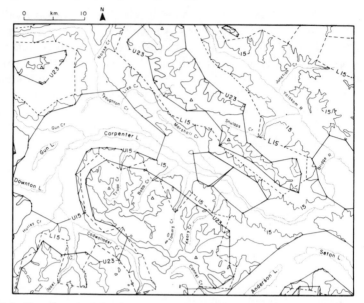

Fig. 2.1 Envelope and Sub-envelope surfaces based on departure from the
real surface over distances of up to 5 km. Based on the 2286 m (7500 ft:
labelled '23') and 1524 m (5000 ft: labelled '15') contours in the middle
Bridge River District (Bendor and Shulaps Ranges), British Columbia,
Canada, the upper (solid) and lower (dashed) envelope surface contours
are labelled U and L respectively.

former should be below, and the latter above, the actual surface at all
points, there is considerable latitude in the drawing of each. It seems
necessary to take as objective as possible a definition of summits (in
terms of relative altitude), and of valleys (in terms of gradient or
stream order). Even so, difficulties occur when, for example, a mountain range has 'valleys' on one side only. Pannekoek (1967) demonstrated the subjectivity of the summit surface concept.

Depending on the magnitude of the summits and valleys used, these
methods can produce anything from very smooth surfaces, to only
slightly generalised versions of the real surface. Such variations are
reflected in the magnitude of the 'relief' estimates derived. Two significant attempts have been made to produce more objective definitions
of summit surfaces. In the first, Fischer (1963; Louis 1963) invoked an
arbitrary constant (the slope of the 'cone of influence' of summits), and
the circular contours of his 'enveloping surface' look unreal.

A more useful but equally arbitrary approach was used by Stearns
(1967). Recognising the large number of alternatives, he rendered the
method reasonably objective by specifying the maximum distance for
which an 'envelope' or 'sub-envelope' surface might depart from the

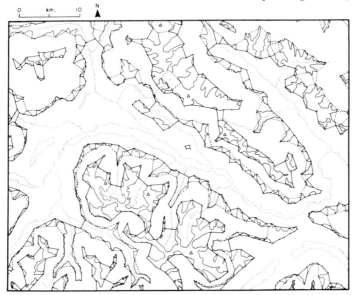

Fig. 2.2 Envelope and Sub-envelope surfaces based on departure from the
real surface over distances of up to 2 km. Area and legend as for fig. 2.1,
but labels were omitted for clarity.

real surface to which it is tangent. The greater the distance, the
smoother the resulting surface and the greater the relief. Figures 2.1
and 2.2 exemplify this for 'departure distances' of 2 and 5 km.

The fact remains that an infinite number of such surfaces may be
drawn, and the resulting values of relief vary widely. Except for simple
situations such as dissected domes, or dip slopes with roughly parallel
valleys, the usefulness of 'stream line surfaces' and 'summit surfaces'
seems to be limited. It is certainly difficult to fit the irregularities of
glaciated mountains into such an approach.

Gutersohn (in Wood and Snell 1960) graphed highest point and
lowest point against radial distance from a randomly selected sample
point; the vertical difference between these two curves is the relief
within that radius. The 'main nick-point' on the difference curve gives
relief around the sample point; this is a highly subjective conclusion
to an otherwise objective method. Thompson (1964) devised an alternative in order to avoid the problem of multiple nick-points. 'Representative' high and low points in the neighbourhood of the sample point
are defined as the first points touched by spheroids respectively lowered
from above and raised from below the sample point. The resulting
maps of the Swiss Alps look interesting and reasonable, but the work
involved is considerable and the critical factor is the radius of the sphere.
In effect, the radii of Thompson's spheroid decide which nick-point is

selected but do this separately for the 'highest point' and the 'lowest point' curves. If the curve of relief plotted against distance tended to some general form, e.g. inverse exponential, it could be represented by one or two parameters; but often it is complex, and thus we are faced once again with either a number of alternative parameters or an arbitrary one.

Another method is to note the range in altitude between the highest and lowest points within a certain radius or grid square. The size of grid square used is arbitrary, varying for example from 1 km² (Chen 1947) to 90 km² (Hammond 1964); areas of about 25 km² have been

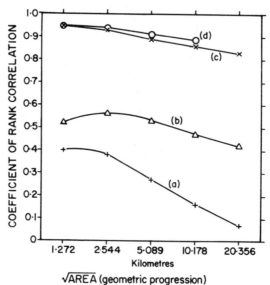

Fig. 2.3 Correlations between pairs of parameters (Data from Wood 1967, Table 2)

+ (a) Relief v. drainage density over same area
△ (b) Mean gradient v. drainage density over same area
× (c) Relief v. mean gradient over same area
○ (d) Relief v. mean gradient over next larger area

favoured (Smith ·1935). If the square is so small (in relation to topographic wavelengths) that it is unlikely to contain a whole slope, 'relief' becomes simply a measure of gradient (indeed, this is the intention of some authors, such as Swan (1967, 1970)). Wood (1967, table 2) showed that the rank correlation between relief and gradient, measured within concentric circles, declines from 0·95 for circles of 1·62 km² to 0·83 for 414 km² (fig. 2.3). For 1 km² squares, Gălăbov (1968) found correlations between range in altitude and mean gradient of 0·92, 0·81, 0·85, 0·83 and 0·73, for regions from the Danubian Platform (highest correlation) to the middle Stara Planina (lowest correlation, highest relief and gradients). Hence in order to make relief as distinct and non-redundant a variable as possible, fairly large areas should be used.

Though there is a need for standard-size squares for inter-regional comparisons, it will probably be necessary to use different sizes in areas of markedly different topography (Hammond 1958). Trewartha and Smith (1941, 31) also suggested that size must be adjusted to texture of relief pattern. The defence of this method is that, since any of the methods discussed so far would be more or less subjective and arbitrary, the simplest and most objective method is preferable. Kaitanen (1969) used a similar method, with squares of 25 km², to construct upper and lower envelope surfaces.

One defect shared by these methods is the use of extreme values. All of them, including those based on enveloping surfaces, use the range in altitude within an area defined in various ways. This is only one aspect of the distribution relating frequency (area of map surface) to altitude, within that area. To describe the dispersion of a distribution, the standard deviation is a more powerful and stable statistic than the range, since it is based on all values and not just the two extreme values. It is therefore logical that relief, defined as dispersion in altitude, should be measured by the standard deviation of altitude. The auto-correlation of altitude admittedly makes range less unreliable than it is for random variables, since on a continuous surface all intermediate values between the extremes must be represented. Yet variance ('power') is commonly used in the analysis of autocorrelated series, and so there is precedence for the use of statistics based on all values. In a large area, maximum altitude may be very unrepresentative.

The 'Kotenstreuung' of Gassman and Gutersohn (1947), to be calculated for regions of homogeneous topography, is a dimensionless version of the standard deviation of altitude. Unfortunately, it is standardised through division by range in altitude; it is intended as a measure of dissection rather than vertical dimension, and is in fact an expression of kurtosis of altitude. 'Relieffaktor', which relates Koten-streuung to half the range in altitude, would appear to be of still more limited value. Péguy (1948) proposed standard deviation in altitude as the most rational index of the distribution of altitudes within a region, an index complementary to mean altitude; but he did not make much use of it, or compare it with other measures of relief. Hobson (1967) used a computer program to measure standard deviation of altitude and standard deviation perpendicular to the best-fit linear trend-surface. These and other surface roughness parameters were based on samples of altitude at sixteen points.

The standard deviation is not used generally because with manual methods it is much more work to establish the altitude frequency distribution and measure its parameters, than to find the highest and lowest points and calculate their difference. The advent of automation, however, makes it reasonable to envisage the more laborious but statistically preferable approach, especially if the digital data is available or will also be used to establish other parameters. Automation reduces the cost of calculating standard deviation much more than that of calculating range.

A case could alternatively be made for use of the inter-quartile range, to allow for skewed or multi-modal distributions of altitude; but since non-normality is rarely so extreme as to render the standard deviation

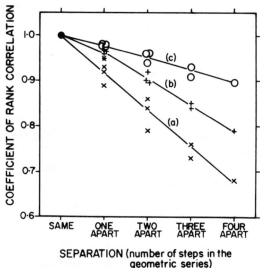

Fig. 2.4 Correlations between same parameter measured over different (concentric) areas (Data from Wood 1967, Table 2)

× (a) Drainage density (changes in sense of gradient, along two diameters)
+ (b) Relief (range in altitude)
○ (c) Mean gradient (contour crossings, along two diameters)

meaningless, I propose that the more powerful statistic be adopted. This does not answer all the difficulties; it remains necessary to define an area within which relief is measured. But change in area has less effect on the standard deviation than on the range. (The rank correlation between range in altitude as measured over concentric areas of $1 \cdot 62$ km^2 and 414 km^2 ($0 \cdot 79$), while greater than that between drainage densities ($0 \cdot 68$), is less than that between gradients ($0 \cdot 9$): see fig. 2.4.) The area used should be sufficiently large to give a representative sample of the topographic surface. With automated methods, there is no case against the replacement of range by standard deviation. The case for use of standard deviation of altitude as the measure of relief is overwhelming when combined with the use of mean and skewness of altitude in an integrated approach to general geomorphometry.

Horizontal scale; texture, wavelength

It is particularly important to review measures of horizontal scale, because they have been used to define the areas within which other geomorphometric parameters were measured. For example, Wood and Snell (1960) derived a 'typical' topographic wavelength from the nick-point in the graph of relief against distance, discussed above; this gives a horizontal dimension equivalent to the vertical relief. Similarly, Thompson (1964) based a measure of crest spacing on his spheroid method. The reciprocal of valley density, or of the drainage density of higher-order streams, might provide a more objective measure. One may question, however, whether such a 'unit cell' approach is really applicable to the varying oscillations present in topography, that is, whether a single wavelength suffices to characterise this attribute of topography.

An alternative approach is to regard topography as the manifestation of a spectrum of various wavelengths: ' . . . most contour maps may be considered to be built up by the superposition of a double infinity of elementary undulating surfaces, each of which has the form of a horizontal sinusoidally corrugated sheet, infinite in extent' (Swartz 1954, 46). Such analysis starts with measurement of the autocorrelation of altitude at various distances, in different directions. Closely spaced points are similar in altitude, since despite its irregularity, the land surface is not a purely random variable. The autocorrelation declines rapidly with distance, but may then rise if periodicity (regular oscillation) is present. Autocorrelation rises to a maximum at the wavelength of periodicity, e.g. the spacing of crests. In lineated topography, autocorrelation declines more rapidly across the trend of lineation than parallel to it.

An early study of autocorrelation along one-dimensional terrain profiles was by Diamantides and Horowitz (1957), who found that autocorrelation declined to 0·87 (Arizona) and 0·72 (Ohio) at a distance of 1·6 km. The first application of two-dimensional spectral techniques to the land surface was by Horton, Hoffman and Hempkins (1962), who used an altitude matrix of 3·81 m mesh for a 122 m × 122 m valley side (now submerged) in Texas. They mapped the autocovariance (autocorrelation × variance), which has contours parallel to the valley side, i.e. autocorrelation is almost constant along the 'strike' of the topography, while it declines monotonically across it ($r = 0·7$ for a lag of 19 m).

Switzer, Mohr and Heitman (1964) presented two-dimensional

correlograms for four adjacent 22×22 matrices of altitude of the sea floor. Each matrix was non-isotropic due to the presence of linear depressions, but the degree and direction of anisotropy differed because each incorporated only one or two depressions. A larger matrix (58×46) gave a smoother correlogram, with elliptical contours portraying a reduced but more stable anisotropy (variation of correlation with azimuth).

McDonald and Katz (1969) measured autocorrelation along radial profiles $30°$ apart, having detrended their bathymetric data by subtraction of the best-fit linear plane. Altitude was read to the nearest 9 m at intervals of 122 m. Though the resulting polar autocorrelation function is not the two-dimensional autocorrelation function, it does reveal lineation. Correlation declined to 0 within 2 km for the $030°–210°$ track, but within 4 km for the others.

The power spectrum (also called energy spectrum or variance spectrum), which represents the importance (amplitude) of oscillations at various wavelengths and orientations, is obtained as the Fourier transform of the autocovariance function. Since this procedure assumes that the series is stationary, it is necessary to remove trends (long-wave variations). Horton *et al.* detrended their altitude matrix by the filter

$$\begin{matrix} 0 \cdot 5 & -2 & 0 \cdot 5 \\ -2 & 6 & -2 \\ 0 \cdot 5 & -2 & 0 \cdot 5 \end{matrix}$$

which eliminates both the mean and the overall gradient by replacing the original altitudes with figures of height relative to the neighbouring altitudes. This filter affects amplitude at all wavelengths, but in a predictable fashion, so that a correction can be applied. The corrected power spectrum is symmetrical, and most of its power comes from wavelengths greater than 12 m.

Rayner (1971, 118–125) gave an example of how two-dimensional spectral analysis clearly demonstrates terrain anisotropy (in this case, structural lineation), and of how an isotropic spectrum should be collapsed to the simpler, one-dimensional case. He also applied cross-spectral analysis to correlate surface altitude with stream pattern, and despite the limitations of the latter as a 'surface' interesting results were obtained.

Fourier analysis (Bhattacharyya 1965) has been applied (after the subtraction of linear trend) to a 49×25 altitude matrix by Preston and Harbaugh (1965), who found that 11×12 Fourier terms accounted for 93% of the original variance. However, there is no reason to expect the land surface to be composed of oscillations at discrete wavelengths, and still less reason for these wavelengths to be integral fractions of a 'fundamental wavelength'. Hence Fourier analysis is too rigid for land surfaces, and there is no reason to prefer it over the more general spectral technique, especially since Esler and Preston (1967) have made available a program for two-dimensional spectral analysis of matrices up to 100 × 100. Bassett and Chorley (1971) also have suggested that spectral analysis might be more suitable than Fourier or polynomial trend analysis, which were not particularly successful in separating topographic components related to different stream orders. Fourier analysis may be more suitable for certain specific landforms, such as those of volcanic or meteoritic origin (Johnson and Vand 1967), or structural domes.

One disadvantage of these sophisticated two-dimensional techniques is that, far from providing a single measure of surface wavelength, they produce a matrix of power at various wavelengths and directions; they replace a map with a map. One-dimensional spectral techniques are therefore of continuing value, for dealing with either the average of the two-dimensional spectrum over all orientations, or the spectrum of profiles across or along the 'grain' of the surface. Carr and van Lopik (1962) reviewed their use for microrelief, arising from work on ocean waves (Pierson and Marks 1952) and runway roughness (Houbolt 1961). Stone and Dugundji (1965) used Fourier analysis with wavelengths of 19·5, 9·8, 4·9, 2·4 and 1·2 m (64, 32, 16, 8 and 4 ft) to derive roughness parameters from radial profiles of microrelief. Jaeger and Schuring (1966) analysed the spectrum of a 150 m random walk on the moon, and Rozema (1969) analysed profiles from four lunar sites, three earth sites, and three artificial 'proving grounds'. He used sample points spaced at 0·611 m and identical techniques throughout, to permit comparisons.

Since power tends to decline exponentially with frequency (which is the reciprocal of wavelength), it is customary to plot the logarithm of 'power' (m^2/cycle/m) against frequency (cycles/m) or its logarithm. The general height of the spectrum ('power spectral density function') measures overall surface roughness. Similar information might be obtained more directly, for randomly varying topography, by the characteristic constant of the exponential decay of autocorrelation with distance (Hayre and Moore 1961).

Rozema showed that, for wavelengths between 1 m and 100 m, the lunar profiles are less rough than the Perth Amboy badlands, the Bonito lava flow, and the Suffield crater. Though all log (power); log (frequency) plots approximate to an exponential decline with a gradient of −2, there are differences in shape between the spectra. The lava flow is very rough throughout, while the badlands are relatively rougher for wavelengths between 3 and 11 m, and the crater is relatively rough at wavelengths greater than 11 m.

None of these microrelief profiles shows evidence of strong periodicity, though one might expect microrelief to be more regularly periodic than macrorelief. Hence it is surprising that spectral techniques, which are especially suited to irregular surfaces, have not been applied more often to macrorelief. On the other hand, Bryson and Dutton (1967) showed definite periodicities in profiles oriented directly across a drumlin field and a series of moraines, and commented on the significance of different exponents for the decline in power (variance). This falls into the domain of specific geomorphometry, where Nordin (1971) has recently applied one-dimensional spectral analysis and other techniques to flume and river beds with ripples and dunes.

Rozema (1969) reviewed the computational problems of spectral analysis. The sampling interval (spacing between points in the series or matrix) must be less than half the shortest wavelength present, i.e. as a necessary but not a sufficient condition the 'lag-one' auto-correlation (between adjacent values) should be high. Green (1967) suggested that this is not the case in the profiles of Stone and Dugundji. When shorter wavelengths are present, they should be smoothed out by an appropriate filter prior to analysis, to prevent aliasing ('folding over' of the shorter wavelengths to affect power around their harmonics within the range of wavelengths analysed). In an appendix, Rozema (1968) showed that the exponentially weighted moving average, which he used for detrending, drastically reduced power at all wavelengths for some profiles but not for others. It appears that subtraction of linear trend would be a safer way of producing stationarity.

Another problem is the possible variability of a spectrum within the profile or area examined. In practice, it is unlikely that even spectral analysis will reveal many of the 'wavelengths' observable on contour maps. This is because (i) though linear, valleys often curve; (ii) valley spacing varies, almost randomly; (iii) there are confluences down-valley, and so ridge height and spacing necessarily vary.

Though they may not solve the 'horizontal dimension' problem, spectral and autocorrelation techniques are extremely important for general geomorphometry, in the analysis of surface roughness. To some extent, they are complementary to the 'altitude derivatives and moments' approach proposed here, since they emphasise the horizontal while the latter emphasises the vertical. In the final analysis, both probably produce similar information, since a 'model' terrain could be reconstituted on the basis of either. They differ in the information which they make explicit, but both can conveniently be calculated from an altitude matrix.

Slope; gradient and aspect

Slope is perhaps the most important aspect of surface form, since surfaces are composed completely of slopes, and slope angles control the gravitational force available for geomorphic work (Strahler 1956).

Slope at a point is defined in terms of a plane tangential to the surface at that point. This plane is characterised by its gradient (maximum inclination to the horizontal, in degrees), and its aspect (direction of a line perpendicular to the contours, measured in degrees clockwise from north, away from the hillside). The tangent of gradient is the first derivative of altitude, i.e. the rate of change of altitude with distance. However, slope frequency distributions cannot be derived reliably from altitude frequency distributions; slope must be obtained separately, from either contours or altitude matrices.

Aspect and gradient are the two components of three-dimensional slope direction. This can be studied integrally, as when poles to the plane of slope are plotted (Chapman 1952), but because gravitational forces are related only to gradient, it is dynamically useful for geomorphologists to separate gradient and aspect.

The measurement of slope involves considerable sampling problems. Gregory and Brown (1966) claim to have obtained a 'complete' sample of slope, by dividing the surface into facets of similar slope, and measuring slope at one or two subjectively chosen points within each facet. However, the delineation of facets in a continuous surface must be rather arbitrary; measurements within the same facet differ by as much as 2° in gradient, and even more in aspect. Taking the area of facets into account may slightly increase the accuracy of estimates of mean gradient, but this method prevents the use of statistical tests. The selection of 'typical' points at which to measure slope probably produces bias, and so the facet method cannot be considered an adequate substitute for a proper sampling design.

Slope frequency, like altitude frequency, is usually expressed in terms of map area, i.e. horizontally projected area. It is more realistic to use surface area, and the conventional gradient frequency distribution itself contains the information for converting to this basis (surface frequency = map frequency/cosine (gradient)). However, it is easier to sample on the basis of equal probability per unit map area.

Hence the sampling of slope may be tackled as a straightforward example of areal sampling, though the measurement of slope from the ensuing systematically or randomly located points is not simple. Starting from a contour map, it is necessary to measure contour spacing, and this is not easily automated. One method is to digitise points on the contours above and below each sample point. Since 'eye-balling' the direction of slope can be quite inaccurate, it is best to take two points to represent each contour, so that the computer program can construct perpendiculars from the sampling point to each contour, and calculate gradient and aspect. The need to digitise five points per slope sample reduces the amount of labour saved, though the result is more accurate than from purely manual methods.

There are several ways of achieving larger slope samples per unit effort. Starting from an altitude matrix of points which are closely spaced relative to the spacing of lines of inflexion, slope may be estimated either (i) for triangles formed by three adjacent points (Hobson 1967); (ii) for best-fit planes through four adjacent points forming a square; or (iii) for each point by taking the differences between neighbouring points north, south, east and west (Tobler 1964, and 1969, 65). Planes formed by these arbitrary facets give both aspect and gradient. Such methods are useful in producing a map of slope based on equal areal units. If a coarser altitude matrix is used in the same way, a crude generalisation of slope is achieved – gradients are reduced by the filtering out of short-wavelength components. Each method produces a systematic sample of surface slope, smoothed according to the resolution of the matrix. While (iii), a finite difference method, conveniently provides a slope measurement for each point at which altitude has been defined, it causes greater smoothing because slope is measured across two intervals of the grid instead of one.

Any matrix or contour representation of the geomorphic surface depends upon the surface being relatively smooth between grid points or contours. Slope is ideally an instantaneous point value, but it must be measured over a small area. Though some field techniques might circumvent this problem by measuring slope tangentially at a point, it would remain necessary to decide what size of feature (e.g. boulders, earth mounds) to ignore (Gerrard and Robinson 1971). An altitude matrix of fine mesh is as good a way as any of measuring slope at the appropriate degree of generalisation.

A method developed by Monmonier, Pfaltz and Rosenfeld (1966) processes digitised contours in matrix fashion, searching from sample points for the nearest contours in four cardinal directions. Some inaccuracy is involved in the low resolution of matrix storage, and the arbitrary distance at which a search is terminated. Though a fair approximation to mean gradient is produced, the distributions of individual gradient and aspect measurements would be much less accurate. Matrix storage of a small set of digitised contours necessitated use of magnetic tape, even for a 360×260 matrix; storage to the 0·1 mm resolution of. the d-mac table could produce a $10{,}000 \times 10{,}000$ matrix. Either resolution, size of area, or computing speed must be sacrificed. For fitting facets in this way, an altitude matrix is probably a better starting point.

Alternatively, starting from digitised contours, it is possible to search from a point on one contour to the nearest point on an adjacent contour. A program to perform this was described by Piper and Evans (1967). This samples all contours at equally spaced points, searching for the nearest points on the contours above and below. It is not easy to overcome the effects of contour divergence (fig. 2.5). In such cases, a straight line can only be normal to one contour; in this program, it is the contour searched rather than the contour sampled. It is hoped that differences between sampling upward and sampling downward will cancel out; but there will inevitably be an increase in dispersion,

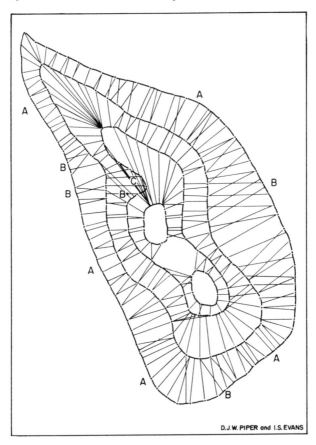

Fig. 2.5 Inter-contour slope lines produced by the program DJWP1045/SA2/SAAD6, written by D. J. W. Piper with I. S. Evans. The slope lines calculated by the program were drawn on a computer-controlled Calcomp plotter. Dots indicate their starting points, the sampled points for which nearest points on the adjacent contour were established. The diagram has been redrafted, with contours added. Where contours are parallel (A), results are satisfactory; but where they are divergent (B), the slope lines resulting from 'upward' searches cross those from 'downward' searches. A contour concavity not accompanied by a similar feature on the adjacent contour is crossed by 'slope lines' (C) which are quite unacceptable. To eliminate this would involve complicated additions to an already long program.

especially of slope aspect. The method is suitable for closely spaced contours. Sampling is proportional in intensity to contour density, and it is necessary to correct for this by dividing slope frequency by the tangent of slope angle.

A quite different automated system, based on triangular facets between labelled points on divides, drainage lines and breaks in slope, has been developed by Hormann (1968, 1969). This is particularly useful for its production of graphs on a line printer, with related statistics printed in the margins. Altimetric, hypsometric and clinographic curves are plotted, average gradient is graphed against aspect for selected classes of gradient and altitude, and various analyses of valley length, depth, gradient and direction are produced. Hormann has also developed a program for analysis of relief and gradient from altitude matrices, but in these papers he emphasises the triangular network because it does not require such a high density of points. On the other hand it involves more labelling, and contains a subjective element which hinders reproducibility: its sampling base is difficult to evaluate and the unequal area of triangular facets causes problems of weighting.

The relationship between slope and altitude is clear; slope is the first derivative of altitude. Hence it is a fundamental property of the geomorphic surface; its definition involves no arbitrary assumption. The frequency distributions of aspect and gradient have an important place in the geomorphometric characterisation of a region. The distribution of aspect reveals the presence of lineation or regional slope (Eyles 1965; Newell, 1970). Because of the equivalence of 0° and 360°, analysis of aspect involves use of circular or periodic statistics (Curray 1956; Watson 1966).

Gradient frequency distributions have been used often by modern geomorphologists, and they permit some inferences about process (Strahler 1950; Tricart and Muslin 1951). Earlier workers estimated mean gradient from contour density (Wentworth 1930), but without information on the statistical or geographic distributions the mean gradient is only suggestive. Evans (1963) modified Wentworth's method to produce estimates of gradient averaged over a kilometre square, so that large-scale maps and frequency distributions could be produced. This method is useful for reconnaissance and for regional studies, but the smoothing involved reduces the dispersion of frequency distributions. The magnitude of this reduction is variable and unknown, since it depends upon the importance of short-wavelength variations in altitude. Compared with a point sample of one slope per square kilometre from the same map, this method improves the quality of the map of gradient and gives a more precise estimate of mean gradient, but biases estimates of standard deviation and higher moments of gradient, and gives no information on aspect. Where automation permits finer sampling of slope, the matrix-based method is clearly superior to the contour density method.

Slope has been expressed in other ways; for example, the ratio of surface area to projected area (Hobson 1967). This ratio is completely determined by the frequency distribution of gradient; it is the mean value of the reciprocal of the cosine of gradient, but is not related to mean gradient as simply as Monmonier *et al.* (1966) suggested. Hobson developed various parameters for surface roughness, and his program for triangular facets in an altitude matrix has been used by Turner and Miles (1968). It provides a measure of surface roughness as the three-dimensional dispersion of vectors normal to the surface facets. This relates principally to gradient and its variability, and though Turner and Miles show that its linear dependence on aspect variability is insignificant, aspect will affect vector dispersion, producing undesirably hybrid results; it is better to separate variability in gradient from variability in aspect.

It is concluded (i) that slope should be measured at a point or over as small an area as possible; (ii) grid-based automated methods provide large samples of slopes without raising serious problems of computer programming or, except for grid mesh size selection, of sampling; (iii) gradient should be expressed in degrees and its frequency distribution should be summarised by mean, standard deviation, skewness and kurtosis. Normality (Strahler 1950) can be assumed only for very restricted conditions, but pronounced bimodality is not as widespread as Calef and Newcomb (1953) imply.

Local convexity

Local convexity is the rate of change of slope, the first derivative of slope or the second derivative of altitude. Just as slope comprises gradient and aspect, convexity downslope can be separated from convexity across slope. To consider downslope convexity in isolation involves the loss of important information (Sparks 1960, 67–8). The convexity, concavity or straightness of slope profiles has long been an object of contention in geomorphology; this cannot be resolved without the analysis of large, representative samples. Downslope (vertical) convexity can be measured from contours in the same way as gradient, but taking the intersection of a slope line with four contours instead of two. Convexity at the sample point is measured by fitting a parabola to points where the slope line crosses contours, and is expressed in degrees/m. This method has the disadvantage of extending over much greater lengths of slope, and being smoothed accordingly, for gentler slopes.

Cross-slope (horizontal) convexity is expressed by contour curvature (degrees/m), which can be measured very efficiently from digitised contours by interpolating short, straight segments of equal length along them. However, it would be inconvenient to have separate bases for the measurement of vertical and horizontal convexity; they would be measured at different locations and could not be interrelated individually. Both of these problems, as well as the sampling problem, could be solved by use of an altitude matrix. Quadratic trend surfaces should be fitted to the neighbourhood (3×3 points) of each sample point. Horizontal and vertical convexity (and aspect and gradient) could be obtained by differentiating the equation for the surface. It would then be particularly easy to interrelate all four derivatives, and altitude itself, for individual points within the matrix.

Regional convexity (dissection, aeration)

Having described how the five basic attributes of a surface may be measured at any point, I will now consider their relevance to a more complex concept. Geomorphometrists have long sought an adequate measure of the 'dissection' or 'aeration' of a landscape, i.e. the extent to which it has been opened up, especially by erosion (Clarke 1966). De Martonne (1940) used the ratio of mean altitude to maximum altitude (= index of articulation, coefficient of massiveness) for natural regions of France; lower ratios imply greater aeration. His formula ignores lowest altitude, so that identical topography is given a lower coefficient if closer to sea level. This error is significant for natural regions, and even more important for smaller areas, such as those used by Loup (1963) in his comparison of Valais and the Pyrenees.

If lowest altitude is subtracted from both mean and maximum (Péguy 1942, 462), de Martonne's formula becomes (mean relative height)/(maximum relative height), where 'relative height' is altitude above the lowest point. Merlin (1965) calls this the relative coefficient of massiveness. This is the same[1] (fig. 2.6) as the area under the dimensionless hypsometric curve (cumulative frequency curve of altitude), which was first developed by Imamura (1937). Proposed by Strahler (1952) as the 'hypsometric integral', this expression is now the most popular coefficient of dissection.

The widespread use of the hypsometric integral, however, should not encourage us to accept it as necessarily the most appropriate expression of convexity or dissection. Like de Martonne's coefficients, it is defined in relation to the altitude frequency distribution, and it is based on the extremes of this distribution as well as on its average. As was shown

[1] Pike and Wilson (1971) have now given two mathematical proofs of the identity of hypsometric integral and the elevation/relief ratio of Wood and Snell (1960), which is another incarnation of Péguy's coefficient.

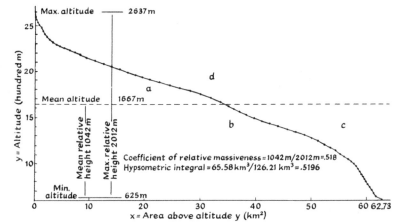

Fig. 2.6 Hypsometric curve of the Keary–Nosebag area, Bridge R. District British Columbia. Areas on this plot of area against altitude represent volumes under the earth's surface; hence the area of the enclosing rectangle, $a + b + c + d$, is the volume (total area under consideration) × (range in altitude within that area). The hypsometric integral is the volume above the minimum altitude and beneath the earth's surface, (represented by the area $a + b$ under the hypsometric curve) expressed as a proportion of the total possible volume (represented by the area $a + b + c + d$) which would be comprised if the whole surface lay at the maximum altitude. This proportion equals the 'coefficient of relative massiveness', the ratio of mean relative height to maximum relative height, i.e. $(b + c)/(a + b + c + d)$; for since mean relative height is the mean height of the hypsometric curve above its base (Péguy 1948), area a must equal area c.

by Péguy (1948, 49), an index based on maximum altitude is very sensitive to small changes in the area measured.

The hypsometric curve for a straight slope of uniform width is straight, and the integral is 0·5; whereas the curve for all the world's land masses is highly concave. De Martonne's comparisons ignore the fact that the larger the region, the more likely it is to approach the concavity of the world hypsometric curve. The boundary conditions are also important; obviously the fact that the whole outline of Corsica is at sea level is likely to produce a greater concentration of area at minimal altitudes than in the Massif Central, the boundary of which varies approximately from 100 m to 500 m in altitude. Use of a single closed contour as the boundary would provide a better basis for comparison of mountain ranges (and islands). But drainage basins are surrounded by higher ground to a variable degree, and could not be included in the comparison because their integrals are on the whole higher.

The main limitation on the use of the hypsometric integral and index of articulation for comparison of regions, however, is heterogeneity within the area measured. This, for example, makes Loup's (1963) comparison of Valais and the Pyrenees extremely dubious. Valais is an homogeneous area with high summits along most of its perimeter; its heavily glaciated valleys have many steps, and these too contribute to a high integral (0·44: index d'articulation 49·5%). The Pyrenees have lower indices (28 to 39%), which Taillefer (1948) nevertheless interprets as massive in comparison to that of the whole French Alps (23%; lack of data on minimum altitudes prevents conversion to hypsometric integrals, but minima are low and so the effect should be small).

Loup disputes this, comparing his figure for Valais and others for French alpine massifs with Taillefer's for four transverse divisions of the Pyrenees, which are comparable in area. The Pyreneean divisions are, however, structurally heterogeneous. They decline northward towards a plain, and have high summits along one border only. Here it is more appropriate to use homogeneous smaller divisions, despite the difference in area, and Taillefer's figures for high Pyreneean massifs show them to be as massive as the alpine massifs and Valais. This emphasises the sensitivity of integrals to area definitions. It is evident that ranges and drainage basins should not be compared, but these examples demonstrate the importance of boundary conditions even for range–range or basin–basin comparisons.

Merlin (1965, 1966) discusses the separation of structural from erosional 'aeration' (dissection) by means of a 'perfect coefficient of massiveness' relating to an hypothetical initial surface (of structural origin, or a peneplain). Its complement is the coefficient of structural aeration, while the difference between the 'perfect' and the actual coefficient of massiveness is the coefficient of erosive aeration. This reduces some of the objections to comparison, but the definition of the initial stage poses difficulty and introduces subjectivity.

Two more complex indices of aeration or dissection have been proposed by Péguy (1942) and de Smet (1951). The former finds the ratio (length of longest contour)/(mean contour length) × tangent (mean gradient) to be useful in comparing drainage basins of tributaries of the upper Durance. Yet while interesting, the contour length ratio could give similar results (as de Smet indicates) for highly concave as for highly convex surfaces; both are likely to have many short and few long contours. This does not appear from the basins studied by Péguy since they are all concave, and his index does reflect the degree of concavity.

Division by mean gradient is intended to deal with the proposition that with steeper slopes, an otherwise similar surface is more compact. Péguy does not attempt to justify his implicit assumption that such a relationship would be linear; this is surely unlikely. His concept of compactness differs from that of de Martonne and of those who use the hypsometric integral (regional convexity) as a measure of compactness. I prefer not to mix the concept of regional convexity with that of mean gradient.

By contrast, de Smet (1951) uses mean gradient (expressed as a percentage, i.e. 100 × tangent (angle)) in the *numerator* of his index of dissection. This index is admittedly not comparable with the others, since it is intended to pass through a maximum when all trace of the 'old surface' is removed. Both mean gradient and the relative increase in surface area compared with map area are presumed to pass through such a maximum, and hence they are multiplied together to give the index. Because a surface encompassing a greater volume is considered less dissected than one otherwise similar, volume becomes (by the same over-simplified reasoning as that of Péguy with respect to slope) the denominator of the index. Incidentally, de Smet's results (p. 132) reveal that his index is not as given, but;

$$\frac{100(S_r - S_p)100 \tan (\text{mean gradient})}{S_p \times \text{volume}}$$

where S_r = real surface area and S_p = projected (map) surface area. The dimension of this index is (length)$^{-3}$, and so it cannot be used to compare areas of different sizes. Moreover, relative increase in surface area depends upon the mean secant of gradient, and so the index is almost (mean gradient)^2/volume. It does not appear to be of any value.

The popularity of the hypsometric integral stems from its relation to the proportionate 'volume to be consumed', i.e. the remaining volume above the lowest point ('base level') of the area considered. Though the tangibility of this concept may be pedagogically useful, both highest and lowest points are transient, and so this relation does not make the hypsometric integral any more meaningful. Whether or not a non-local base level such as sea level is used in derivation of a volumetric integral, the base level itself will change considerably before erosion proceeds far. The decline of the hypsometric integral has been used as a measure of 'stage' of dissection (Strahler 1957), but it does not measure the variation through time of the 'unconsumed volume', because both highest and lowest points change with time. Eyles (1969) has shown the complexity of the resulting relationship between hypsometric integral and range in altitude; in Malaya the latter appears to be a better measure of 'stage'.

Hence there is no special case for the use of extreme values, which can be misleading for either large or fairly small areas. It seems appropriate to turn once more to a parametric statistic derived from the altitude frequency distribution. The skewness of this distribution contains all the information sought from the hypsometric integral, without resorting to extreme values. Tanner (1959, 1960) used skewness (with signs opposite to convention) and kurtosis to express the shape of hypsometric curves. Skewness is the expected cubed deviation from the mean. From sample data, it is best estimated by Fisher's (1954) statistic:

$$k_3 = \frac{\Sigma (y - \bar{y})^3 N}{(N - 1)(N - 2)}$$

Since this is in units of the original variate, cubed, it is standardised by dividing by the cube of the standard deviation:

$$g_1 = \frac{k_3}{(k_2)^{3/2}} = k_3 \left(\frac{\Sigma (y - \bar{y})^2}{(N - 1)} \right)^{-3/2}$$

where k_2 is the variance, k_3 and therefore g_1 are zero for symmetrical distributions. Estimates of g_1 from different samples are distributed normally with variance

$$V(g_1) = \frac{6N(N - 1)}{(N - 2)(N + 1)(N + 3)}$$

A g_1 of 0 is equivalent to a probable hypsometric integral of 0·5, i.e. a

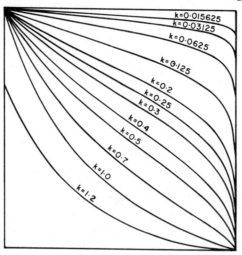

Fig. 2.7 Lubkin–Strahler model hypsometric curves, defined by the equation $y = \left(\frac{(d - x)a}{x(d - a)} \right)^k$, for the family $r = 0·3$, giving $a = 0·4286$, $d = 1 + a$ $= 1·4286$. Curves for 12 values of k are shown, but that for $k = 1·2$ is not properly computed; in fact, curves for all values of k greater than unity are unmanageable. k controls curve height and therefore hypsometric integral, while r controls curve sinuosity. Curve families have been calculated and plotted for $r = 0·01, 0·05, 0·1, 0·15, 0·25, 0·3, 0·5, 0·7$ and $0·9$.

symmetrical frequency distribution of altitude. A lower integral relates to positive skewness ('tail' towards higher values). However, it is not otherwise known which values of g_1 correlate with which hypsometric integrals.

A program was written to calibrate skewness and hypsometric integrals against each other, using curves (fig. 2.7) generated from a model

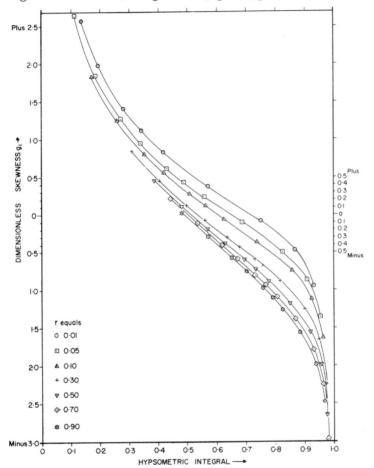

Fig. 2.8 Relationships between altitude skewness and hypsometric integral for different families of model hypsometric curves. Though non-linear, the relationship between skewness and hypsometric integral is perfectly regular within each family of altitude frequency distributions. The 'families' are characterised by r, a sinuosity parameter. The lower the value of r, the more sinuous is the hypsometric curve, i.e. the higher is the peak frequency of altitude for a given hypsometric integral. For low values of r, zero skewness corresponds to hypsometric integrals greater than 0·5.

function developed by Lubkin for Strahler (1952). The use of this model, which provides a fit for most hypsometric curves, smooths out small perturbations and avoids the problem of aberrant extreme values which could have an excessive influence on the hypsometric integral.

Figure 2.8 shows that, given this smoothing, the correlation with skewness is good. It also shows that the lower r (the more sinuous the curve), the higher the skewness for an integral of 0·5, and the higher the integral for 0 skewness. This effect may be produced by curves which are generally symmetric, but about altitudes above 0·5. As Strahler suggests, the value of r for an empirical curve may be estimated by visual comparison with model curves. It is then possible to estimate skewness from the hypsometric integral and vice versa.

It is concluded that the hypsometric integral, and all the variant measures of 'dissection' or 'aeration' discussed above, may usefully be replaced by the skewness of the altitude frequency distribution. This does not avoid the effects of boundary conditions or of internal heterogeneity, which influence any parameter based on the frequency distribution. However, a statistic based on all values is less sensitive to small changes in area boundaries. If skewness is matrix-based, the flexibility of the automated system permits its calculation for over-lapping and different-sized areas, so that the effect of changed areal definition is readily perceived.

Relationships between variables

The procedures of measurement described above provide, at any point on a surface, five measures of important geometric properties. These are (i) altitude, z; (ii) gradient, z_v'; (iii) aspect, z_h'; (iv) vertical convexity, z_v'' (v) horizontal convexity, z_h''. Any area is described by their frequency distributions which may be summarised by statistical moments. Some of these moments are appropriate parameters for other established geomorphometric concepts. Specifically, standard deviation of altitude is a measure of vertical dimension (relief), and skewness of altitude is an inverse measure of regional convexity (dissection, aeration). Standard deviation of gradient is another expression of degree of curvature, but does not distinguish convexity from concavity. These moments, then, provide areal parameters of surface geometry.

There are a number of inbuilt relationships between the moments of vertical derivatives; these are shown in table 2.1.

This model suggests some degree of correspondence between the nth moment of one of these distributions, the $(n-1)$th moment of its first derivative, and the $(n-2)$th moment of its second derivative. This generalisation applies only to the moments and derivatives shown in table 2.1; no extension to higher moments or derivatives is implied.

Table 2.1   *Inbuilt relationships between areal parameters; moments of vertical derivatives*

Columns represent distributions of derivatives, and rows represent concepts. The first four moments are represented by $\bar{Z}$, S, g_1 and g_2. Higher derivatives tend to have more purely local significance, and hence they define rougher surfaces. Kurtosis, $g_2 = k_4/k_2{}^2$, where

$$k_4 = \frac{N(N+1)\sum(y-\bar{y})^4}{(N-1)(N-2)(N-3)} - \frac{3\sum^2(y-\bar{y})^2}{(N-2)(N-3)}$$

	Z *Altitude*	Z'_v *Gradient*	Z''_v *Vertical curvature*
(Altitude)	$\bar{Z}$ $+\updownarrow$		
(Gradient)	$S(z)$ $\xrightarrow{\ +\ }$	$\bar{Z}'_v$ $\updownarrow +$	
(Curvature)	$\lvert g_1(z)\rvert$ $\xrightarrow{\ +\ }$	$S(z')$ $+$	$\xleftarrow{\ +\ }$ $\lvert Z''_v\rvert$ $\updownarrow$
(Heterogeneity)	$-g_2(z)$	$g_1(z')$ $\xrightarrow{\ +\ }$	$S(z'')$

Regional ←⟶ Local

Smooth ←⟶ Rough

Vertical and horizontal second derivatives take positive and negative values, and for a trendless surface their means would be zero (see below, p. 67). Hence their magnitude is best described by mean modulus, which is closely related to standard deviation.

Apart from relationships between mean and standard deviation of the same distribution, five specific interrelationships are hypothesised, for areas of equal extent.

(i) Increasing dispersion in altitude (relief) is likely to increase mean gradient, by increasing slope height. Any increase in slope height, for constant slope length, necessarily increases gradient; hence the high correlations between relief and gradient noted by Wood (1967). (ii) Increasingly positive or negative skewness in altitude (regional concavity or convexity) is likely to be associated with increased dispersion in gradient, and with (iii) increasingly positive or negative downslope convexity. This is because a landscape which is more concave or convex, either locally or regionally, necessarily implies contrasting gradients. Since on the other hand gradient or its dispersion can change without influencing the altitude or concavity distributions, dispersion in gradient is regarded as the 'dependent' variable of these three.

Increasing skewness of gradient is likely to be associated (iv) with variability in convexity; both these parameters may perhaps be labelled as representing surface 'heterogeneity'. (v) Kurtosis of altitude is a regional measure of homogeneity. Bimodality (negative kurtosis, heterogeneity) is likely to produce variation in gradient with altitude (cf. two gently sloping surfaces separated by a scarp), increasing $S(z')$ and, by producing more convexity and concavity, increasing both $S(z'')$ and $|z'|$. These relationships between parameters are unlikely to be the only ones, but they are those which may be expected *a priori* from the definitions adopted.

This model also comprises the trend from characteristics of regional significance on the left, to increasingly local characteristics on the right. For example, regional convexity (cf the hypsometric integral) is measured by the third moment of the original distribution, while local convexity is measured by the first moment of the second derivative. It is hypothesised that the degree of autocorrelation diminishes as successive derivatives are taken. The rougher surfaces formed by higher derivatives require increasingly fine-meshed altitude matrices for adequate definition. The matrix of any of these derivatives can be smoothed to produce a more autocorrelated surface, and the statistical moments of the resulting frequency distributions may be of interest, though the degree of smoothing must be arbitrary.

Geomorphologists have long been interested in the relationship of gradient to altitude (Kaitanen 1969) and to aspect (Gregory and Brown 1966). Too often, however, they have expressed these in terms of percentages in arbitrary classes (an unnecessary descent from ratio to nominal scales of measurement), or have dealt with mean values and ignored dispersion. Empirical studies may be based either on point measures, or on areal means and other moments. There is some difficulty in synthesising the available evidence, because of differences in scale of study and in definition of variables.

Peltier (1954) found that, for 70 maps, $\bar{Z}' = 0.079R - 0.66$, where $\bar{Z}'$ = mean gradient (in degrees) and R = average range in altitude (m) per square of 2.6 km². $\bar{Z}'$ also increases with drainage density, but at a rate varying with climate and lithology. In 1955, Peltier added the relationship;

$$\tan \bar{Z}' = \tan 60° \, (2\bar{h}/\bar{w})$$

where h = hill height and w = distance between ridge crests, at a random orientation with respect to the axis of the valley. He suggested that other landform parameters could be estimated from such relationships, but unfortunately he did not give the strengths of the relationships. Correlations between relief and gradient have however been given by Wood (fig. 2.3) and Gălăbov (above), as well as by Tobler and by King (below).

Hobson (1967) correlated parameters for 45 contiguous 1·58 × 1·58 km squares, each based on altitudes at 16 points. He found strong linear correlations (>0·7) between standard deviation of gradient, vector strength (an inverse measure of slope dispersion in both gradient and aspect), proportionate surface area, and various measures of dispersion in altitude (including standard deviation); all of these measure surface roughness. Mean gradient ($0·25 < r < 0·35$) and mean altitude ($0·2 < r < 0·55$) were much more weakly correlated with this set, but it should be noted that Hobson did not test for possible curvilinearity in the relationships. For example, both vector strength and vector dispersion are completely specified by sample size and length of resultant vector. Hence their (curvilinear) correlation, apart from several presumed data errors, is 1·0, and their mapped patterns are almost identical. But the linear correlation coefficient used by Hobson is 0·1235.

A study by Tobler (1969) is particularly relevant here because he computed many different point measures, including derivatives, from an altitude matrix. The 635 m mesh (interval between grid points) of his 60 × 60 matrix of Ogemaw County, Michigan, U.S.A., is coarse, since the 1/250,000 map (contour interval 15 m) on which it is based shows variation over shorter wavelengths. Hence the matrix would not be suitable for spectral analysis, and Tobler's first derivative (gradient) values are local trends rather than estimates of true point values. However, this is a deficiency of data rather than method.

Table 2.2 shows how the variables are derived from each other, and fig. 2.9, based on Tobler's correlation matrix for 'point' values, portrays all correlations greater than 0·30. Except for contrast enhancement, the size of the local neighbourhood used is 3 × 3 matrix positions, i.e. 1905 × 1905 m = 3·629 km².

The set of six interrelated variables is composed of three pairs, each of which contains an unsmoothed variable and its smoothed equivalent. However, all the smoothed variables are related to each other more closely than to their corresponding unsmoothed variables. Local range and first derivative (gradient) are closely related ($r = 0·83$) to each other, but second derivative correlates poorly with the other two unsmoothed variables. In general, the unsmoothed variables correlate more strongly with the smoothed than with the unsmoothed versions of other variables. Table 2.3 summarises these distinctions.

Hence relations with smoothed variables, which except for variance are measured effectively over larger areas, are stronger than relations with other variables at fine scale. The main part of this anomaly might be explained by the size of the area over which second derivative is measured, even before smoothing; this is another disadvantage of the coarseness of the altitude matrix. Alternatively, the weak relationships between unsmoothed variables may be due to random local 'noise'

Table 2.2   *Derivation of Tobler's variables*
Number in brackets indicates area over which the variable is measured.

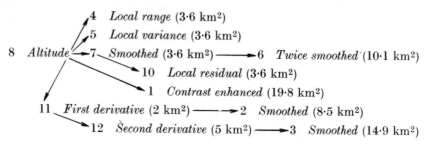

4   *Local range* = maximum − minimum (feet), within 3 × 3 unit neighbourhood

5   *Local variance* = mean square variance (feet²), over 3 × 3 unit neighbourhood

7   *Smoothed* = low pass filter = $\frac{1}{16} Z_{ij}$ $\begin{Bmatrix} 1 & 2 & 1 \\ 2 & 4 & 2 \\ 1 & 2 & 1 \end{Bmatrix}$ $\begin{matrix} j+1 \\ j \\ j-1 \end{matrix}$ Feet.
$\begin{matrix} i-1 & i & i+1 \end{matrix}$

6   *Twice smoothed* = low low pass filter = repeat of operation 7. Feet.

10   *Local residual* = high pass filter = $Z_{ij} + Z_{ij}$ − (low pass filter). Feet.

1   *Contrast enhanced* = 'local dodging' = constant + (Z_{ij} − local minimum) (regional range/local range), where local neighbourhood
= 7 × 7 units. Feet.

11   *First derivative* = gradient =
$(1/2d) \sqrt{(Z_{i,\,j+1} - Z_{i,\,j-1})^2 + (Z_{i-1,\,j} - Z_{i+1,\,j})^2}$
where d = mesh of matrix (635 m here). Feet/mile.

12   *Second derivative* = convexity = rate of change of gradient = repeat of
operation 11. Feet/mile².

3   *Second derivative smoothed* = operation of low pass filter on variable 12.
Feet/mile².

---

which is of no value in interrelating these properties. This is supported
by the lack of relationships of the two highly local variables (contrast
enhancement and local residual), except to each other.

For the scale studied by Tobler, it is evident that smoothed variables
are the more useful measures of the three properties concerned. A
further conclusion is that the second derivative of altitude is less stable
(more improved by smoothing) than is the first, and it is less interrelated with other variables. All of these results support the model of
general geomorphometry proposed above.

Further valuable information on relations between morphometric

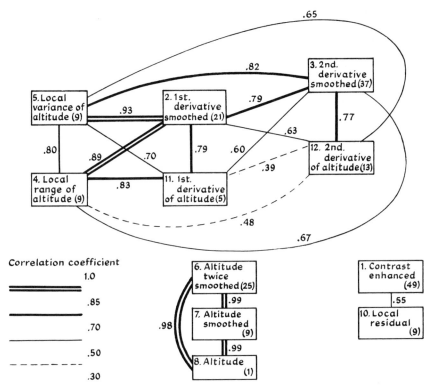

Fig. 2.9 Structure of site correlations between different transformations of an altitude matrix, based on a sample of 180 points drawn from a 60 × 60 matrix of Ogemaw County, U.S.A. (Tobler 1969). Altitudes were interpolated at points 2·54 mm apart on a 1/250,000 map with a 15 m (50 ft) contour interval, giving a spacing of 635 m between grid points in both *x* and *y*. Correlations not portrayed here range between −0·17 and +0·05. The eleven variables fall into three isolated groups.

Table 2.3   *Correlations between the six variables in the 'altitude derivatives and relief' group*

	correlations						average
Smoothed v. smoothed		0·93	0·82	0·79			0·85
Smoothed v. corresponding unsmoothed		0·80	0·79	0·77			0·79
Smoothed v. non-corresponding unsmoothed	0·89	0·70	0·67	0·65	0·63	0·60	0·69
Unsmoothed v. unsmoothed		0·83	0·48	0·39			0·57

variables was provided by C. A. M. King (1968). The basis of her study was the 1/63,360 map, divided into 5 × 5 km units; since initial units were areal, derivatives were not used. However, altitude, relief, mean gradient and drainage density were among the variables, whose inter-correlation is shown in table 2.4. The sample of 120 units was equally divided between six regions of northern England, and areas transitional between regions were not studied.

King found that altitude, range in altitude and mean gradient were closely intercorrelated. The highest correlation was 0·91, between the two latter: this is very close to that found for areas of 25 km² by Wood (1967). Drainage density and number of stream sources correlated highly with each other and with altitude. A particularly interesting result is the extent to which correlations vary between regions. Taking the subset of four variables in Table 2.4 (b), intercorrelations are high within the Lake District and Solway Lowland, and low within the Alston and Askrigg blocks of the Pennines. The low correlations are related to low coefficients of variation of altitude and gradient in the Askrigg and Alston blocks.

Strahler (1950) suggested that mean gradient increases with relief and with drainage density (see also fig. 2.3). Though there is abundant evidence for the former, the latter relationship is much weaker than expected, perhaps because (due to glaciation and human interference) drainage density is a poor measure of valley density in northern England, or because it varies over only a small part of its world-wide range. Mean gradient should increase also with summit density and magnitude. King's results confirm this for number of upland summits. The negative correlation between number of lowland summits and mean gradient is probably through there being more lowland summits where lowland is more extensive and mean gradient is therefore less; it would be interesting to replace summit number with summit density.

King's correlations are low probably because they involve parameters measured over sizeable areas, instead of at anything approaching a point; this is an important distinction. The varying correlations are themselves characteristics of the areas involved; King points out that where slopes are longer, mean gradient is correlated more highly with range in altitude.

It is important to remember that, for measurements which approach the zero of a ratio scale, standard deviation is rarely independent of mean: if negative values are impossible, zero mean must be accompanied by zero standard deviation. Hence this relationship should be studied along with other relationships between areal parameters. Analysis of data (fig. 2.10) collected for the study reported in Evans (1963) shows that for gradient, the relationship is not removed through the simple expedient of dividing standard deviation by mean to give

Table 2.4   *Correlations between morphometric variables for 5 × 5 km squares in northern England, based on a cluster sample of 120 squares located within six regional cores* (Based on C. A. M. King (1968))

a) Northern England

	1	2	3	4	5	7	8
1) Mean of maximum altitude per 1 km^2							
2) Range in altitude	·81						
3) Number of upland summits	·45	·63					
4) Number of lowland summits	−·45	−·35	−·35				
5) Tangent of mean gradient	·75	·91	·74	−·38			
7) Number of stream sources	·80	·44	·26	−·40	·43		
8) Drainage density	·69	·42	·22	−·42	·37	·85	
10) Bifurcation ratio first/second order	·32	·24	·22	−·11	·25	·33	·14

b) Within six regions

	1) Mean max. altitude						2) Range in altitude						5) Mean gradient					
	Askrigg	Alston	Cheviot	Lake District	Solway Lowland	Northumberland coastal	Askrigg	Alston	Cheviot	Lake District	Solway Lowland	Northumberland coastal	Askrigg	Alston	Cheviot	Lake District	Solway Lowland	Northumberland coastal
1) Mean max. altitude																		
2) Range in altitude	−·00	·02	·60	·77	·90	−·69												
5) Mean gradient	·03	·08	·46	·81	·89	·65	−·08	−·24	·44	·68	·88	·75						
8) Drainage density	·29	·54	·46	·33	·27	·33	−·11	·52	·20	·21	·29	·27	·18	·23	−·11	·17	·28	·31

coefficient of variation. Instead, standard deviation increases much more rapidly over low mean gradients. Variability should be characterised as high or low in relation to the best-fit line, and not in relation to the straight line representing a given coefficient of variation. Use of the latter would classify all low-gradient areas as highly variable, just as use of standard deviation would classify them all as little variable.

In conclusion, the available empirical evidence does not contradict the proposed model (table 2.1). Use of derivatives (at points) and moments (over areas) would help to clarify the distinction between

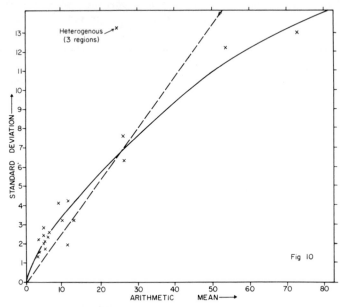

Fig. 2.10 Relationship between mean and standard deviation for estimates of mean gradient per 1 kilometre square. Each point is based on a contiguous sample of 100, 200 or 600 kilometre squares. The points define a curvilinear relationship, which is only coarsely approximated by the straight-line comparison (dashed line) on which coefficient of variation is based.

point correlations and areal correlations, which is blurred in Tobler's (1969) study. Much more investigation along these lines is required.

A trial analysis; the Keary–Nosebag area

The proposed system of derivatives and moments has not yet been implemented in full. However, it was desirable to check some of the proposals, especially where changes from previous practice were suggested. Hence a pilot study of altitude frequency distributions and their variation with scale was mounted. This permitted comparison of standard deviation with range, and of skewness with hypsometric integral, for areas between 475·2 m square and 7920 m square. Correlations between different parameters, based on areas of various sizes, were also obtained.

A suitable data set was already available, following a project which evaluated different sampling patterns. A 50 × 50 altitude matrix based on a square grid interval of 158·4 m had been created for an area around Nosebag Creek and lower Keary Creek, on the eastern slope of the Bendor Range, British Columbia. The westernmost column of point altitudes lies along 122° 30′ W, cutting through the middle of Keary Lake, and the southernmost row lies along 50° 45′ N, from an unnamed glacier to Nosebag Mountain. The area represented, 7·92 km square, is referred to hereafter as the 'Keary–Nosebag area' (figs. 2.11 to 2.15).

The topography included is varied, with cirques and arêtes in the south-west, fluvial topography east of Nosebag Creek, and the steep-sided trench of the Bridge River with a flat alluvial floor in the north-east. The hypsometric curve of the area is given in fig. 2.6. The integral shows a slight overall convexity, due largely to the areas of abnormally gentle slope high on either side of the trough of lower Keary Creek; in this respect the area is not typical of the Bendor Range.

The matrix was created from the south-west corner of a 1/31,680 dye-line pre-print of B.C. sheet 92 J/16 West. This had been compiled by Young in 1948, from air photography at the same scale, and it has a 30·48 m (100 ft) contour interval. The altitude of each grid point was recorded manually in contour interval units, to the nearest contour below. This method is faster than manual interpolation between con-tours, and involves little loss of information on the steep slopes which cover most of the map.

Metric graph paper was used, and the altitude of the south-west corner of each 5 mm square was recorded within the square. While the resulting 158·4 m horizontal interval produced some smoothing of the topography, comparison of fig. 2.12 with fig. 2.11 shows that no im-portant features were filtered out. A 100 m interval would be desirable (especially for measurement of slope), while a 50 m interval, combined with estimation of altitude to the nearest 10 ft (3 m), would certainly retain almost all the information present in the contours, and avoid any aliasing problem in spectral analysis.

In discussing U.S. Army Map Service altitude matrices, derived by interpolation from digitised contours, Connelly (1968, 107) recom-mended an horizontal interval of 122 m (0·02 in on the 1/250,000 maps used as the data source), and a vertical resolution of 10 ft (3 m). Taking into account the great relief of the Keary–Nosebag area, the altitude matrix used here does not depart excessively from this overall speci-fication. It is necessary to balance a geometrical increase in work and computation as the grid interval is decreased, against a diminishing increase in precision, the benefit of which is further diminished by any inaccuracy in the source data (the contour map). The use of matrices created in this way is only a stop-gap, pending the availability of matrices created directly from air photograph stereo models.

Geometric properties of the Keary–Nosebag 50 × 50 altitude matrix were studied by means of an autocode program (ISE 1174/15P5/SUBMAT), for *submat*rix analysis. For squares of various widths (P), which are odd multiples of the grid mesh (from 3 × to 49 × 158·4 m),

c

this program calculates eight parameters of altitude. Numbered according to their output streams, these are:

 ø. Estimated population standard deviation. (The $P \times P$ points in the altitude sub-matrix are regarded as a systematic sample of points in the related $(158 \cdot 4P)^2$ m² area.)

1. Hypsometric integral ((mean altitude − minimum altitude)/range in altitude).
2. Local convexity (altitude of central point − mean altitude of peripheral points).
4. Minimum altitude.
5. Range in altitude (maximum − minimum).
6. Skewness k_3 (the expected cubed deviation from the local mean altitude).
7. Dimensionless skewness, g_1 (see p. 46).

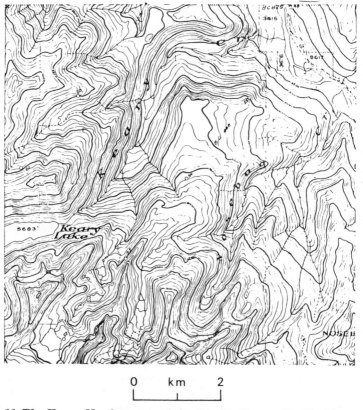

0   km   2

Fig. 2.11 The Keary–Nosebag area of the Bendor Range, Pacific Ranges, south-western British Columbia. 1/50,000 map with 30 m (100 ft) contour interval, reproduced by permission of the British Columbia Department of Lands, Forests and Water Resources.

8. Maximum altitude.
9. Local height (altitude of central point — local mean altitude).

Parameters (2) and (9) relate to central points; the others relate to the whole area over which they are calculated. Output is in the form of a 'numerical map' of the geographic variability of each parameter, at each scale (P). The necessary square grid of numbers is produced by using five line printer positions horizontally and three vertically for each square, but this limits map width to 24 squares (sub-matrices).

For parameters (ø) standard deviation, (2) local convexity, (5) range, and (6) g_1 only, class frequencies, mean, standard deviation and co-efficient of variation are produced. The scaling factors used for classes, for the map, and for the mean etc., usually differ, and are fixed for each parameter by the program. Note that in dealing with these frequency distributions, there is occasion to discuss the standard deviation and

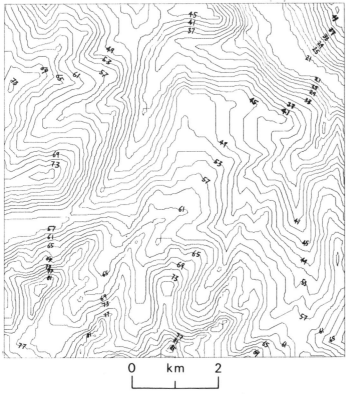

Fig. 2.12 Contour map of Keary–Nosebag altitude matrix. Contours (interval 60 m or 200 ft; labelled in hundreds of feet) interpolated by Titan library routines 23 and 24, called by program ISE1174/15P4/MATCON, and plotted on a Calcomp drum plotter in 15 minutes. The computer 'charging time' for production of this and fig. 2.13 was 36 seconds. Comparison with fig. 2.11 shows how little information has been lost by taking approximate altitudes at discrete points spaced 158·4 m apart.

skewness (over the Keary–Nosebag matrix) of standard deviation and of skewness (of altitude, over sub-matrices).

As the size (P^2) of square (submatrix) is increased, there is an increasing number of points near the periphery of the matrix which are the centres of squares for which parameters cannot be evaluated, because data on altitude for the outer parts of such squares is lacking. This increases the difficulty of comparing the mean and dispersion of

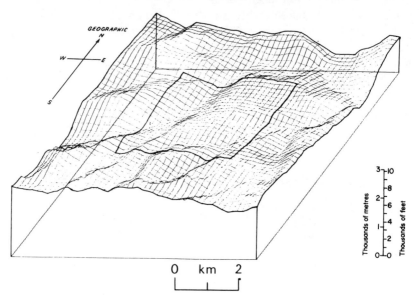

Fig. 2.13 Block diagram of Keary–Nosebag altitude matrix, viewed from the south-east. Profiles drawn north–south and east–west by program ISE1174/15P4/MATCON, in parallel perspective. Vertical scale 1/60,000; east–west horizontal scale 1/51,968. The central 24 × 24 units of the matrix are outlined heavily.

parameters for different scales, since peripheral parts of the matrix are given less weight overall in the sets of parameters based on larger areas. The main comparison here uses areas centred on the central 24 × 24 points in the matrix; this gives 576 squares for values of P between 3 and 27, but smaller samples of larger squares centred on increasingly central matrix positions only.

The sample squares overlap, to an increasing extent as square size is increased. Hence the number of independent samples is much less than the sample size. The procedure is one of repeated sampling with replacement, from a finite subset of an infinite population. Overlap could be avoided with the initial version of the program (15P2/SUBMAT), which permitted the spacing of centres of sample squares to be stipulated in grid mesh units. With a spacing of 25 units, non-overlapping squares up to $P = 25$ units wide could be generated, but the sample size would be only 4. Nor would each sample be completely independent of the others, especially for larger squares, because of the autocorrelation of altitude.

A third option, embodied in an intermediate version of the program (15P2), is to take a systematic sample of all possible square centres. For small squares, this differs from the final version in sampling peripheral areas (results for large squares are similar in both versions, except for sampling density). Since there is little difference in the results,

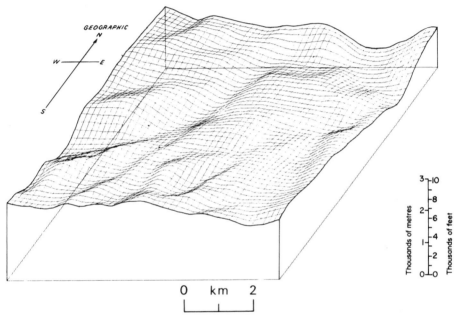

Fig. 2.14 Block diagram of Keary–Nosebag matrix as smoothed by the operator. Viewed from the south-east; vertical scale 1/60,000; east–west horizontal scale 1/51,968.

1	2	1
2	4	2
1	2	1

it seems that in this case the peripheral areas are not greatly different from the central portion of the matrix, in their local properties.

It is indeed difficult to decide what is the proper basis for comparison of variability at different scales. Wood's (1967) use of physically separated sample areas is perhaps ideal, though involving much more work for results such as those presented here. However, even this procedure is dubious in the presence of heterogeneity, for the larger areas cover ground not dealt with by the smaller. The present approach is valid for comparisons of parameters at one scale and for estimating their means, but because of overlapping their spatial variability is considerably underestimated, especially for large squares. This defect could be overcome only by greatly increasing the size of the matrix. Finally, a rough check of extreme 'corner' values of range and standard deviation, measured over squares which do not overlap for P smaller than 25,

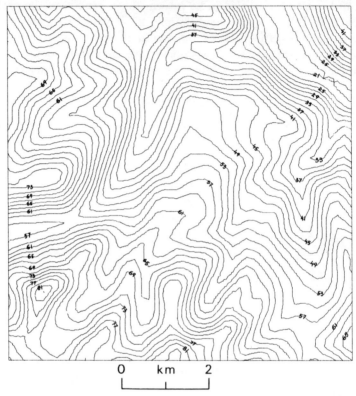

Fig. 2.15 Contour map of Keary–Nosebag matrix as smoothed by the operator. Contours at 61m (200 ft) interval, labelled in hundreds of feet. Scale 1/84,280.

1	2	1
2	4	2
1	2	1

showed that their mean and variability did not differ consistently from those of the whole matrix of parameters.

Since it was not easy to interpret the numerical maps without contouring them, I decided that visual displays of the results were essential. After an abortive attempt to manipulate paper tapes of output from 15P5, for input to a program for plotting block diagrams on a Calcomp drum plotter, the latter program was incorporated into 15P5. The resulting program (ISE 1174/15P6/SUBMATBD) produced figs. 2.21 to 2.43, plotted on a Calcomp drum plotter at Cambridge University Computing Laboratory.

Theoretically the program could have plotted all those diagrams in one run, but limitations on plotter output made it necessary to produce each set for a given value of P in a separate run, five in all.

The block diagrams are drawn in parallel perspective, without removal of hidden lines. Each parameter in turn is read into the subscripted variable Z, and the subscript denotes position in the matrix. The matrix is plotted as a skewed grid, and a height proportional to Z is added to the Y co-ordinate of each point. Lines are drawn connecting points first down columns and then along rows, producing a 'draped fish-net' representation of the surface. The datum and corners are drawn in, to complete the block, and a Z scale with intervals of 40 units is drawn to the right. The Z scaling factor is constant for each parameter, and is more suitable at some scales than at others; some of the block diagrams require greater vertical exaggeration, but comparison over different scales is facilitated.

Range and standard deviation of altitude

The assumption that standard deviation is more stable than range was tested with reference to (a) scale dependence, (b) geographic variability, and (c) smoothness of this variability. Figures 2.16 and 2.17 show the variation of both parameters with square size. Range increases rather more rapidly than standard deviation, for example the ratio between mean range for $P = 27$ and for $P = 7$ is 3·25, whereas that for standard deviation is 2·53; but both increase throughout the range of sizes investigated here.

Standard deviation is somewhat less scale-dependent, especially for areas greater than 1·5 km². For P greater than 27, the results are increasingly affected by the limited size of the whole matrix, and are in effect based on a very small independent sample. Comparison of range with Wood's (1967) sample of 204 areas in the U.S.A. is interesting; mean range in altitude for the Keary–Nosebag matrix rises from the 94 percentile for areas of 0·627 km², to the 99 percentile for 13·28 km², and for the whole 62·73 km² the range is about 300 m above Wood's 100 percentile.

The expected ratio between the standard deviation and the range where both are based on the same independent sample from a normally distributed population has been tabulated as a function of sample size. The ratio between mean standard deviation and mean range observed here is considerably greater than this expectation (fig. 2.18): the discrepancy is slight for small sub-matrices, but increases rapidly up to samples of 9×9, for which a ratio of 0·249 compares with an expectation of 0·206. This ratio is appropriate to a sample of 27 independent observations, rather than 81.

However, since the observed ratio for 50×50 is 0·22 (corresponding to 54 independent observations), compared with 0·21 for 23×23

(corresponding to 70), it does not seem appropriate to base an 'effective number of independent observations' on this ratio. The discrepancy is probably too great to be explained by the difference between mean ratio and ratio of means.

One possible explanation of the deviation from expectation is that, due to skewness or kurtosis, standard deviations are greater than ranges

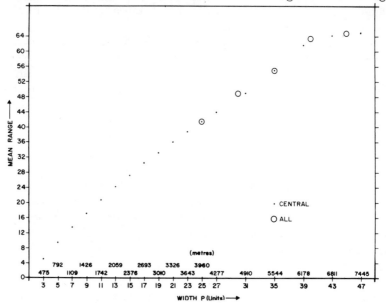

Fig. 2.16 Variation of range with square size (P)
(a) for squares centred on points among the central 24 × 24 of the Keary–Nosebag matrix;
(b) for a 1 in 4 sample of all possible centres within the matrix.

would suggest. This must be rejected because moduli of skewness and kurtosis are greater for small squares, where the discrepancy is slight. Rather, it seems that range is curtailed by autocorrelation, and is always smaller than expected from the standard deviation.

The geographic variability (b) of the two parameters is represented by fig. 2.19, as a function of scale. The curves for range are particularly interesting; two maxima are separated by a shallow minimum, but these are differently located on the curve based on all possible squares and on that for squares around central positions only. In both cases, the variability of the standard deviation is greatest when the latter is measured for 2 km² squares. The relative variability of the standard deviation, compared with the range, is least between 4 and 11 km². Wood's results (1967, fig. 1) imply that, between independent areas, the variability of range continues to increase with area, though less rapidly than the mean range increases. Levelling off of variability in the present study presumably relates to increased overlap and non-independence; it is based on very small samples.

Because of the difference in absolute magnitudes, a comparison of geographic variability should be based on coefficients of variation (standard deviation/mean). Though the disadvantages discussed on

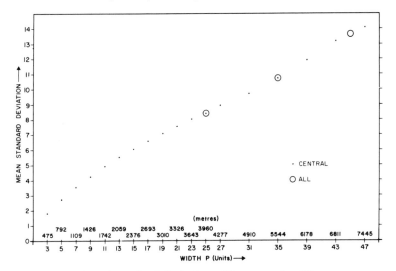

Fig. 2.17 Variation of standard deviation with square size (P)
    (a) for squares centred on points among the central 24 × 24 of
    the Keary–Nosebag matrix;
    (b) for a 1 in 4 sample of all possible centres within the matrix.

p. 55 apply, and a plot such as fig. 2.10 cannot be constructed because standard deviations of parameters for large squares are biased by over-lapping, coefficient of variation may be valid for comparison of range and standard deviation at the same scale. Figure 2.20 shows that, some-what surprisingly, standard deviation is more variable over space than is range, at all scales greater than 0·4 km². The difference is most marked between 2 and 6 km².

A different conclusion follows from study of (c) the smoothness of geographic variability. For $P = 3$, there is very little difference in the form of surface representing the two statistics (figs. 2.21 and 2.22). Both are angular, since statistics based on samples of 9 vary abruptly. However, for larger areas (figs. 2.23 to 2.30), surfaces for standard deviation are much smoother than those for range. This is the sense in which standard deviation is more stable than range. Its high geographic variability measures the real variability of the land surface. In its local variability, standard deviation is less affected by accidents of precise sample location than is range. The abruptness of variation of range is not simply due to the coarse increment of the initial data; the mean height of the surface in fig. 2.23 is 24 units. Range changes suddenly where a particular high or low point is embraced. Similarly, maxima and minima (fig. 2.31) are not suitable for contouring as 'summit' or 'streamline' surfaces, without first being smoothed.

There is precedent for the use of range to describe other autocorrelated

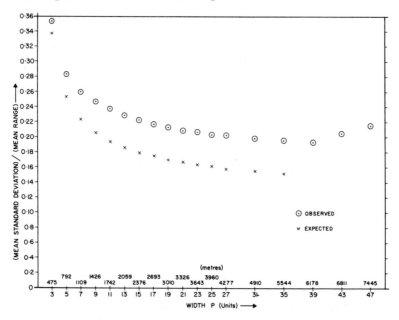

Fig. 2.18 Observed (i) and expected (ii) ratios of mean standard deviation to mean range of altitude for sets of areas of different sizes (P), from the Keary–Nosebag altitude matrix. The expectations are those for (standard deviation)/(range) of samples of P^2 independent variates.

distributions, especially the periodic time-variations of climatic parameters. Terrain, however, is not a regularly periodic distribution, especially when the second autocorrelated dimension is considered. If it was, relief would increase with scale up to one half wavelength, and then remain stationary; fig. 2.16 shows that this is not the case. The results of this trial support the recommendation that standard deviation of altitude is a better measure of relief than is range of altitude. Compared with the range, the standard deviation removes a lot of unwanted 'noise'.

Local convexity, hypsometric integral, and skewness of altitude

Initial attempts to measure 'convexity' from an altitude matrix were based on deviation of central point altitudes from surrounding altitudes.

'Local height' is defined as deviation from the local mean altitude over the $P \times P$ neighbourhood, and 'local convexity' is deviation from the mean of altitudes at the edge of the neighbourhood, i.e. at a distance of $(P - 1)/2$ grid units from the central point in four cardinal directions, and $(P - 1)/\sqrt{2}$ diagonally. Both parameters pool the vertical and

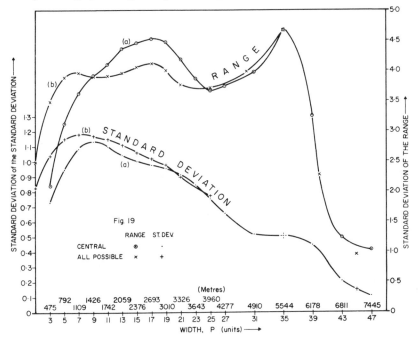

Fig. 2.19 Standard deviations of local estimates of range and of standard deviation of altitude, for areas of different sizes (P), from the Keary–Nosebag altitude matrix. (a) for squares centred on points among the central 24 $\times$ 24 of the matrix; (b) for a 1 in 4 sample of all possible squares within the matrix. The scale for range (right) is 2·5 $\times$ the scale for standard deviation (left).

horizontal components of convexity, and provide a 'contrast enhancement' of the altitude surface.

A disadvantage of these parameters is that for a stationary (trendless) surface, the expectation of mean convexity or height is zero. Hence the character of a surface – whether it is multiconcave, with sharp convexities, or multiconvex, with sharp concavities, is apparent not in the position but in the shape of the frequency distribution of local convexity or of local height. Local convexity has been studied in more detail; local height is very similar in pattern and slightly smaller in magnitude, at all scales.

Skewness of distributions of the parameters was not programmed, but local convexity can be measured by its median. The median local convexity is positive for all values of P, indicating that the surface at most points is convex; this is balanced by a tail of highly negative convexities (the distributions are positively skewed). Since the mean is not in fact zero, but because of trends in altitude with distance is increasingly negative as P increases (i.e. the central part of the matrix

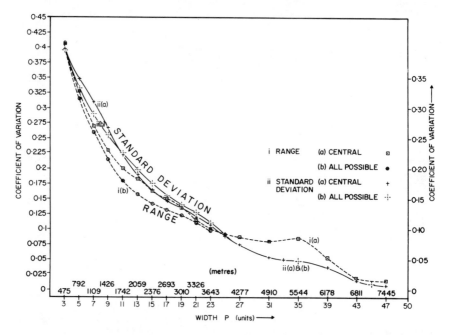

Fig. 2.20 Coefficients of variation (standard deviation/mean) of local estimates of range and of standard deviation of altitude, for areas of different sizes (P), from the Keary–Nosebag altitude matrix.

is lower than its surroundings), it is necessary to correct the median by subtracting the mean. The results are shown in table 2.5, which compares local convexity with skewness and hypsometric integral.

Frequency distributions of local convexity are unimodal and symmetrical about 0 for $P = 3$ and 5, but local convexities measured over larger areas produce platykurtic distributions with little difference in frequency between $+180$ m and -110 m ($+600$ and -350 ft). Consequently the distributions are polymodal, but the modes seem to be random and their positions are not stable over different scales. Standard deviation of local convexity increases steadily in proportion to P (width of area) until it is 185 m (605 ft) for $P = 23$; it then levels off, and declines due to increased overlapping and decreased sample size.

Maps of local convexity (or of local height) are very closely related to

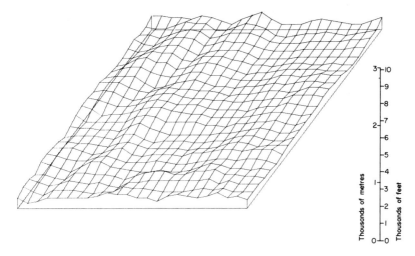

Fig. 2.21 Range in altitude over squares of 3×3 units (225,815 m^2) in the central portion of the Keary–Nosebag altitude matrix. The two ridges of high range in the north-west are the walls of the trough of Keary Creek.

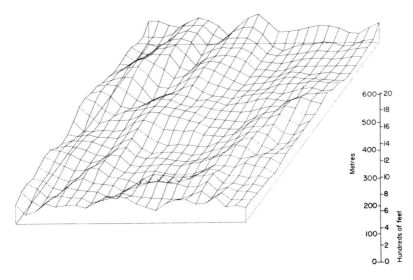

Fig. 2.22 Standard deviation of altitude over squares of 3×3 units (225, 815 m^2) in the central portion of the Keary–Nosebag altitude matrix. Note the similarity to fig. 2.21.

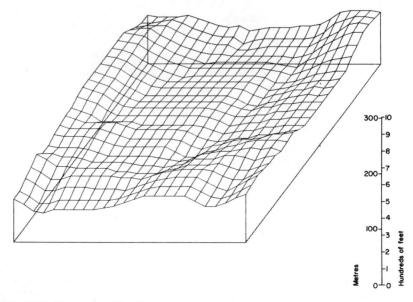

Fig. 2.23 Range in altitude over squares of 13 × 13 units (4,240,305 m²) in the central portion of the Keary–Nosebag altitude matrix. The pattern is no longer related to adjacent topography.

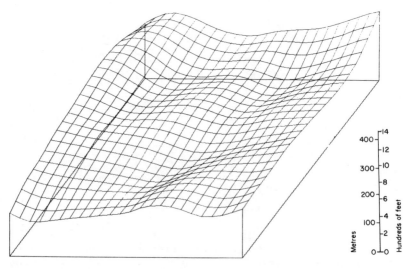

Fig. 2.24 Standard deviation of altitude over squares of 13 × 13 units (4,240, 305 m²) in the central portion of the Keary–Nosebag altitude matrix. Note the similarity in pattern to fig. 2.23, but the greater smoothness.

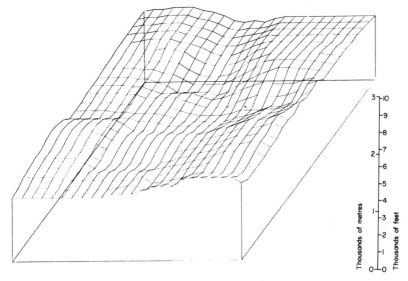

Fig. 2.25 Range in altitude over squares of 23 × 23 units (13,272,906 m²) in the central portion of the Keary–Nosebag altitude matrix.

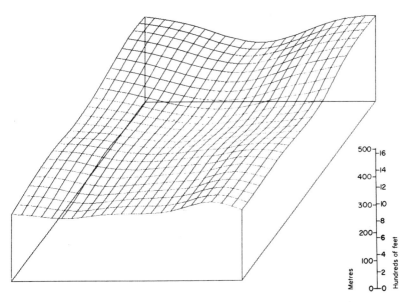

Fig. 2.26 Standard deviation of altitude over squares of 23 × 23 units (13,272,906 m²) in the central portion of the Keary–Nosebag altitude matrix. The lowest values are for the plateau area, which is farthest from the mountains to the south-west and the Bridge River trough to the north-east.

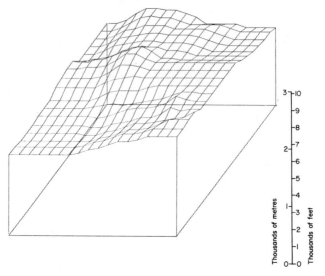

Fig. 2.27 Range in altitude over squares of 33 × 33 units (27,323,620 m²) in the central portion of the Keary–Nosebag altitude matrix. Note the reversal compared with range over smaller squares (fig. 2.25)

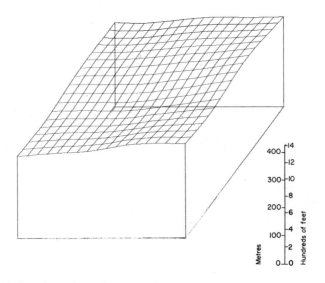

Fig. 2.28 Standard deviation of altitude over squares of 33 × 33 units (27,323,620 m²) in the central portion of the Keary–Nosebag altitude matrix. Compared with that for smaller squares (fig. 2.26), the pattern has not changed so drastically as has that of range. There is now great consistency in estimates, with a gentle geographic variation.

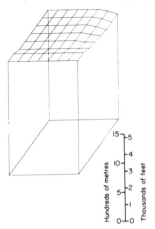

Fig. 2.29 *Range in altitude* over squares of 43 × 43 units (46,392,445 m²) in the central portion of the Keary–Nosebag altitude matrix.

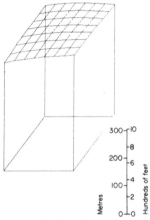

Fig. 2.30 *Standard deviation of altitude* over squares of 43 × 43 units (46,392,445 m²) in the central portion of the Keary–Nosebag altitude matrix.

altitude, with negative values in valleys and positive ones on ridges and plateau edges. With increasing scale, i.e. as altitude is related to points at increasing distance, the surface is simplified but exaggerated. Minor valleys are 'filled in', and only major features remain. For $P = 13$ or larger areas, the local convexity maps (fig. 2.33) are very similar to each other and to altitude. Changes for P greater than 27 are of very little interest, since they reflect principally the narrowing towards the centre of the matrix which is due to the reduced number of possible centres for large areas.

The latter point indicates a further disadvantage of local convexity (and of local height), namely an emphasis on the value of altitude at

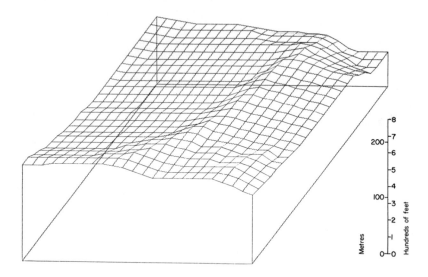

Fig. 2.31 Minimum altitude within squares of 13 × 13 units (4,240,305m²) in the central portion of the Keary–Nosebag altitude matrix. Most of the map is controlled by the Keary Creek valley, but the effect of the Bridge River trough is felt in the north-east.

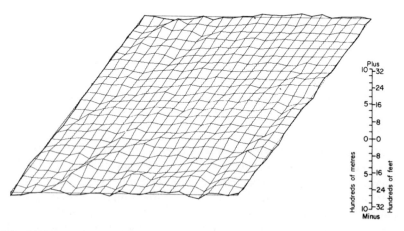

Fig. 2.32 Local convexity within squares of 3 × 3 units (225,815 m²) in the central portion of the Keary–Nosebag altitude matrix. These values are of local significance only.

the central point. This does not affect two regional measures of convexity, the hypsometric integral and the skewness of the altitude frequence distribution. Though the values of these two parameters are more meaningful for large areas, while those of local convexity and local height are more meaningful for small areas, they have all been measured over the whole range from $P = 3$ to $P = 47$.

While the local convexity of a multiconcave (scalloped) surface would have large positive values for the sharp crests, to balance the more

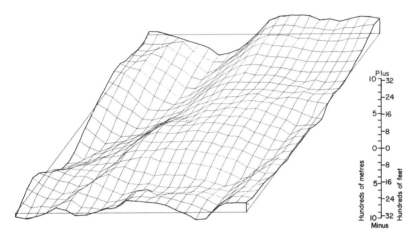

Fig. 2.33 Local convexity within squares of 13×13 units (4,240,305 m²) in the central portion of the Keary–Nosebag altitude matrix. The surface produced by these values is very closely related to the altitude surface; note the Keary Creek trough and the plateau to the east.

general negative values in the hollows, the skewness could be positive and the hypsometric integral could be below 0·5 around all points, indicating the concave curvature throughout. If the concavities were superimposed on broader trends, some parts of the surface might show as convex by any parameter, even over the smallest areas. On the whole, hypsometric integral and skewness are more sensitive to multi-concavity or multi-convexity at any scale, and less sensitive to particular values of altitude, than are local convexity or local height.

The program provides frequency distributions for altitude skewness (g_1). There is a marked negative skewness of these distributions for small areas (up to $P = 11$), but otherwise they are fairly symmetrical. The variability (standard deviation) of skewness declines steadily as P increases (table 2.5). The frequency distribution of hypsometric integral for $P = 3$ is platykurtic, with 83% of integrals evenly spread between 0·40 and 0·60.

Table 2.5

P	SKEWNESS			HYPSOMETRIC INTEGRAL	CONVEXITY (metres)	
	median	*mean*	*standard deviation*	*median*	*(median − mean)*	*standard deviation*
3	·0138	—	—	·5010	2·5	24·1
5	·0422	·0076	·4157	·4900	4·8	49·2
7	·0480	−·0400	·4093	·4949	6·3	73·3
9	·0146	−·0784	·3897	·4951	9·2	95·9
11	−·0285	−·1101	·3566	·4983	16·3	116·6
13	−·1085	−·1297	·3047	·5119	16·5	135·0
15	−·1685*	−·1387	·2497	·5196	20·4	150·7
17	−·1572*	−·1419	·2086		21·9	163·4
19	−·1309	−·1439	·1859	·5092	26·2	173·3
21	−·1319	−·1453	·1807		26·8	180·3
23	−·1401	−·1478	·1901		24·1	184·5
25	−·1401	−·1529	·2034	·5014	23·5	185·6
27	−·1368	−·1609	·2097		23·2	183·3
31	−·1439	−·1732	·1886	·5112	39·9	182·0
35	−·244*	−·213	·1642	·5196	64·9	179·9
39	−·334*	−·302	·1070	·5316	41·1	147·3
43	−·365	−·379	·0497	·5224	19·2	90·9
47	−·385	−·386	·0313	·5208	11·0	52·4

* skewness of skewness positive. Otherwise it is negative.

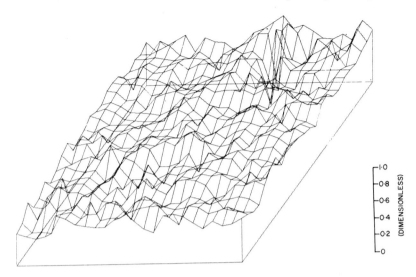

(DIMENSIONLESS)

Fig. 2.34 *Hypsometric integral* over squares of 3 × 3 units (225,815 m²) in the central portion of the Keary–Nosebag altitude matrix. The integral is infinite (shown here as zero) if all nine points have the same altitude.

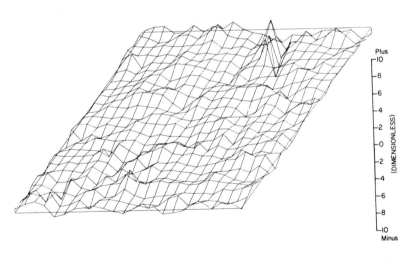

Plus

(DIMENSIONLESS)

Minus

Fig. 2.35 *Dimensionless skewness*, g_1, over squares of 3 × 3 units (225,815 m²) in the central portion of the Keary–Nosebag altitude matrix.

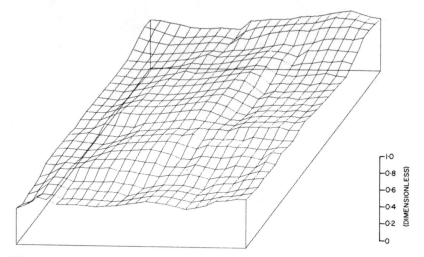

Fig. 2.36 Hypsometric integral over squares of 13 × 13 units (4,240,305 m²) in the central portion of the Keary–Nosebag altitude matrix. Low integrals (concavity) are found along Keary Creek in the north-west. The main 'ridge' of high integrals (convexity) follows the Keary–Nosebag divide, north-east–south-west.

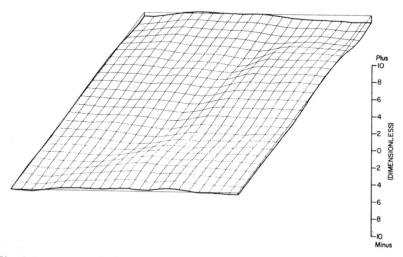

Fig. 2.37 Dimensionless skewness, g_1, over squares of 13 × 13 units (4,240,305 m²) in the central portion of the Keary–Nosebag altitude matrix. The spatial pattern is very close to that for hypsometric integral, but inverted (positive skewness of altitude = concavity).

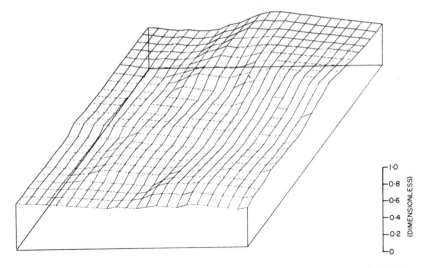

Fig. 2.38 Hypsometric integral over squares of 23 × 23 units (13,272,906 m²) in the central portion of the Keary–Nosebag altitude matrix.

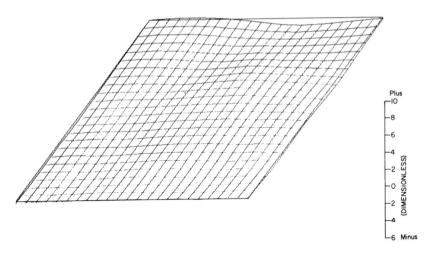

Fig. 2.39 Dimensionless skewness, g_1, over squares of 23 × 23 units (13,272,906 m²) in the central portion of the Keary–Nosebag altitude matrix. The surface is much smoother than that for hypsometric integral, even allowing for the difference in vertical scale.

Contrary to local convexity, both hypsometric integral and median skewness indicate a net concavity for small areas ($P = 5$ to 9). For large areas ($P = 13$ to 47) there is a net convexity, as there is for the matrix as a whole. The surface must therefore be multiconcave, with concavities less than $13 \times 158 \cdot 4 = 2060$ m across. However, the effect is not so strong that the balance could not be shifted easily from net concavity to net convexity by a change in area definition.

The geographic patterns of skewness and of hypsometric integral correlate well, though comparison of the block diagrams (figs. 2.34 to 2.43) is difficult because the correlation is negative. For $P = 13$ and smaller areas, the pattern is similar to that of local convexity, with a

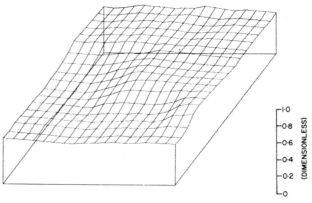

Fig. 2.40 Hypsometric integral over squares of 33×33 units (27,323,620 m²) in the central portion of the Keary–Nosebag altitude matrix. The 'ridge' of high integrals now trends north-west–south-east.

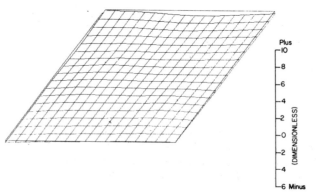

Fig. 2.41 Dimensionless skewness, g_1, over squares of 33×33 units (27,323,620 m²) in the central portion of the Keary–Nosebag altitude matrix. Skewness is now dominantly negative. The vertical scale of this diagram is inadequate.

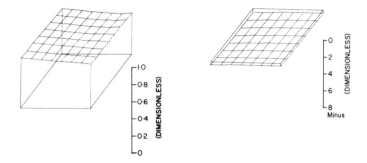

Fig. 2.42 Hypsometric integral over squares of 43 × 43 units (46,392,445 m²) in the central portion of the Keary–Nosebag altitude matrix.

Fig. 2.43 Dimensionless skewness, g_1, over squares of 43 × 43 units (46,392,445 m²) in the central portion of the Keary–Nosebag altitude matrix.

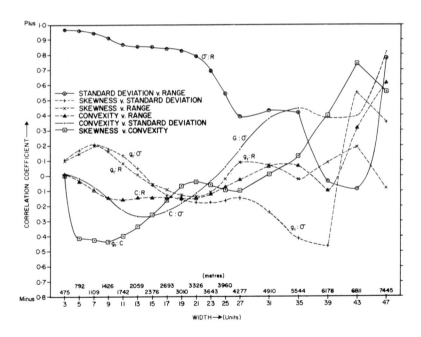

Fig. 2.44 Simple correlations between range, standard deviation, convexity and dimensionless skewness, for different sizes of square (P) within the Keary–Nosebag altitude matrix; based on squares in the central portion.

progressive simplification of the altitude surface as P increases. For larger areas there is a progressive change, so that the main concave zone trends north-west – south-east instead of south-west – north-east. Both surfaces are very rough for $P = 3$, because of the small sample size. Skewness becomes smooth more rapidly than hypsometric integral, as might be expected from its higher information content. Hence some support is provided for the use of skewness rather than hypsometric integral, while both are definitely to be preferred to 'local convexity' and 'local height'.

Interrelationships

Program 15P5 provides also a small matrix of correlations between range, standard deviation, local convexity and skewness (g_1). The variation of these correlations with size of square (P) is shown in fig. 2.44. These results suggest caution in comparing correlations at different scales, as well as in comparing correlations of point values with correlations of area values.

It is not surprising that correlations change in magnitude, for example the decline of correlation between standard deviation and range as P increases is in line with Wood's (1967) results. It is, however, disturbing that correlations change sign. For P greater than 27 changes are particularly abrupt, as the number of square centres shrinks. These results are shown to indicate the sudden changes at $P = 27$, but no other significance is attached to them.

The negative correlation between skewness and local convexity is lower than expected, and declines further as the spatial pattern of skewness changes with increasing P. Near-zero correlation for $P = 3$ presumably reflects the capricious nature of parameters from these small samples. Taking all correlations together, they seem to be expressed best at $P = 7$ or $P = 9$ ($1 \cdot 2$ or $2 \cdot 0$ km²). It is interesting that variability of altitude correlates with concavity or positive skewness, but the correlations are too low for general conclusions to be drawn.

Conclusions

Results reported in the three latter sections demonstrate how much information can be obtained by program from even a small altitude matrix. Though currently limited to intensive studies, this form of geomorphometry is extremely efficient in relation to the effort involved. Fuller implementation of the proposed system is recommended, including calculation of gradient and aspect, their relations to each other and to altitude, and the separation of vertical and horizontal convexity. Though parametric statistics are recommended, there is room for debate, and for further investigation, of their merits relative to percentile statistics. Rosenfeld (1969) has suggested a large variety of 'picture processing' operations for matrices.

The system of derivatives of altitude at a point, and moments of their distributions over an area, directly covers all geomorphometric concepts except for horizontal dimension. The latter is implicit in horizontal convexity and in the parameters taken in combination: there are probably fewer degrees of freedom of variation of surface form than there are parameters. More information about horizontal geographic distributions may be obtained by study of the autocorrelation properties of altitude and its derivatives, leading on to spectral analysis. It is also interesting to study the variation of the summary statistics (moments) with scale, for square areas of different sizes. Finally, the initial altitude field may be smoothed (figs. 2.14 and 2.15) and re-analysed. There is a considerable overlap between these three approaches.

One very good reason why this simple approach has not been attempted previously is that the measurement and calculation involved is extremely onerous if manual methods are used. This is no longer a good reason, for most geomorphologists have access to powerful computers. It is best, however, that computer processing should not be simply a final stage, tagged on to a process of data definition and measurement which is essentially manual. It is more efficient to involve the computer at a much earlier stage, in fact to input altitude of the land surface to the computer and to derive all other variables and parameters by program.

Data availability is still a problem, but the advantages of altitude matrices are such that they are much in demand by engineers, geophysicists and military officers. Hence their supply will probably increase rapidly, especially with their direct generation from air photos in stereoplotters, and direct recording on to magnetic tape. Digitised contours are becoming available with the automation of cartography. Thus there is every prospect of the onerous digitisation stage being performed by others, and the results made available to geomorphologists.

The use of statistical frequency distributions is a more abstract form of analysis than the detailed study of slope profiles. The two are essentially complementary, especially since the latter is usually performed to a higher degree of resolution, and over small areas. Pitty (1968) pointed out that 'the shape of a slope–angle histogram, and hence sample statistics summarising the distribution, give little indication of the land form of the area in which the observations were made'. It is true that very different profiles may produce similar gradient frequency distributions, but their distributions of altitude would differ. For example, variability of altitude is greater for a convex upper slope and concave lower than for a concave upper and convex lower. Hence when distributions of altitude, gradient, aspect, and vertical and horizontal convexity are taken together, the surface form is rather well specified.

Morphometric mapping (Savigear 1965), the subdivision of the land surface into relatively homogeneous facets, is another complementary approach which is of practical value, especially in reconnaissance studies. It is however less suitable as a basis for statistical analysis and for interrelating altitude, gradient, aspect and convexity. Although it recognises the continuous nature of spatial variation in altitude, it essentially presupposes that variation in gradient is discontinuous, and it ignores the undeniably continuous nature of variation in aspect. It leads on to the selection of 'typical' sites for each facet, which is a very dangerous form of sampling.

Square grid sampling is also suspect, but in practice natural period-icities seem to be too irregular to produce bias, and the results are generally more precise and accurate than those from random sampling (Evans 1969B). Certainly the square grid method is safe if, in accord with the sampling theorem, the grid mesh is less than half the shortest wavelength of variability present. Grid mesh sizes of 20, 50 or 100 m are suitable for study of mesorelief. If altitude is sampled with a mesh of 1 km or coarser, slope and convexity should be measured separately at each point, and not derived from the altitude matrix.

A further advantage of the matrix approach is that while a map of facets and their gradients could be obtained from an altitude matrix, the reverse is not usually possible with any accuracy. The facet approach has been quantified in the NORDISK ADB digital terrain model (Hallmen 1968). This is an empirical, flexible system which defines terrain by arbitrary lines which join up to form irregular polygons; contours and cross-sections may be included as special cases of such lines, but 'break-lines' are usually preferred for simple features such as tip heaps. Data redundancy is thus reduced, and the altitude of any point can be interpolated fairly accurately. However, parallel cross-sections (forming a matrix) are used for the less well-defined terrain, which is more common in nature. This emphasises the flexibility of the matrix approach, which also facilitates more rapid retrieval and interpolation.

Acknowledgements

I would like to thank R. Webster, R. J. Chorley, D. Connelly and A. Goudie for commenting on early partial drafts; M. Evans, G. Brown and P. Howe for plotting graphs and figures; U.S.A.F. Cambridge Research Labs., U.S. Army Engineer Waterways Experiment Station,

U.S. Office of Naval Research (Geography Branch), W. Tobler and E. Hammond for providing copies of reports and unpublished work; P. Wharton, M. Hunter and S. Eckford for typing; and the Director of the Computing Laboratory, Cambridge University, for making computer time available on TITAN.

References

ARMSTRONG, J. S. (1967) Derivation of theory by means of factor analysis or Tom Swift and His Electric Factor Analysis Machine; *The American Statistician* 21(5), 17–21.

BASSETT, K. and CHORLEY, R. J. (1971) An experiment in terrain filtering; *Area* 3(2), 78–91.

BHATTACHARYYA, B. K. (1965) Two-dimensional harmonic analysis as a tool for magnetic interpretation; *Geophysics* 30(5), 829–57.

BOEHM, B. W. (1967) Tabular representations of multivariate functions – with applications to topographic modelling; *Proceedings, Association for Computing Machinery National Meeting*, 403–15.

BRYSON, R. A. and DUTTON, J. A., (1967) The variance spectra of certain natural series; In W. L. Garrison and D. F. Marble (Eds.), *Northwestern University Studies in Geography*, 14, Quantitative Geogr., Part 2; 1–24.

CALEF, W. and NEWCOMB, R. (1953) An average slope map of Illinois; *Annals of the Association of American Geographers* 43(4), 305–16.

CARR, D. D. and VAN LOPIK, J. R. (1962) Terrain quantification, Phase 1; Surface geometry measurements; *U.S.A.F. Cambridge Research Laboratories, Contract AF 19(628)–481, Project 7628, Report AFCRL 63–208*, 85p., 348 annoted ref.

CHAPMAN, C. A. (1952) A new quantitative method of topographic analysis; *American Journal of Science* 250, 428–52.

CHEN, S. P. (1947) The relative relief of Tsunyi, Kweichow; *Journal, Geographical Society of China* 14(2), 5.

CHORLEY, R. J. (1969) The drainage basin as the fundamental geomorphic unit; In R. J. Chorley (Ed.), *Water, Earth and Man* (Methuen, London), 77–99.

CLARKE, J. I. (1966) Morphometry from maps; In G. H. Dury (Ed.), *Essays in Geomorphology* (Heinemann, London), 235–74.

CONNELLY, D. S. (1968) *The Coding and Storage of Terrain Height Data: An Introduction to Numerical Cartography* (M.Sc. Thesis, Cornell University), 141p.

CURRAY, J. (1956) The analysis of two-dimensional orientation data; *Journal of Geology* 64(2), 117–31.

DIAMANTIDES, N. D. and HOROWITZ, M. (1957) Autocorrelation of the earth's crust with analog computers; *Review of Scientific Instruments* 28(5), 353–60.

DURY, G. H. (1951) Quantitative measurement of available relief and of depth of dissection; *Geological Magazine* 88, 339–43.

ESLER, J. E. and PRESTON, F. W. (1967) FORTRAN IV program for the

GE 625 to compute the power spectrum of geological surfaces; *Kansas Computer Contribution* No. 16, 22p.

EVANS, I. S. (1963) Map analysis of mean slope; *Geographical Articles* (Cambridge) 2, 34–40.

EVANS, I. S. (1969A) The geomorphology and morphometry of glaciated mountains; In Chorley, R. J. (Ed.), *Water, Earth and Man* (Methuen, London), 369–80.

EVANS, I. S. (1969B) Some problems of sampling spatial distributions; reported by B. J. Garner, *Area* 1(1), 40.

EVANS, I. S. (1970) The implementation of an automated cartography system; In Cutbill, J. L. (Ed.), *Data Processing in Biology and Geology, Systematics Association Special Volume* No. 3 (Academic Press, London), 39–55.

EYLES, R. J. (1965) Slope studies on the Wellington Peninsula; *New Zealand Geographer* 21(2), 133–44.

EYLES, R. J. (1969) Depth of dissection of the West Malaysian landscape; *Journal of Tropical Geography* 28, 23–31.

FISCHER, R. K. (1963) Hüllfläche und Sockelfläche des Reliefs, dargestellt am Beispiel der Schweizer und Salzburger Alpen; *Abhandlungen der Bayerischen Akademie der Wissenschaften, Math.-nat. Klasse* n.f. 113, 38p., München.

FISHER, R. A. (1954) *Statistical Methods for Research Workers*, 5th Edn, (Oliver & Boyd, Edinburgh), 319p.

FREY, C. (1965) Morphometrische Untersuchung der Vogesen; *Basler Beiträge zur Geographie und Ethnologie, Geographische Reihe*, heft 6, 150p.

GĂLĂBOV, Z. (1968) On the application of morphometric analysis in geomorphological investigations (results of a geomorphological study of the middle Stara Planina), p. 7–40; In *Problemina na geografijata na Bălgariya* v. 2, Nauka i izkustvo, Sofia (in Bulgarian).

GASSMAN, F. and GUTERSOHN, H. (1947) Kotenstreuung und Relieffaktor; *Geographica Helvetica* 2, 122–39.

GERRARD, A. J. W. and ROBINSON, D. A. (1971) Variability in slope measurements: a discussion of the effects of different recording intervals and micro-relief in slope studies; *Transactions of the Institute of British Geographers* 54, 45–54.

GOLDBERG, G. M. (1962) The derivation of quantitative surface data from gross sources; *Surveying and Mapping* 22, 537–48.

GREEN, R. (1967) The spectrum of a set of measurements along a profile; *Engineering Geology* (Elsevier) 2(3), 163–8.

GREGORY, K. J. and BROWN, E. H. (1966) Data processing and the study of land form; *Zeitschrift für Geomorphologie* n.f. 10(3), 237–63.

HALLMEN, B. (1968) In Blaschke, W. (Ed.), Digital terrain models; a presentation of the practical application of existing D.T.M.s; *International Society of Photogrammetry, Intercommissional Working Group* IV/V.

HAMMOND, E. H. (1958) Procedures in the descriptive analysis of terrain; *U.S. Office Naval Research, Geography Branch, Contract Nonr 1202(01), Project NR 387–015, Final Report*, 85p. + Appendices (72p.).

HAMMOND, E. H. (1964) Analysis of properties in land form geography; an application to broad-scale land form mapping, *Annals of the Association of American Geographers* 54, 11–19.

HAYRE, H. S. and MOORE, R. K. (1961) Theoretical scattering coefficient for near vertical incidence from contour maps; *Journal of Research, National Bureau of Standards* 65 D (5), 427–32.

HOBSON, R. D. (1967) FORTRAN IV programs to determine surface roughness in topography for the CDC 3400 computer; *Kansas Computer Contribution* No. 14, 27p.

HORMANN, K. (1968) Rechenprogramme zur morphometrischen Kartenauswertung (Computer programs for the morphometric analysis of maps); *Schriften des Geographischen Instituts der Universität Kiel* 29(2), 154p.

HORMANN, K. (1969) Geomorphologische Kartenanalyse mit Hilfe elektronischer Rechenanlagen (Geomorphological analysis of maps with the aid of electronic computers); *Zeitschrift für Geomorphologie* 13(1), 75–98.

HORTON, C. W., HOFFMAN, A. A. J. and HEMPKINS, W. B. (1962) Mathematical analysis of the microstructure of an area of the bottom of Lake Travis; *Texas Journal of Science* 14, 131–42.

HOUBOLT, J. C. (1961) Runway roughness studies in the aeronautical field; *Journal of the Air Transport Division, Proceedings of the American Society of Civil Engineers*, 87 (AT 1), 11–31.

IMAMURA, G. (1937) Past glaciers and the present topography of the Japanese Alps; *Science Reports of Tokyo Bunrika Daiguku*, C. 7, 61p.

JAEGER, R. M. and SCHURING, D. J. (1966) Spectrum analysis of terrain of Mare Cognitum; *Journal of Geophysical Research* 71(8), 2023–8.

JOHNSON, G. G. and VAND, V. (1967) Application of a Fourier data smoothing technique to the meteoritic crater Ries Kessel; *Journal of Geophysical Research* 72(6), 1741–50.

KAITANEN, V. (1969) A geographical study of the morphogenesis of Northern Lapland; *Fennia* 99(5), 85p.

KING, C. A. M. (1968) An example of factor analysis applied to geomorphological data for six areas in Northern England; In Cole, J. P. and King, C. A. M. *Quantitative Geography* (Wiley, London), 319–34.

LEOPOLD, L. B., WOLMAN, M. G. and MILLER, J. P. (1964) *Fluvial Processes in Geomorphology* (W. H. Freeman & Co, San Francisco), 522p.

LEWIS, L. A. (1969) Analysis of surficial landform properties: the regionalization of Indiana into units of morphometric similarity; *Proceedings, Indiana Academy of Science* 78, 317–28.

LOUIS, H. (1963) Über Sockelfläche und Hüllfläche des Reliefs; *Zeitschrift für Geomorphologie* n.f. 7(4), 355–66.

LOUP, J. (1963) Altitudes moyennes et coefficients d'aération dans la Valais; *Revue de Géographie Alpine* 51(1), 5–18.

DE MARTONNE, E. (1940) Interprétation géographique de l'hypsométrie française; *Comptes Rendus hebdomadaires des Séances de l'Académie des Sciences* 211, 426–8.

MATALAS, N. C. and REIHER, B. J. (1967) Some comments on the use of factor analyses; *Water Resources Research* 3(1), 213–23.

MCDONALD, M. F. and KATZ, E. J. (1969) Quantitative method for describing the regional topography of the ocean floor; *Journal of Geophysical Research* 74(10), 2597–607.

MERLIN, P. (1965) A propos des méthodes de morphométrie; *Acta Geographica* (Paris) 56, 14–20.

MERLIN, P. (1966) Résultats d'une analyse morphométrique de quelques massifs montagneux nord-africains; *Acta Geographica* (Paris) 61, 14–15.

MIESCH, A. T. (1969) Critical review of some multivariate procedures in the analysis of geochemical data; *Journal, International Association for Mathematical Geology* 1(2), 171–84.

MONMONIER, M. S., PFALTZ, J. L. and ROSENFELD, A. (1966) Surface area from contour maps; *Photogrammetric Engineering* 32(3), 476–82.

NEWELL, W. L. (1970) Factors influencing the grain of the topography along the Willoughby Arch in northeastern Vermont; *Geografiska Annaler* 52A(2), 103–12.

NORDIN, C. F. (1971) Statistical properties of dune profiles; *United States Geological Survey, Professional Paper* 562-F, 41p.

PANNEKOEK, A. J. (1967) Generalized contour maps, summit level maps, and streamline surface maps as geomorphological tools; *Zeitschrift für Geomorphologie* n.f. 11, 169–82.

PÉGUY, CH. P. (1942) Principes de morphométrie alpine, *Revue de Géographie Alpine* 30, 453–86.

PÉGUY, CH. P. (1948) Introduction à l'emploi des méthodes statistiques en géographie physique; *Revue de Géographie Alpine* 36(1), 5–101.

PELTIER, L. C. (1954) Some properties of the average topographic slope (abstract); *Annals of the Association of American Geographers*, 44(2), 229–30.

PELTIER, L. C. (1955) Landform analysis in operational research (abstract); *Bulletin of the Geological Society of America* 66, 1716–17.

PIERSON, W. J. and MARKS, W. (1952) The power spectrum analysis of ocean-wave records; *Transactions of the American Geophysical Union* 33(6), 834–44.

PIKE, R. J. and WILSON, S. E. (1971) Elevation–relief ratio, hypsometric integral, and geomorphic area–altitude analysis; *Bulletin of the Geological Society of America* 82, 1079–84.

PIPER, D. J. W. and EVANS, I. S. (1967) Computer analysis of maps using a pencil follower; *Geographical Articles* (Cambridge) 9, 21–5.

PITTY, A. F. (1968) Some comments on the scope of slope analysis based on frequency distributions; *Zeitschrift für Geomorphologie* 12(3), 350–5.

PRESTON, F. W. and HARBAUGH, J. W. (1965) BALGOL programs and geologic application for single and double Fourier series using IBM 7090/7094 computers; *Kansas Geological Survey, Special Distribution Publication* 24, 72p. (Summarised in p. 143–6 of Harbaugh, J. W. and Merriam, D. F., *Computer Applications in Stratigraphic Analysis*, Wiley New York).

RAYNER, J. N. (1971) *An Introduction to Spectral Analysis* (Pion, London), 174p.

ROSENFELD, A. (1969) *Picture Processing by Computer* (Academic Press, New York), 196p.

ROWAN, L. C., MCCAULEY, J. F. and HOLM, E. A. (1971) Lunar terrain mapping and relative-roughness analysis; *United States Geological Survey Professional Paper* 599–G, 32p.

ROZEMA, W. J. (1968) The use of spectral analysis in describing lunar surface roughness; *National Aeronautics and Space Administration, Contract Report* CR106336 (C.F.S.T.I. Acception Number N69–40421), 34p. (includes Appendix on effects of detrending).

ROZEMA, W. J. (1969) The use of spectral analysis in describing lunar surface roughness; *United States Geological Survey Professional Paper* 650-D, 180–8.

SAVIGEAR, R. (1965) A technique of morphological mapping; *Annals of the Association of American Geographers* 55, 514–38.

DE SMET, R. (1951) Problèmes de morphométrie; *Bulletin Société Belge d'Études Géographiques* 20, 111–32.

SMITH, G.-H. (1935) The relative relief of Ohio; *Geographical Review* 25, 272–84.

SPARKS, B. W. (1960) *Geomorphology* (Longmans, London), 371p.

SPEIGHT, J. G. (1968) Parametric description of land form; In Stewart, G. A. (Ed.), '*Land Evaluation; Proceedings of a C.S.I.R.O. Symposium*', p. 239–50.

STEARNS, R. G. (1967) Warping of the Western Highland Rim Peneplain in Tennessee by ground-water sapping; *Bulletin of the Geological Society of America*, 78, 1111–24.

STONE, R. O. and DUGUNDJI, J. (1965) A study of microrelief – its mapping, classification and quantification by means of Fourier analysis; *Engineering Geology* (Elsevier) 1(2), 89–167.

STRAHLER, A. N. (1950) Equilibrium theory of erosional slopes approached by frequency distribution analysis; *American Journal of Science* 248, 673–96 and 800–14.

STRAHLER, A. N. (1952) Hypsometric (area–altitude) analysis of erosional topography; *Bulletin of the Geological Society of America* 63, 1117–41.

STRAHLER, A. N. (1956) Quantitative slope analysis; *Bulletin of the Geological Society of America* 67(5), 571–96.

STRAHLER, A. N. (1957) Quantitative analysis of watershed geomorphology; *Transactions of the American Geophysical Union* 38(6), 913–20.

SWAN, S.B.St C. (1967) Maps of two indices of terrain, Johor, Malaya; *Journal of Tropical Geography* 25, 48–57.

SWAN, S.B.St C. (1970) Land surface mapping, Johor, West Malaysia; *Journal of Tropical Geography* 31, 91–103.

SWARTZ, C. A. (1954) Some geometrical properties of residual maps; *Geophysics* 19(1), 46–70.

SWITZER, P., MOHR, C. M. and HEITMAN, R. E. (1964) Statistical analyses of ocean terrain and contour plotting procedures; *Project Trident, Technical Report, (U.S.) Department of Navy, Bureau of Ships* Nobsr–8154 SS-050, ix + 79p. Ref. A.D. 601538.

TAILLEFER, F. (1948) L'altitude moyenne des régions naturelles des Pyrénées Françaises. Essai d'interprétation morphologique; *Revue de Géographie alpine* 36, 145–60.

D

TANNER, W. F. (1959) Examples of departure from the Gaussian in geomorphic analysis; *American Journal of Science* 257(6), 458–60.

TANNER, W. F. (1960) Numerical comparison of geomorphic samples; *Science* 131(3412), 1525–6.

THOMPSON, W. (1964) Determination of the spatial relationships of locally dominant topographic features; *U.S. Department of Army, Natick, Massachusetts*, 14p + 10p. of figs.

TOBLER, W. R. (1964) Computation of 'slope' maps; 2p., mimeographed, for *Michigan Inter-University Community of Mathematical · Geographers. Department of Geography, University of Michigan, Ann Arbor.*

TOBLER, W. R. (1969) An analysis of a digitalized surface; In Davis, C. M. (Ed.) *A Study of the Land Type* (U.S. Army Research Office, Durham, North Carolina, Contract DA–31–124–ARO–D–456), p. 59–76 and 86.

TREWARTHA, G. T. and SMITH, G.-H. (1941) Surface configuration of the driftless cuestaform hill land; *Annals of the Association of American Geographers* 31, 25–45.

TRICART, J. (1947) Sur quelques indices géomorphométriques; *Comptes Rendus hebdomadaires des Séances de l'Académie des Sciences* 225, 747–749.

TRICART, J. (1965) *Principes et méthodes de la géomorphologie* (Masson, Paris), 496p. (see p. 159–82).

TRICART, J. and MUSLIN, J. (1951) L'étude statistique des versants; *Revue de Géomorphologie dynamique* 2, 173–82.

TURNER, A. K. and MILES, R. D. (1968) Terrain analysis by computer; *Proceedings of the Indiana Academy of Science* 77, 256–70.

WATSON, G. S. (1966) The statistics of orientation data; *Journal of Geology* 74, 786–97.

WENTWORTH, C. (1930) A simplified method of determining the average slope of land surfaces; *American Journal of Science*, Series 5, vol. 20, no. 117, 184–94.

WOOD, W. F. (1967) Qualitative considerations in quantitative physical geography; In Garrison, W. L. and Marble, D. F. (Eds.), *Northwestern University Studies in Geography*, 14; Quantitative Geography, Part 2, 227-42.

WOOD W. F. and SNELL, J. B. (1960) A quantitative system for classifying landforms; *U.S. Department of Army, Natick, Massachusetts, Technical Report* EP–124, 20 p.

3 Geomorphology and information theory

DANIEL S. CONNELLY

AutoMap Consultants, Whitesboro, New York

Automated cartographic compilation, automated radar simulation and automated placement of communications facilities are some of the new, rapidly growing research areas which require numerically coded 'messages' representing the earth's terrain. Many geomorphologists may be aware that there are also computer-based systems in use which are capable of simple, automatic interpretation of numerically coded terrain features. In each of these automated processes the first step is the encoding of landforms into a machine-readable format. Of course, the encoding of terrain information, the rendering of it into digits, is a purely mechanical operation, and of little direct interest to geomorphologists. Much more significant to geomorphologists is the coding of landforms, the creation of a numerical 'language' for efficiently forming a message which accurately represents the terrain.

Landforms as messages

Messages, and their transmission from source to destination, may be analysed with provocative insight by a body of techniques known as information theory. By considering the terrain as a source for coded messages, these techniques and insights become available to the geomorphologist in his analysis of landforms and the processes which shape them. To appreciate landforms as messages, consider the task of transmitting, over a telegraph line, a terrain description of sufficient accuracy to make a plastic relief map of the area described. The telegraph signals (dots and dashes) can be transmitted at only a fixed rate, so many signals per second. Since shorter transmission times mean lower transmission costs, one of the first questions to be asked is this: How fast can we transmit a landform by telegraph?

Clearly, if we are not very clever in the way we translate the landform into telegraphic code, it could take a very long time to transmit the message. We could, on the other hand, hope to speed the transmission by carefully choosing the code that we use. It was reasoning such as this which led Samuel Morse, in devising his telegraphic code, to choose short code words for the frequently used letters E and T and longer code words for the infrequently used letters Z and Q.

Unfortunately, there is a limit to our cleverness. We come sooner or later to a point where no improvement in transmission speed is possible. That point reflects the 'information content' of the message, the information which must be transmitted to differentiate this particular message from all other possible messages. Therefore, if we are to seek the fastest method of transmitting descriptions of landforms by telegraph, we must first discover by information theoretic analysis the information content of landforms.

Textual messages are analysed for information content by subdividing them into their component parts, letters, numbers and punctuation. To analyse landforms in a similar fashion, they too must be subdivided into a set of component features. These component features must be so simple that they can be easily coded, and yet they must be descriptive so that, when re-assembled, they accurately reconstruct the landform.

Let us, then, consider a landform to be made up of a large number of spot elevations. Spot elevations are easily coded as three numbers representing the spatial co-ordinates of the sampled point. Furthermore, when the spot elevations are sampled with sufficient density, they will represent the landform accurately.

If a gridded height sampling were used to collect spot elevations, the coding could be simplified. The horizontal co-ordinates of each point may be specified by recording the limits of the sampling grid, its mesh size and an implicit scanning sequence. The heights may then be sequentially listed. The position of each point is determined by its place in the sequence of listed heights. Consequently the horizontal co-ordinates of each sample point need not be explicitly coded for storage and transmission.

Such a gridded height sample is used in many computer programs processing terrain data and is known as both a digital terrain model and a numerical map. It is, in fact, an array of heights representing some real topographic shape. The mesh size used in practice varies considerably, depending on the needed resolution. For construction surveys it is sometimes as small as 4 m. In aircraft navigation studies it can be as large as 100 m. The elevations are usually recorded to the nearest metre, although decimetre intervals are sometimes used.[1]

The shape of the land is, then, described by a regular array of heights. Each location in the array exhibits a single height and, taken as an entity in itself, the collection of heights describes the land's form. It is the message we are investigating.

Microstates and macrostates of an array

Before continuing this discussion of the representation of terrain by an array of heights, let us consider the ways in which we might convey information about a general array of objects. One method of array specification is to look at each array element individually. If each such element is fully described, then the array specification is equivalent to a message representing the array in coded form. A second method of specifying an array is simply to examine it superficially, noting in a quick 'glance' characteristic features of the whole array. While a cursory overview does not constitute a complete description of an array, it does give us some idea of what to expect when we begin examining each individual element. Since a generalised description of an array is often available *a priori* (e.g. the array represents a fluvial landform), we should know, roughly, what to expect when we examine each individual member of the array. This should make the task of transmitting the complete specification somewhat easier.

Jagjit Singh has summarised these two methods of array specification:

> Now whenever we are confronted with crowds of entities, whether of men, messages, or molecules, there are two ways of dealing with them. *Either* we specify the attribute(s) under consideration of each and every individual in the crowd, *or* we specify the over-all statistical average(s) of their individual attribute(s). The former is said to define the internal structure or *microstate* of the crowd and the latter its outer façade or *macrostate* (Singh 1966, 73). (Singh's emphasis.)

The task of specifying the microstate is precisely the task of transmitting a message representing the array. The specification of the macrostate, however, will not, as a rule, convey all the information about the array. Singh defines the array macrostate as 'the over-all statistical averages' of the properties of the array elements. More usefully, we may restate the definition of the array macrostate as follows: An array macrostate consists of the specification of the

---

[1] In fact, many systems in use are based on Imperial units. The author prefers to describe these techniques as though metric measurements were universally used.

number of occurrences of the different properties exhibited by the array elements.

As an example illustrating the relationship between microstates and macrostates of an array, consider the familiar array of red, amber and green lights known as a traffic signal. Each light may exhibit the property of being either on or off. A listing of all the possible microstates and macrostates of this array (possible, that is, only after considerable modifications in the Highway Code) is given in the following table (table 3.1).

Table 3.1   *States of a traffic signal array*

Elements of the array: red light, amber light, green light
Attributes: ON, OFF (lighted or not lighted)

| *Microstates – array elements distinct* | | | |
microstate label	*red light*	*amber light*	*green light*
1	ON	ON	ON
2	ON	ON	OFF
3	ON	OFF	ON
4	ON	OFF	OFF
5	OFF	ON	ON
6	OFF	ON	OFF
7	OFF	OFF	ON
8	OFF	OFF	OFF

| *Macrostates – array elements indistinct* | | |
macrostate label	*number of occurrences of ON*	*number of occurrences of OFF*
I	3	0
II	2	1
III	1	2
IV	0	3

There are eight ($2^3 = 8$) different microstates but only four macrostates. Corresponding to macrostate III there are three microstates, 4, 6 and 7. Consequently, knowledge that the traffic signal configuration is characterised by macrostate III limits the possibly occurring microstates to three out of the eight. Even more limiting is macrostate IV which has only one corresponding microstate. Therefore, specification of this macrostate uniquely determines the corresponding microstate.

The number of microstates equivalent to any one macrostate is called the probability number of that macrostate. The probability number indicates two characteristics of the macrostate. First, assuming all microstates are equally probable, a large probability number signifies a likely macrostate. This macrostate will have a high probability of occurrence since it may be fulfilled by a large number of equally probable microstates. Secondly, knowledge of the occurrence of such a likely macrostate conveys very little information about the specific microstate which is, in fact, giving rise to the observed macrostate. That is, a high probability number signifies a large uncertainty about the actual microstate. A low probability number, on the other hand, reduces the uncertainty since there are only a few possibilities from which to choose.

The descriptive capacity of an array

McLachlan (1958, 255–6) has used the concept of array states to investigate two-dimensional pictorial arrays such as photographs. A photograph may be considered as a lattice of grid squares, each square being 0·02 cm on a side. This 0·02-cm mesh size approximates the resolving power of the human eye. On a photograph measuring 20 cm^2 there will be 1,000,000 of these grid squares. It is possible to distinguish among ten shades of grey in each grid square. Consequently, the total number of possible microstates of this pictorial array is $10^{1,000,000}$. McLachlan calls this number the descriptive capacity of the photograph. It is the number of different pictures which could be imaged by a 20-cm^2 photograph. Of course the vast majority of these images would appear to be only a random pattern of grey dots and would be unintelligible to visual interpretation. In fact, $10^{1,000,000}$ is such a huge number that there are not enough atoms in the universe to produce more than a negligible quantity of the many possible photographs (see Ashby, 1964, p. 167).

In general, the descriptive capacity of an array may be defined as the total number of possible microstates of that array. Whenever external constraints or physical limitations placed on the array limit the number of microstates actually possible, then the descriptive capacity of the array is similarly decreased. In an array which is restricted to a particular macrostate, the descriptive capacity is then just the probability number of the macrostate.

If it were possible to determine the relative frequencies of the ten grey shades in the photograph, that is to determine the macrostate of the photograph, then this *a priori* knowledge would decrease the number of possible microstates of the pictorial array. The descriptive capacity of the photograph is also decreased when the photograph is restricted to a particular macrostate. Consequently, it is theoretically possible to transmit this photograph in less time than would be required if nothing were known of the actual macrostate.

The equilibrium macrostate of causal arrays

Unfortunately the macrostate of an arbitrary descriptive array is not always easy to determine in advance. Furthermore, there is no guarantee that an arbitrary array will maintain its macrostate over any appreciable length of time. However, it may be possible to determine a steady-state macrostate of an array if the array is causal rather than arbitrary. That is, it may be possible to determine a stable macrostate of an array if the array consists of real objects which are interrelated by deterministic laws of mechanics. The work of Bertalanffy in general systems theory is largely directed towards this problem. To Bertalanffy (1950, 146) a system is a set of simultaneous differential equations which are, in effect, the mechanical laws governing the physics of real bodies. These equations have the following form:

$$\frac{dQ_1}{dt} = f_1(Q_1, Q_2, \ldots, Q_n)$$

$$\cdot \qquad\qquad \cdot$$
$$\cdot \qquad\qquad \cdot$$
$$\cdot \qquad\qquad \cdot$$

$$\frac{dQ_n}{dt} = f_n(Q_1, Q_2, \ldots, Q_n)$$

where Q_i is an attribute of the ith element of the array. This set of equations is a 'black box' which accepts as input the array $(Q_1, Q_2, \ldots, Q_n)$ and yields the same array as it appears after time dt. This system relates every member of the array $(Q_1, Q_2, \ldots, Q_n)$ to every other member. Thus, there is a feedback network within the system. A change in one member of the array produces compensating changes throughout the array over time.[1]

In statistical mechanics it has been observed that energy tends to become distributed throughout an array of molecules in such a manner that the total amount of work needed to maintain this distribution is a minimum. For example, consider a vessel which contains a gas at constant temperature. The gas molecules are moving about and colliding with each other in a random fashion. Although it is possible that this random motion could lead to a segregation of the molecules in one section of the vessel, this eventuality is highly unlikely. That is, the probability number of this segregated array is low since there is only a limited number of ways in which the individual molecules may occupy

---

[1] For an interesting discussion of feedback theory see Maruyama (1963).

a small area of the vessel. It is more likely that the molecules will be evenly distributed throughout the vessel. This macrostate has the greatest probability number since any one molecule might be located anywhere within the vessel. If the molecules are evenly distributed within the vessel, then any one molecule will have the same number of collisions, on the average, as any other molecule in the vessel. There will be no net change of kinetic energy of the average molecule, and consequently there can be no net work performed within the system. Thus this macrostate consisting of an even distribution of molecules within the vessel is not only the most probable spatial distribution, it is also the macrostate which distributes the energy so as to fulfil the least-work condition. This distribution is in mechanical equilibrium and will maintain itself indefinitely over time.

Now consider that the vessel is not kept at constant temperature throughout but is heated at some points and cooled at others. Then there will be an uneven distribution of energy within the vessel. Under such conditions the warmer sections will be continually heating the cooler sections, thus performing work on them. Even in this case the principle of least-work may be applied. The spatial distribution of molecules will be such that the total work performed in maintaining this arrangement will be minimised. There will be fewer molecules within the warm regions and a greater number in the cool regions. This macrostate will be the most probable condition allowed under the given constraints. Since this macrostate is mechanically stable, it is often called a steady-state condition. It will not change over time so long as the constraints remain constant.

These statistical mechanical concepts have been transplanted by general systems theorists to other disciplines such as biology, ecology, sociology, geography and geomorphology (e.g. Chorley 1962). In most cases these analogies to the behaviour of gas molecules have been used in order to derive the equilibrium macrostates or steady-state conditions. Such arguments are based on the general principle that the distribution of energy among the members of the array leads to the macrostate with minimum possible internal work. In descriptive arrays where this notion is applicable, the task of finding a valid macrostate may be greatly simplified. Note, however, that nature does not co-operate as well as she might. The principle of least-work invariably leads to a steady state whose probability number is the largest possible under the given constraints. Nevertheless, if the system is controlled by persistent, external constraints, then the array may be forced to maintain a macrostate of relatively low probability. Such systems are intrinsically easier to describe since they are capable of only a restricted number of microstates.

Limitations on the descriptive capacity of terrain

A numerical topographic map is a descriptive array of topographic heights. The descriptive capacity of this array, as that of any other causal array, is limited by its macrostate. The number of possible microstates may be calculated from the statistics for the occurrence of the various terrain heights. If there is great variation in the possible terrain heights in an area, then there is great uncertainty as to the actual terrain (microstate) stored in the numerical map. If, on the other hand, there is little variation in the possible terrain heights, then this terrain will be easier to describe since there are fewer possible microstates.

The most important consideration in the determination of the equilibrium macrostate of terrain is the land-shaping process of erosion by running water seen as a causal system operating under deterministic laws. Leopold and Langbein (1962) have considered topography as a device which drains water from the land and carries it to the ocean. This device is subject to the same least-work principles that are observed in statistical mechanics. On the basis of these principles Langbein and Leopold (1964) have deduced an equilibrium morphology for river channels that is adjusted to the most efficient flow of water. As a result of their work, it may be expected that the equilibrium shape of the earth's surface, in general and in particular, may be determined and utilised in the coding and the transmitting of numerical terrain data.

As an initial approach to the descriptive capacity of terrain, let us consider that the shaping of landforms by fluvial erosion and deposition is a process of uniform intensity operating under uniform constraints. Strahler (1950, 673) notes that a uniformity of process produces a marked uniformity in the landscape. He states:

> In a mature stage of development the landscape consists of an intricate combination of channels, slopes and divides. Although the same unit form is not exactly repeated throughout this type of topography, an observer cannot fail to recognize that the forms are approximately the same throughout an area having a uniform lithology, geologic history, climate, soil and vegetation.

Since the landscape is the source for the storage and transmission of terrain height information, the observed regularity of the landscape may well be used to optimise the coding of this information. This regularity of the terrain may be regarded as a restriction on the descriptive ability of the land's surface. When the landscape is viewed as a descriptive array of topographic heights, it is seen that this array is not arbitrary but is constrained by the laws of fluvial mechanics. It is constrained to a morphology which is characterised by a particular macrostate of the height array. The descriptive capacity of this macro-state is its probability number, the number of microstates whose form and regularity are compatible with fluvial erosion and deposition.

The average information of a macrostate

The information content of a macrostate is measured as the logarithm to the base two of the probability number of the macrostate. The logarithm is used as an information measure so that the information content of two combined arrays is given as the sum of their separate information contents. The base two is used so that the information content becomes equivalent to the number of binary digits needed to create a binary array with the same descriptive capacity. That will be just the number of binary digits needed to store or transmit a message completely describing the array.

Since the information content of an array macrostate is dependent on the number of elements in that array, it is difficult to compare the information contents of two arrays of different size. The most useful measure of the information conveyed by an array constrained to a particular macrostate is, therefore, the average information of the array macrostate. It is the number of binary digits needed, on average, to transmit one element of the array. Although the average information could be calculated directly from the macrostate specification by means of combinatorial mathematics, it is far simpler to use Shannon's formula to approximate the average information of an array element.[1] Shannon's formula gives the average information H as

$$H = \sum_{i=1}^{N} P_i \log_2 \frac{1}{P_i}$$

where N is the number of different attributes which each element might adopt and P_i is the probability of the occurrence of the ith attribute at any element throughout the array.

A first approximation to the average information of topography may be obtained from the probability distribution of the heights as they would occur on a numerical map data grid. From these probabilities the average information may be calculated according to Shannon's formula. It is important to note, however, that the array macrostate is then specified by the probabilities of the *independent* occurrences of the heights on the data grid. Conditional probabilities based on the values of neighbouring heights are not considered. Thus, the regularity inherent in the smoothness and continuity of the landscape is neglected. The probability distribution of the heights as independent events provides

[1] For a derivation of Shannon's formula from the combinatorial expression for the average information of an array see either Brillouin (1962) or Connelly (1968).

only a bare outline of the actual regularity of the earth. Much theoretical work remains to be done before the earth's true regularity can be specified statistically and the true average information determined accurately.

Equilibrium states of the terrain

Papers by Hack (1960) and Chorley (1962) have argued that the persistence of certain regular landforms is due to an equilibrium distribution of hydraulic energy over the land's surface. The idea of energy equilibrium in stream channels is much older. It was first proposed by the Italian engineer Domenico Guglielmini in 1697. The interdependence of stream equilibrium and landscape geometry was later stressed by G. K. Gilbert in his 'Report on the Geology of the Henry Mountains' (1877). Around 1902, William Morris Davis popularised the term 'grade' as a synonym for the equilibrium profile of a stream. In extending the principle of grade to the whole landscape Chorley (1962) borrowed the general systems theory terminology of 'steady state' to describe the equilibrium shape of the land.

Leopold and Langbein (1962) view the flow of water over the terrain as a vast system which rapidly adjusts to the equilibrium condition of least-work. They argue that the natural feedback interaction between erosion and deposition imposes a stability on the landscape. This stability is characterised by a continual tendency of the land to minimise the work expenditure of the water running over its surface. Thus, the fluvial landforms fluctuate about, but perhaps never actually attain, the ideal condition of least-work.

However, this quasi-stability should not be interpreted as leading to a permanence of the terrain features. Although it maintains the form which, under the given constraints, minimises the expenditure of work, the land will continue to degrade gradually thus reducing the available relief. Other constraints such as vegetation, climate and rock structure may also change through time and produce changes in the equilibrium configuration. Leopold and Langbein (1962, A19) conclude that 'Landscape evolution is an evolution in the nature of constraints in time, maintaining meanwhile a dynamic equilibrium or quasi-equilibrium.'

Changes in the constraints take place very slowly. By comparison the modelling of the landscape by fluvial erosion and deposition is very rapid indeed. Thus, the landscape will usually maintain itself constantly in dynamic equilibrium despite the evolution of the constraints. Inequilibrium caused by catastrophic changes in the constraints will be short-lived. It follows that landforms are primarily the result of the constraints imposed at the present time. To a large extent, the observed landscape is independent of its previous evolution. This minimises any disorder in the landscape resulting from differences in development. Instead, the uniformity of the present-day constraints leads to the order and regularity observed in fluvial landforms.

Much work on determining the equilibrium form of the terrain remains to be done. It is possible now to explain only a few topographic forms directly on the basis of the minimum work configuration of running water. Taking this approach, Langbein (1964) and Langbein and Leopold (1964) have studied and described the least-work configuration of river channels. While it is true that only a small fraction of the earth's surface is directly shaped by channelised flow, it should be realised that the erosional slopes in a drainage basin are adjusted to and often conform with the dominant stream channel. For the purposes of this investigation it will be sufficient to consider a stream profile as a typical terrain profile. Thus, the occurrence of heights along a river at grade will be representative of the occurrence of heights throughout the steady-state configuration of the terrain.

The equilibrium profile of a river

The equilibrium profile of a river is determined by the distribution of energy in the river. In general, the energy E of flowing water in a reach along the channel of a stream is given by Bernoulli's equation,[1]

$$E = \gamma V h + \frac{V \rho v^2}{2}.$$

The first term $\gamma V h$ is the potential energy of the water in the reach in terms of its volume V, its height h and the specific weight of water γ. In the second term, which represents the kinetic energy, v is the velocity of the water in the reach and ρ is the density of water. The loss of energy ΔE over the length of the reach ΔL represents the work expended by the stream per unit of stream length, $\Delta W / \Delta L$. Thus,

$$\frac{\Delta W}{\Delta L} = \frac{\Delta E}{\Delta L} = V \gamma \frac{\Delta h}{\Delta L} + \frac{V \rho}{2} \frac{\Delta v^2}{\Delta L}.$$

Leopold (1953) has shown that a stream will adjust its depth so as to maintain a constant velocity throughout its length at any one time. This is particularly true at the time of bankfull discharge, the time when the stream exerts a critical force on the shape of the land. Consequently,

$$\frac{\Delta v^2}{\Delta L} = 0 \quad \text{and} \quad \frac{\Delta W}{\Delta L} = V \gamma \frac{\Delta h}{\Delta L}.$$

Since the stream is flowing, water is continually passing by a given

[1] Bernoulli's equation is discussed by Streeter (1958), p. 86–103.

point and work is continually expended in the reach. The work expenditure per unit of time $\Delta W/\Delta t$ is termed the power expenditure ΔP. Thus,

$$\frac{\Delta W}{\Delta L\,\Delta t} = \frac{\Delta P}{\Delta L} = \frac{V}{\Delta t}\gamma\frac{\Delta h}{\Delta L} = Q\gamma\frac{\Delta h}{\Delta L}.$$

where Q is the discharge of the stream measured in volume of stream flow per unit of time.

The above formula may be simplified by the assumption that the stream runs flat enough so that ΔL may be approximated by Δx, the change in horizontal distance. Under this assumption

$$\frac{\Delta h}{\Delta L} = \frac{\Delta h}{\Delta x} = S,$$

the slope of the stream. Consequently,

$$\frac{\Delta P}{\Delta L} = \gamma QS;$$

that is, the power expenditure per unit length is proportional to QS, the discharge times the slope.

Langbein and Leopold (1964) have used this relationship to deduce a condition for the least-work stream profile. In doing so they implicitly use the empirical relationship $S = -kQ^z$ where $k > 0$ and z are constants determined by observation. Then

$$\frac{\Delta P}{\Delta L} = -\gamma kQ^{z+1} \quad \text{and} \quad P = \Sigma -\gamma kQ^{z+1}\,\Delta L$$

is the total power expended by the river. P is minimised as z goes to negative infinity. In practice, however, the minimum value of z is found to be about $-1\cdot 0$. This appears to be its physical limit due to the interrelationships with the other factors of hydraulic geometry. It is concluded, therefore, that in streams with the minimum allowable power expenditure $z = -1\cdot 0$. Since

$$\frac{\Delta P}{\Delta L} = -\gamma kQ^{z+1},$$

$$\frac{\Delta P}{\Delta L} = -\gamma kQ^0 = -\gamma k \text{ (constant)}.$$

Therefore $\quad \gamma QS = \dfrac{\Delta P}{\Delta L} = -\gamma k$ (constant)

and $\qquad\qquad\qquad QS = -k$ (constant).

This result is summarised by Leopold, Wolman and Miller (1964, 270), who state, 'Thus constant power expenditure per unit of stream length can be shown to be equivalent to minimum rate of work in the river system.'

The slope of a river is, then, dependent on its discharge. The discharge is in turn dependent on the area and the shape of the drainage basin. Empirical relationships show that the discharge is, on the average, proportional to the length of the stream measured horizontally from the drainage divide. That is, $Q \approx cx$ where c is a constant determined by observation and x is the horizontal distance along the stream. For simplicity this approximation will be assumed to be a true equality.

Substituting this relationship into the previous expression,

$$cxS = cx\frac{dh}{dx} = -k \text{ (constant).}$$

Therefore
$$x\frac{dh}{dx} = \frac{-k}{c} = -K \text{ (constant).}$$

Integrating this differential equation gives the following relationship between the topographic height h and the horizontal distance x:

$$-K \ln x = h + C$$

where C is the constant of integration. This constant is determined by the slope of the river at its mouth. This equation leads to the longitudinal profile shown in fig. 3.1.

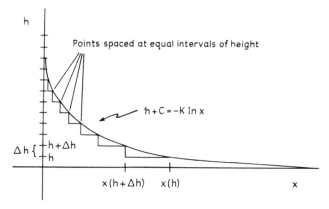

Fig. 3.1 Stream profile of minimum work.

The probability distribution of heights

The profile shown in fig. 3.1 may be approximated by a set of points spaced at equal intervals of height. That is, this profile may be given as a step function whose steps occur at height intervals of Δh, the unit height. The widths of these steps may vary from very narrow at the highest elevations to very broad at the river's mouth. The ratio of the step widths to the total horizontal length of the stream gives the probability $p(h)$ of the occurrence of the height h.

Since $h + C = -K \ln x$, $x = e^{-\left(\frac{h+C}{K}\right)}$ or, in function notation, $x(h) = e^{-\left(\frac{h+C}{K}\right)}$. Using this formula for the horizontal distance $x(h)$, the probability $p(h)$ is derived as follows:

$$p(h) = \frac{x(h) - x(h + \Delta h)}{x(0)}$$

$$p(h) = \frac{e^{-\left(\frac{h+C}{K}\right)} - e^{-\left(\frac{h+\Delta h+C}{K}\right)}}{e^{-\left(\frac{C}{K}\right)}}$$

$$p(h) = \frac{e^{-\left(\frac{h+C}{K}\right)}\left(1 - e^{-\left(\frac{\Delta h}{K}\right)}\right)}{e^{-\left(\frac{C}{K}\right)}}$$

$$p(h) = e^{-\left(\frac{h}{K}\right)}\left(1 - e^{-\left(\frac{\Delta h}{K}\right)}\right).$$

The term $\left(1 - e^{-\left(\frac{\Delta h}{K}\right)}\right)$ is a constant dependent on Δh and K.

Let $\left(1 - e^{-\left(\frac{\Delta h}{K}\right)}\right) = T$ (constant), then

$$p(h) = Te^{-\left(\frac{h}{K}\right)}.$$

Since all the probabilities must sum to one,

$$\sum_{h=0}^{\infty} p(h)\, \Delta h = \sum_{h=0}^{\infty} Te^{-\left(\frac{h}{K}\right)}\, \Delta h = 1.$$

Since Δh is very small compared with the upper limit of h this summation is equivalent to an integration over all h.

$$\int_{0}^{\infty} Te^{-\left(\frac{h}{K}\right)}\, dh = 1$$

$$-TKe^{-\left(\frac{h}{K}\right)}\Big]_{0}^{\infty} = 1$$

$$TK = 1$$

$$T = \frac{1}{K}.$$

Therefore
$$p(h) = Te^{-hT}.[1]$$

The probability distribution of heights along a stream in equilibrium is thus determined except for the value of T. Although T might well

[1] The same formula is used, but not derived, by V. I. Sukhov in his investigations of the information content of maps.

be determined by theoretical considerations, it is much simpler to compute T empirically from observation.

Since the terrain is largely controlled by the profiles of the major streams, a similar probability distribution is to be expected for all points on the land's surface.

Let us make the outrageous assumption that the probability distribution for the dry land heights over the globe is of the form $p(h) = Te^{-hT}$ as derived above. Figures given by Wagner (1912, 279) show that about 73% of the earth's dry land lies between sea level and 1000 m in elevation. Therefore, taking $\Delta h = 1$ m,

$$0{\cdot}73 = \int_{0\,\text{m}/\Delta h}^{1000\,\text{m}/\Delta h} p(h)\,\mathrm{d}h = \int_0^{1000} Te^{-hT}\,\mathrm{d}h = -e^{-hT}\Big]_0^{1000}$$

$$= -e^{-1000T} + 1.$$

$$-e^{-1000T} = -0{\cdot}27$$

$$-1000T = \ln 0{\cdot}27$$

$$T = 0{\cdot}0013.$$

Consequently, the probability $p(h)$ of the occurrence of the height h over the entire globe is given by the formula,

$$p(h) = 0{\cdot}0013e^{-(0{\cdot}0013)(h)}.$$

The average information of the terrain

On this basis the array macrostate of the earth's terrain heights may be specified in terms of the probability distribution of those heights, $p(h) = 0{\cdot}0013e^{-(0{\cdot}0013)(h)}$. With this distribution it is possible to compute the average information of this macrostate according to Shannon's equation,

$$H_{\text{heights}} = \sum_{h=0}^{N} p(h) \log_2 \frac{1}{p(h)}.$$

Although the maximum difference in relief of the earth's land is about 8850 m (elevation of Mt Everest), it suffices to take $N = 8192 = 2^{13}$. The average information of the world's heights may then be evaluated by direct summation on a digital computer. It is found that H_{heights} is very nearly 11 bits per spot elevation.[1] Since it would require 13 bits

---

[1] The unit length Δh is equivalent to the contour interval on the numerical map, the smallest difference possible between two recorded elevations. The average information H_{heights} as computed here is dependent on the value of Δh and, similarly, the average information conveyed by a spot height on the numerical map is dependent on the contour interval used. Values of the average information (in bits per spot elevation) corresponding to various values of Δh are given below:

$$\Delta h = \quad 1\text{ ft}, \quad H_{\text{heights}} = 12{\cdot}714 \text{ bits};$$
$$\Delta h = \quad 2\text{ ft}, \quad H_{\text{heights}} = 11{\cdot}725 \text{ bits};$$

(Footnote continued on next page)

per recorded elevation to differentiate these 8192 heights using a fixed length code, the efficiency of that code would be $11/13 = 85\%$. The redundancy is 15%.

The computed average information, $H_{\text{heights}} = 11$ bits, is only a first approximation to the actual average information of landforms. A second, more accurate approximation could be obtained by considering the terrain as a grid of height changes. Then, the probabilities for the transition from one height to another between adjacent grid cells would be used to calculate the average information conveyed by a spot height. The difficulty here is the determination of the probabilities for the height transitions. The use of the height grid suggests the application of cellular automata theory to the determination of the least-work configuration for slopes and, hence, the probability distribution for height transitions. Until such work is done, the probabilities for the occurrences of heights alone, and not height transitions, will remain the most accurate method of specifying the macrostate of the earth's surface.

It is significant that the analysis of landforms as messages is based on a realisation of the remarkable regularity of landforms. If we think of terrain as being the output from, rather than the input to, a numerical map, we begin to appreciate how great is the number of logically possible landforms and, in comparison, how small is the number of actually occurring landforms. The United States Topographic Command numerical maps, for example, are able to distinguish $10^{25,000,000}$ logically possible map sheets. The number of conceivable topographic patterns, though large in absolute terms, is but a small fraction of this total number. This is, of course, significant if one is trying to encode landforms for digital storage and transmission. It is also significant for the geomorphologist. The principal idea which he should gain from the study of landforms as messages is this: It is no longer the complexity of landforms which confounds us. It is their unexplained simplicity.

References

ABRAMSON, N. (1963) *Information Theory and Coding* (McGraw-Hill, New York).

ASHBY, W. ROSS (1964) Introductory remarks at panel discussion; In Mesarovic, M. D. (Ed.), *Views on General Systems Theory* (Wiley, New York), 165–9.

$$\Delta h = \quad 5 \text{ ft}, \quad H_{\text{heights}} = 10{\cdot}413 \text{ bits};$$
$$\Delta h = \quad 10 \text{ ft}, \quad H_{\text{heights}} = 9{\cdot}422 \text{ bits};$$
$$\Delta h = \quad 20 \text{ ft}, \quad H_{\text{heights}} = 8{\cdot}435 \text{ bits};$$
$$\Delta h = \quad 50 \text{ ft}, \quad H_{\text{heights}} = 7{\cdot}143 \text{ bits};$$
$$\Delta h = 100 \text{ ft}, \quad H_{\text{heights}} = 6{\cdot}180 \text{ bits};$$
$$\Delta h = 200 \text{ ft}, \quad H_{\text{heights}} = 5{\cdot}233 \text{ bits}.$$

BERTALANFFY, L. VON (1950) An outline of general systems theory; *British Journal for the Philosophy of Science* 1, No. 2, 134–65.

BERTALANFFY, L. VON (1956) General systems theory; *General System Yearbook* 1, 1–10.

BRILLOUIN, L. (1962) *Science and Information Theory*; 2nd Edn (Academic Press, New York).

CHORLEY, R. J. (1962) Geomorphology and general systems theory; *U.S. Geological Survey Professional Paper* 500-B, Washington, D.C.

CHORLEY, R. J. (1965) A re-evaluation of the geomorphic system of Davis, W. M.; In Chorley, R. J. and Haggett, P. (Eds.), *Frontiers in Geographical Teaching* (Methuen, London), 21–38.

CHORLEY, R. J. (1967) The application of statistical methods to geomorphology; In Dury, G. H. (Ed.), *Essays in Geomorphology* (American Elsevier, New York), 275–387.

CONNELLY, D. S. (1968) *The Coding and Storage of Terrain Height Data: An Introduction to Numerical Cartography* (M.Sc. Thesis, Cornell University, Ithaca, New York).

CULLING, W. E. H. (1957A) Multicycle streams and the equilibrium theory of grade; *Journal of Geology* 65, 259–74.

CULLING, W. E. H. (1957B) Equilibrium states in multicycle streams and the analysis of river-terrace profiles; *Journal of Geology* 65, 451–67.

CURRY, L. (1966) Chance and landscape; In House, J. W. (Ed.), *Northern Geographical Essays* (Oriel Press, Newcastle-upon-Tyne), 40–55.

HACK, J. T. (1960) Interpretation of erosional topography in humid temperate regions; *American Journal of Science* Vol. 258-A, Bradley Volume, 80–97.

HOWARD, A. D. (1965) Geomorphological systems – equilibrium and dynamics; *American Journal of Science* 263, 302–12.

LANGBEIN, W. B. (1964) Geometry of river channels; *Proceedings, Journal of the Hydraulics Division, American Society of Civil Engineers* 90, No. HY2, 301–12.

LANGBEIN, W. B. and LEOPOLD, L. B. (1964) Quasi-equilibrium states in channel morphology; *American Journal of Science* 262, 782–94.

LEOPOLD, L. B. (1953) Downstream change of velocity in rivers; *American Journal of Science* 251, 606–24.

LEOPOLD, L. B. and LANGBEIN, W. B. (1962) The concept of entropy in landscape evolution; *U.S. Geological Survey Professional Paper* 500-A, Washington, D.C.

LEOPOLD, L. B. and MADDOCK, T., JR (1953) The hydraulic geometry of stream channels and some physiographic implications; *U.S. Geological Survey Professional Paper* 252, Washington, D.C.

LEOPOLD, L. B., WOLMAN, M. G. and MILLER, J. P. (1964) *Fluvial Processes in Geomorphology* (W. H. Freeman, San Francisco).

MARUYAMA, M. (1963) The second cybernetics: Deviation amplifying mutual causal processes; *American Scientist* 51, 167–79.

McLACHLAN, D., JR (1958) Description mechanics; *Information and Control* 1, No. 3, 240–66.

MILLER, C. L. and LAFLAMME, R. A. (1958) The digital terrain model – Theory and applications; *Photogrammetric Engineering* 24, 433–442.

RBZA, F. M. (1961) *An Introduction to Information Theory* (McGraw-Hill, New York).

SCHEIDEGGER, A. E. (1964) Some implications of statistical mechanics in geomorphology; '*Bulletin of the International Association of Scientific Hydrology*, 9, 12–16.

SCHUMM, S. A. and LICHTY, R. W. (1965) Time, space and causality in geomorphology; *American Journal of Science* 263, 110–19.

SHANNON, C. E. (1948) A mathematical theory of communication; Parts I and II, *Bell System Technical Journal* 27, 379–423.

SINGH, J. (1966) *Great Ideas in Information Theory, Language and Cybernetics* (Dover, New York).

STRAHLER, A. N. (1950) Equilibrium theory of erosional slopes approached by frequency distribution analysis; *American Journal of Science* 248, 672–96 and 800–14.

STRAHLER, A. N. (1952) Dynamic basis of geomorphology; *Bulletin of the Geological Society of America* 63, 923–38.

STREETER, V. L. (1958) Fluid-flow concepts and basic equations; *Fluid Mechanics* (McGraw-Hill, New York), 78–136.

SUKHOV, V. I. (1967) Information capacity of a map: Entropy; *Geodesy and Aerophotography* (in translation from the Russian), No. 4, 212–15.

TANNER, W. F. (1959) Examples of departure from the gaussian in geomorphic analysis; *American Journal of Science* 257, 458–60.

TANNER, W. F. (1962) Geomorphology and the sediment transport system; *Southeast Geology* 4, No. 2, 113–25.

TOBLER, W. R. (1965) Computation of the correspondence of geographical patterns; *Papers of the Regional Science Association* 15, 131–9.

WAGNER, H. (1912) *Lehrbuch der Geographie*; 1 (Hahnsche Buchhandlung, Hannover and Leipzig.)

WHOLEY, J. S. (1961) The coding of pictorial data; *Transactions on Information Theory, Institute of Radio Engineers*, Vol. IT-7, No. 2, 99–104.

PART II
Point systems

4 Probabilities of surface karst

H. McCONNELL and J. M. HORN

Department of Geography, Northern Illinois University

Introduction

This paper considers spatial variation of surface karst within the context of macroscopic randomness of geomorphic process.[1] In it, we assess the utility of several alternative hypotheses, each containing one or more random components, about the broadscale processes responsible for the spatial distribution of karst depressions in the unglaciated portion of the Mitchell Plain of southern Indiana.

The analysis comes under the rubric of quadrat methods.[2] Individual mappable depressions on 7·5 minute series U.S. Geological Survey topographic quadrangles are observed as points on a plane. Frequencies of depressions are enumerated for an area sample and compared statistically with expected frequencies of the Poisson, Negative Binomial, and Mixed Poisson probability distributions. The expectation of each distribution is set equal to the mean cell count and each probability law has a spatial analog which operationalises one of the hypotheses in the abstract.

[1] The notion of randomness of process as it applies to karst geomorphology was evidently first employed by Curl (1959, 1966). In the latter, it was used to develop a stochastic model of the number and length distributions of entranceless caves and the length distribution of all caves in a region.

[2] The single example of this technique in geomorphology is the paper by LaValle (1967). Quadrat methods have been used to some degree in palaeontology (see Miller and Kahn 1962) and are well established in quantitative plant ecology (see Greig-Smith 1964, for a comprehensive treatment). King (1969) and McConnell (1967) discuss this technique within the context of geographic research.

The conceptual framework: macroscopic randomness

Much of the recent literature of geomorphology indicates that the application of principles of statistical mechanics is a viable approach to understanding the development of landforms. (See, for example, works by Chorley (1962), Dacey (1968A, 1968B), Leopold and Langbein (1962), Scheidegger (1966), Scheidegger and Langbein (1966), Shreve (1967) and Werner (1970).) This approach, which may be termed the conceptual framework of macroscopic randomness, is heuristic because landform assemblages are considered as the most probable outcome of a collection of random or quasi-random processes. As noted by Scheidegger and Langbein (1966, C1), microscale processes are ignored because:

> To understand the development of a landscape in terms of fundamental mechanical principles is a problem of extreme complexity. . . . The processes that are operative represent the cumulative effect of many small-scale events which are impossible to follow in detail. . . .
>
> . . . many direct approaches at explanation of landscape evolution have been tried . . . in view of the great complexity of the phenomena involved, these approaches may lead to as many explanations as there are cases, meaning that there is no general theory. A hypothesis based upon the concepts of statistical mechanics (that is, a probabilistic rather than a direct approach) . . . may yield a satisfactory and practical theory.
>
> One of the common assumptions that are introduced in statistical mechanics is the randomness of some of the processes. Under certain circumstances, this randomness can lead to logistic difficulties because in geomorphology nothing is really random; when it rains, the fall of each raindrop is governed by Newton's law of motion and by atmospheric friction. When the drop hits the ground and erodes some of the soil particles, again the appropriate laws of mechanics determine the initiation of events. The same is true for the formation and deepening of small gullies that form, say, on a slope and ultimately cause its decay. Yet, it would be quite hopeless to try to account for these processes in detail. . . .
>
> Our knowledge of the individual events can never be complete enough to deduce therefrom, say, the recession of a slope, the juncture of channels, or the velocity of flow that caused them. Nevertheless, certain average relationships can be deduced from our incomplete knowledge of the individual processes, simply because the net effect of the many individual events is the same as if the individual events were to occur at random, although the events are, strictly speaking, predetermined. . . . The random walk generation of a

drainage basin (Leopold and Langbein, 1962), verified as realistic (Schenck, 1963), does not belie the fact that rivers join at specific places for specific reasons. But the whole result is as if the process were random.

Each process is deterministic; however, the rates and periods differ, and the result may be indistinguishable from random.

This conceptual base is convenient and necessary for development of theory in geomorphology, keeping in mind that a landscape or stream network, for example, which may in the aggregate be considered the most probable assemblage for a set of macroscopically random processes is in actuality the predetermined expression of a collection of interacting but 'homogeneously complex' mechanical and chemical processes as noted by Strahler (1952). However, it is impossible to observe the detail of the individual processes at work in time and space.

In an operational sense, the presence of random properties in a landscape is accepted as evidence of macroscopically random components in the processes which formed it.

Purpose of the analysis

The present status of karst geomorphology is one of minimal sophistication when compared with modern studies in fluvial geomorphology. Only recently have karst phenomena been considered as components of dynamic systems (White *et al.* 1970). To this we add an element of randomness: it is the aim of this analysis to treat the utility of an overall hypothesis that the spatial distribution of karst depressions results from geomorphic processes which contain random components. A conclusion that such a hypothesis is tenable should provide a reference point for more detailed analyses.

Such a conceptual framework of karst development has been introduced by Curl (1959, 1966) and LaValle (1967). However, it was done in an entirely different context by the former. The paper by LaValle appears to contain several conceptual and methodological shortcomings. One is the conclusion that a randomly located depression has the propensity to generate another in close proximity and that the growth rate is logarithmic; a second is the use of a small level of significance in conducting statistical tests in what was evidently a Type II error situation.

Karst development and the Mitchell Plain

Selection of study region

Karst development is evidently enhanced in the presence of four primary conditions.

The first is a pure, soluble carbonate unit at or near the surface (Dicken 1935). Secondly, the unit should be highly jointed, dense and thin bedded, leading to infiltration of precipitation into zones of concentration along joints and bedding planes (Woodward 1961). Conversely, if the unit were permeable, precipitation would be absorbed

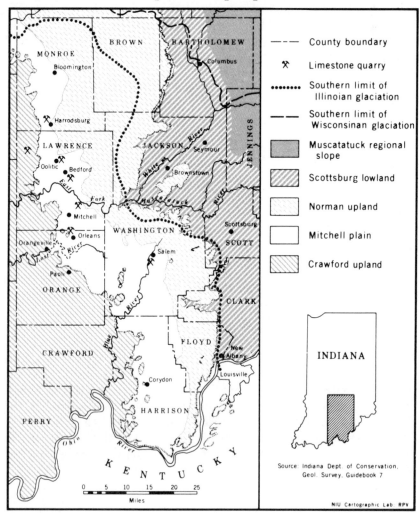

Fig. 4.1 Generalised physiographic map of Mitchell Plain and surrounding area.

uniformly, leading to uniform downwasting. A third condition is entrenchment of valleys (Thornbury 1969). This produces a subsurface conduit network which is the primary system for transporting materials removed by solution. Lastly, solution is intensified by an abundance of organic matter for the formation of organic acids (McGregor and Rarick 1962). It may be noted that classic areas of ongoing karst development have sufficient rainfall for forest growth with occasional high intensity precipitation.

All of these conditions are present in the Mitchell Plain. Moreover, it has not been the subject of modern quantitative analysis in geomorphology.

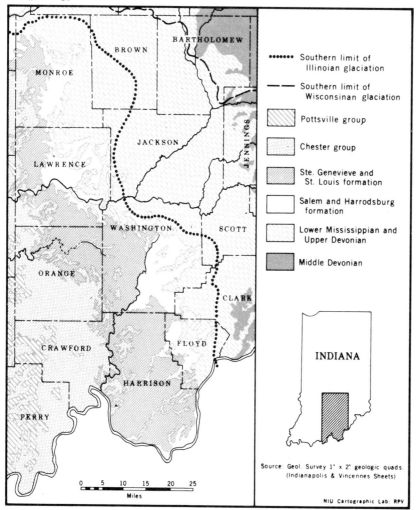

Fig. 4.2 Generalised geologic map of Mitchell Plain and surrounding area.

Topographic setting

The cuesta-like Mitchell Plain (fig. 4.1), which extends into Kentucky as the Pennyroyal Plateau, encompasses approximately 1125 square miles (Frey and Lane 1966) and lies on the western flank of the Cincinnati Arch. The regional dip is into the Illinois Basin at the rate of approximately 20 ft per mile. The northern portion is covered with till deposited during the Illinoian (Riss) glacial stage, obscuring much of the surface karst. The southern portion, or study region for this analysis, is unglaciated. The Mitchell Plain is bounded on the west and east by the Crawford and Norman Uplands, respectively (see fig. 4.1). These three topographic units are separated by two sandstone escarpments; the Mitchell Plain from the Crawford Upland by the Chester Escarp-

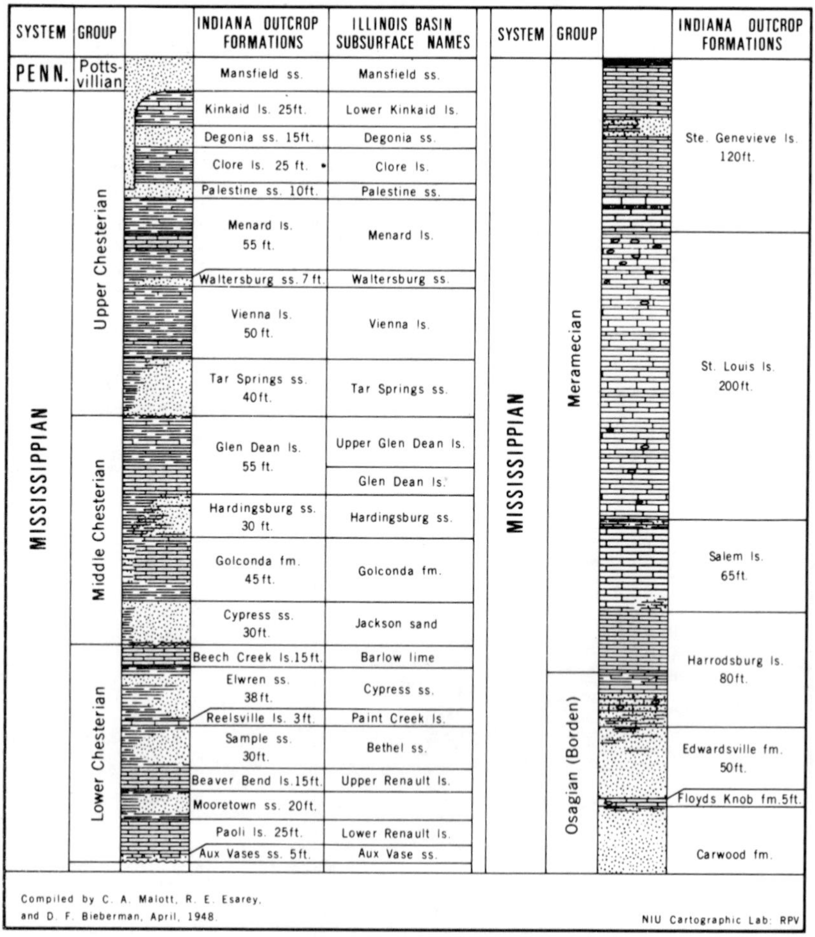

SYSTEM	GROUP		INDIANA OUTCROP FORMATIONS	ILLINOIS BASIN SUBSURFACE NAMES	SYSTEM	GROUP		INDIANA OUTCROP FORMATIONS
PENN.	Potts- villian		Mansfield ss.	Mansfield ss.				
MISSISSIPPIAN	Upper Chesterian		Kinkaid ls. 25ft.	Lower Kinkaid ls.	MISSISSIPPIAN	Meramecian		Ste. Genevieve ls. 120ft.
			Degonia ss. 15ft.	Degonia ss.				
			Clore ls. 25 ft.	Clore ls.				
			Palestine ss. 10ft.	Palestine ss.				
			Menard ls. 55 ft.	Menard ls.				
			Waltersburg ss. 7 ft.	Waltersburg ss.				St. Louis ls. 200ft.
			Vienna ls. 50 ft.	Vienna ls.				
			Tar Springs ss. 40ft.	Tar Springs ss.				
	Middle Chesterian		Glen Dean ls. 55 ft.	Upper Glen Dean ls.				
				Glen Dean ls.				
			Hardingsburg ss. 30 ft.	Hardingsburg ss.				Salem ls. 65ft.
			Golconda fm. 45ft.	Golconda fm.				
			Cypress ss. 30ft.	Jackson sand		Osagian (Borden)		Harrodsburg ls. 80ft.
	Lower Chesterian		Beech Creek ls.15ft.	Barlow lime				
			Elwren ss. 38ft.	Cypress ss.				
			Reelsville ls. 3ft.	Paint Creek ls.				Edwardsville fm. 50ft.
			Sample ss. 30ft.	Bethel ss.				
			Beaver Bend ls.15ft.	Upper Renault ls.				Floyds Knob fm.5ft.
			Mooretown ss. 20ft.					
			Paoli ls. 25ft.	Lower Renault ls.				Carwood fm.
			Aux Vases ss. 5ft.	Aux Vase ss.				

Compiled by C. A. Malott, R. E. Esarey, and D. F. Bieberman, April, 1948

NIU Cartographic Lab: RPV

Fig. 4.3 Mississippian stratigraphic column.

ment and from the Norman Upland by the Knobstone Escarpment. But a small number of major surface streams traverse the Mitchell Plain (fig. 4.1), pointing to the dominance of subsurface drainage. The most intensive karst development is found immediately updip of the Chester Escarpment.

Lithologic units

The Mitchell Plain is comprised essentially of the outcrop pattern of Middle Mississippian (Meramecian) limestone formations (fig. 4.2). The

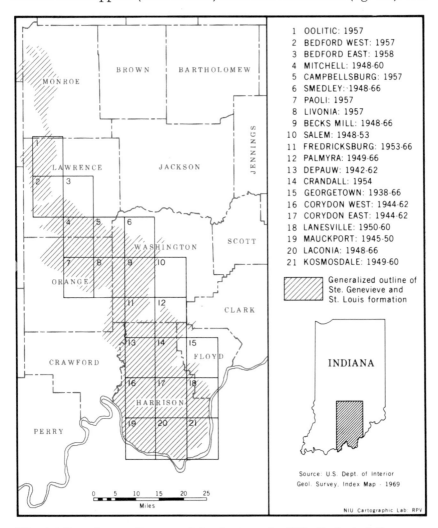

1	OOLITIC: 1957
2	BEDFORD WEST: 1957
3	BEDFORD EAST: 1958
4	MITCHELL: 1948-60
5	CAMPBELLSBURG: 1957
6	SMEDLEY: 1948-66
7	PAOLI: 1957
8	LIVONIA: 1957
9	BECKS MILL: 1948-66
10	SALEM: 1948-53
11	FREDRICKSBURG: 1953-66
12	PALMYRA: 1949-66
13	DEPAUW: 1942-62
14	CRANDALL: 1954
15	GEORGETOWN: 1938-66
16	CORYDON WEST: 1944-62
17	CORYDON EAST: 1944-62
18	LANESVILLE: 1950-60
19	MAUCKPORT: 1945-50
20	LACONIA: 1948-66
21	KOSMOSDALE: 1949-60

Generalized outline of Ste. Genevieve and St. Louis formation

INDIANA

Source: U.S. Dept. of Interior Geol. Survey, Index Map - 1969

NIU Cartographic Lab: RPV

Fig. 4.4 Locations and names, 7·5 minute series U.S. Geological Survey topographic quadrangles.

Meramecian series is subdivided into two groups: (1) the Sanders group, including the Harrodsburg and Salem limestones, and (2) the Blue River group which includes the Ste Genevieve and St Louis limestones. Classic karst development is limited to the Blue River formations (Perry *et al.* 1954). Thus, the analysis pertains only to the areal extent of the Ste Genevieve and St Louis limestones (see fig. 4.4 for a list of topographic quadrangles). The following descriptions are adapted from Malott *et al.* (1948) and Frey and Lane (1966).

The Ste Genevieve formation (fig. 4.3) consists of thin- to thick-bedded limestones of a lithographic to crystalline texture and is locally oolitic. It ranges from 75 to 170 ft in thickness and is somewhat purer than the underlying St Louis formation with insoluble residue content varying from 2 to 3% (Gray 1970, personal communication).

The St Louis formation consists of granular to lithographic limestone which is thin-bedded, hard, compact, and locally argillaceous, both in the limestone and in basal shale layers. Chert nodules and lenses are present locally in the middle and upper layers. Insoluble residue content varies from 12 to 15% (Gray 1970, personal communication).

Principal karst features

Surface phenomena

The discussion will be limited to negative features. The most common of these are sinkholes, varying in depth from barely perceptible indentations of several feet to a maximum of 100 ft or more. Most range from 10 to 30 ft in depth. Their horizontal extent ranges from a few square feet to several acres. The great number of sinkholes in the Mitchell Plain is exemplified by a field study of a single square mile near Orleans, Indiana (fig. 4.1) by Malott and Shrock (Powell 1961) in which 1022 sinkholes were discovered. In the same study, it was estimated that there may be as many as 300,000 individual sinkholes in the Mitchell Plain. Although the number observable on the 7·5 minute series topographic quadrangles is necessarily smaller (contour interval – 10 feet), we counted 6479 individual depressions in a random sample of 80 square miles with mean and sample variance approximately 81 and 2980, respectively.

The sinkholes are of two major types. The first and most frequently encountered are small and have moderately sloping sides. These are called dolines or corrosion sinks and are formed above the water table by solution along zones of weakness and by the ponding of runoff. They are enlarged by solution from water moving downward into the subterranean conduit system. Some corrosion sinks, known as swallow holes, provide underground entry for low order surface streams.

The second major type, collapse sinks, result from roof collapse above subsurface caverns and have steep sides immediately after formation.

However, creep and surface wash tend to obscure the flanks, leading to broad topographic lows which are often shallower and have even more gentle slopes than corrosion sinks. The spatial distribution of collapse sinks is dependent upon the subsurface distribution of caverns, although corrosion sinks form independently of the presence of caverns and, as Howard (1963) has noted, presuppose cavern formation.

We noted no systematic variation in number of sinks per square mile between the Ste Genevieve and St Louis formations, despite differences in bedding thickness and insoluble residue content.

Another negative surface phenomenon which has been described by Howard (1963) in the Mitchell Plain is the solution enlarged vertical joint or 'cutter' in which the soil mantle has been removed by erosion.

Subsurface phenomena

The primary subsurface form is the subterranean conduit. A conduit is a subsurface water passageway which usually develops along joints and bedding planes within the limestone and is part of a network which either drains into an incised stream or joins the regional ground water system. Development of the network is thus constrained by the joint system, which may be postulated as evenly spaced (Rhoades and Sinacori 1941). Enlargements within the conduit system are known as caverns (or caves if they have surface entrances). Cavern development may be considered as occurring more or less randomly along the conduit system, although it is noted that maximum enlargement obtains where two conduits intersect in the upper portion of the saturated zone (White *et al.*, 1970). Cavern development is also enhanced by ground water movement. As Swinnerton (1929, 1932), Rhoades and Sinacori (1941) and Kaye (1957) have noted, bedrock solution and cavern development appear to vary directly with velocity of solvent flow.

More than 700 caves have been discovered in the Mitchell Plain (Powell 1961, 1966). The number of caverns is unknown, but is presumed to be enormous.[1]

Mode of analysis: quadrat methods

Quadrat analysis is a statistical technique for directly evaluating the arrangement of points in a region (map) which has been partitioned into quadrats. In a more interesting corollary, it is an indirect method for assessing the utility of hypotheses concerning the nature of the

[1] We were unable to obtain any estimate of such from the Indiana Geological Survey, Bloomington, nor was that agency able to provide us with water table information. Otherwise, the co-operation of the Indiana Geological Survey is gratefully acknowledged.

process or processes accounting for the phenomena which are abstracted as points. It is assumed that each process contains a random component.

The number of points per cell is enumerated. Then, the similarity of frequencies of points per cell with theoretical frequencies is evaluated. This enables a researcher to identify an array as (1) random, (2) more regular than random or (3) more clustered than random with respect to quadrat size. The quadrat count mean for a random point set is equal to its variance. Thus, it is sufficient to regard an array as more regular than random if the variance is smaller than the mean and more clustered than random if the variance is larger than the mean. A more regular than random array may give the impression of a lattice if the points are widely dispersed. Points occur as clumps in highly clustered distributions.

Spatial processes

Most point sets which are identified as being more regular than random result from a mix of random and spatial competition processes with a net decrease in number of points through time. More clustered than random arrays seem to best reflect such quasi-random processes as colonisation and diffusion, although multiple random processes tend to generate spatial distributions which, based upon the variance–mean ratio, may be described as more clustered than random. Random point sets are the outcomes of single stochastic processes. See Hudson (1969) for an illuminating discussion of such spatial processes as colonisation and competition and the relationship between process and arrangement within the context of settlement geography.)

Hypotheses involving process

Quadrat methods are unable to inferentially identify processes of generation for two reasons: (1) several unlike processes may produce point sets which conform with a single probability law; (2) a single array may be described adequately by several probability laws. Thus, as Harvey (1966) and Rogers (1965) have indicated, the use of quadrat methods is appropriate only when a deductive hypothesis suggests a particular arrangement of points; the mode is one of substantiating or rejecting the plausibility of a hypothesised process rather than one of postdiction.

Processes and probability laws

Since the sample variance (V) is larger than the mean $(\bar{X})$, there is empirical evidence that the spatial distribution of karst depressions in the unglaciated Mitchell Plain is more clustered than random. Thus, only random, quasi-random and multiple random processes and appropriate probability laws are discussed.

Random distribution of points on a plane

A succinct description of a random spatial process is given by Miller and Kahn (1962, 365):

1. In a random distribution of points over a plane, any point has equal probability of occurring at any position on that plane.
2. A small subdivision of the plane has as much chance of containing a point as any other small subdivision of the same size.
3. The position of a point in the plane is in no way influenced by the position of any other point.[1]

The probability of occurrence is uniform over a study region. While this is tantamount to the geographic primitive, the uniform plane (and is equally unattainable), it is necessary. In a real sense, it seems that apparently random point sets might best be expected on surfaces imposing no more than minimal constraints on resultant point locations.

Poisson probability law

If a study region containing a random array is partitioned into conterminous quadrats or sampled in some manner, the number of points per cell ($x = 0, 1, \ldots$) is distributed as the Poisson limit of the binomial probability law with expectation μ. Estimation of μ is by

$$\hat{\mu} = n/Q \qquad (1)$$

where n is the number of points and Q is the number of cells. Thus, $\hat{\mu}$ is the quadrat count mean and is dependent upon quadrat size. If n is invariant, $\hat{\mu}$ varies directly with quadrat size, indicating that an array may be considered random only with respect to quadrat size.

The probability that a quadrat contains exactly x points is given by the density function

$$P(x) = \hat{\mu}^x e^{-\mu}/x'. \qquad (2)$$

Fitting the Poisson distribution to a spatial series involves the implicit hypothesis that the points are distributed in accord with a single random spatial process, which precludes the presence of locational constraints. The assumption of independence among point locations also raises some question about its utility in situations where the points are logically distributed as the result of a quasi-random or Markov process (stream intersections in a dendritic network provide such an example). The Poisson model also assumes that while the probability of occurrence is uniform, it is unknown and 'small' with respect to size of unit area, nor is the possible number of points per unit

[1] See Dacey (1964) for a rigorous derivation.

E

area known. If the probability of occurrence and maximum number of points per unit area can be determined, the binomial probability law is the appropriate model for assessing randomness.

Probability models for more clustered than random arrays

Two probability models for more clustered than random point sets are discussed. Neither is appropriate if the variance is smaller than the mean. The first is the Negative Binomial distribution, which may be derived from either the binomial or Poisson series and is an independent events model to the extent that the number of points in any cell is indeterminate based upon the number of points in any other cell.

The density function is

$$P(x) = \binom{k + x - 1}{k - 1} p^k q^x \tag{3}$$

with expectation

$$\begin{aligned} E(x) &= kq/p \\ &= \bar{X} \end{aligned} \tag{4}$$

where p and k are the parameters.

p is a measure of randomness which imparts an evenness bias to the distribution. Skewness and spatial clustering increase as p converges on zero and decrease as p approaches 1·00. If $V = \bar{X}$, $p = 1·00$, and the series reduces to the Poisson model.

k is an index of clustering. The smaller its value, the greater the skewness and the greater the spatial clustering. If $V = \bar{X}$, $k = \infty$, and the distribution is simply the Poisson ($p = 1·00$ and $k = \infty$ are accompanying conditions).

The method of moments estimation of parameters is (Bliss 1953; Williamson and Bretherton 1963):

$$\begin{aligned} \hat{p} &= \bar{X}/V & 0 < p \leqslant 1 & \tag{5} \\ \hat{k} &= \bar{X}^2/(v - \bar{X}), & 0 < k & \tag{6} \\ & & q = 1 - p & \end{aligned}$$

If k is not a positive integer

$$P(0) = p^k \tag{7}$$

and

$$P(x > 1) = \frac{q}{x}(x + k - 1)\,P(x - 1) \tag{8}$$

may be used as an algorithm for computing probabilities.

Several interpretations (see Harvey 1966) have been given for spatial distributions for whom the frequency of points per quadrat is adequately described by the Negative Binomial distribution:

1. The points are randomly distributed, but the density and Poisson

expectation vary over space as the result of minor inhomogeneities (some sites are more conducive to the random process than others).

2. The array consists of randomly distributed clumps.

3. An initial random array generates new points of the same type at a logarithmic rate. Diffusion is a function of proximity. Thus, the process is quasi-random and may be described as contagious.[1]

In any event, to paraphrase Williamson and Bretherton (1963, 9), any variation in probability of occurrence, in particular, any tendency for one event to increase the probability of another, will increase the variance relative to the mean and the Negative Binomial distribution will invariably describe the data better than the Poisson.

As noted earlier, multiple random spatial processes tend to generate spatial distributions which are, based upon the variance–mean ratio, more clustered than random. An appropriate model for a combination of two random processes is the Mixed or Double Poisson distribution.

$$P(x) = k_1\left(\frac{e^{\gamma}{}_1\gamma_1{}^x}{x!}\right) + k_2\left(\frac{e^{-\gamma}{}_2\gamma_2{}^x}{x!}\right) \tag{9}$$

with expectation

$$E(x) = k_1\gamma_1 + k_2\gamma_2$$
$$= \bar{X} \tag{10}$$

This, the general case (Schilling 1947), may be used to describe an array in which the points are distributed as the sum of two mutually independent random spatial processes of variable intensity. k_1 is the 'weight' of the first process, or the likelihood that any point in any quadrat is the result of the first process, k_2 is the probability of the second process, and γ_1 and γ_2 are the parameters for the first and second processes, respectively, or the expectations if a quadrat is affected exclusively by a single process.

This model may be applied meaningfully if it is logical to hypothesise that each point is the result of a single random process, but that the study region is affected by two random processes because of either (1) areal exclusiveness or (2) inhomogeneity beyond that encompassed by the Negative Binomial model. In the former, it is sufficient to regard the study region as consisting of two distinct parcels. In the latter, the random processes are mutually independent but are distributed as an irregular mosaic, generating an array which is in fact a mixture of two

[1] In this respect, the Negative Binomial model is one of a family of contagious probability distributions. A characteristic of these distributions is that the presence of an event increases the likelihood of a subsequent or nearby event of the same kind. Other such models include the Polya-Aeppli, Thomas's Poisson, and Neyman Type-A distributions. They have been widely used in plant ecology and epidemiology. See Dacey (1967), Harvey (1966, 1968), and Hudson (1969) for applications of the contagious probability distributions in geographic research.

point sets, one the outcome of the first process, the other the outcome of the second process.[1]

Estimation of parameters by the method of moments is as follows:

$$\hat{\gamma}_1 = 0 \cdot 5(j - d) \qquad 0 < \gamma_1 \tag{11}$$
$$\hat{\gamma}_2 = 0 \cdot 5(j + d) \qquad 0 < \gamma_2 \tag{12}$$
$$k_1 = (\gamma_2 - \bar{X})/d \qquad 0 < k_1 < 1 \tag{13}$$
$$k_2 = 1 - k_1 \qquad 0 < k_2 < 1 \tag{14}$$

where
$$j = 2\bar{X} + g \tag{15}$$
$$d = (g^2 + h)^{\frac{1}{2}} \tag{16}$$
$$h = 4(V - \bar{X}) \tag{17}$$
$$g = \{(\mu_3 - \bar{X})/(V - \bar{X})\} - 3 \tag{18}$$

and
$$\mu_3 = (\textstyle\sum Fx^3/Q) - 3\bar{X}V - \bar{X}^3 \tag{19}$$

Test for goodness-of-fit

Observed frequencies of x are assessed for correspondence with expected frequencies where

$$E(Fx) = P(x)\,Q \tag{20}$$

The Kolmogorov–Smirnov one-sample statistic is undoubtedly the most powerful means of assessing goodness-of-fit of discrete events. Denote the absolute difference between cumulative frequencies by K, the maximum absolute difference by K, and the critical value for level of significance α by K_α. If $K_0 < K_\alpha$, one accepts the null hypothesis (of no difference between observed and expected frequencies) and concludes that the distribution is random or other than random with respect to quadrat size. Rejection and the conclusion that the spatial distribution is not the outcome of the hypothesised process are indicated if $K_0 \geqslant K_\alpha$.[2]

Hypotheses of karst depression development

Three alternative hypotheses are assessed for $\alpha = 0 \cdot 20$ regarding the broadscale processes responsible for the spatial distribution of karst depressions in the unglaciated Mitchell Plain.

1. The distribution is the outcome of a single random process.

---

[1] The Mixed Poisson model is incapable of discerning whether the condition is one of areal exclusiveness or inhomogeneity. This can only come from knowledge of the study region and the processes affecting it.
[2] Evaluating goodness-of-fit is a curve-fitting procedure. Thus, minimising the probability of committing a Type II error is a major consideration in the inferential phase of a quadrat analysis.

Acceptance is indicated if the Poisson model adequately describes the frequency distribution of quadrat counts.

2. The distribution is the result of a contagious process. Acceptance is indicated if the frequency distribution is adequately described by the Negative Binomial model.

3. The distribution is the result of two mutually independent random processes. Acceptance is indicated if the distribution is adequately described by the Mixed Poisson model.

The first hypothesis is the least tenable and is assessed merely as a point of departure for the other two. It involves the assumption that all depressions in the study region are formed by either (1) cavern roof collapse exclusively or (2) near-surface corrosion exclusively. We have already noted that at least two types of sinks occur in the Mitchell Plain. Further evidence of the tenuous nature of this assumption is given by Howard (1968) and White *et al.* (1930), who note that depressions on the Ste Genevieve and St Louis formations in the Kentucky Pennyroyal include both collapse sinks and dolines.[1]

Although it has been accepted previously (La Valle 1967), the hypothesis of a contagious process seems nearly as untenable as that of a single random process. First, it considers only one type of depression. Secondly, regardless of the depression-forming process, the depressions serve to divert surface runoff into the subsurface conduit system in the most efficient manner and therefore enlarge rather than multiply as drainage is accelerated. Thus, it is reasonable to assume that once a depression is formed in an arbitrarily small area, the likelihood of another depression forming in an adjacent area of the same size is decreased, rather than increased. Furthermore, the Negative Binomial model contains a logarithmic growth component. In contrast, it has been noted that karst depressions tend to elongate and increase in size through time, such that nearly constant density per unit area may be assumed until such time as the sinks coalesce.

Rejection of this hypothesis leads to the third and most tenable. It may be described by a logical set of events which includes elements of Rhoades and Sinacori (1941), Howard (1963) and White *et al.* (1970): caverns form randomly in a conduit system which is a more or less random network in three dimensions, given the constraint of an essentially evenly spaced pattern of joints as postulated by Rhoades and Sinacori (1941). Subsequent collapse of an arbitrary number of cavern roofs accounts for a random distribution of collapse sinks.

[1] Richard Karsten of the Department of Geography at Northern Illinois University has suggested that if the solution probabilities were completely uniform as is the assumption of the Poisson model, this would lead to a lack of surface karst. Rather, the surface would downwaste uniformly.

Doline formation is, to the extent of the minor constraint noted below, independent of this process and is attributable to corrosion from intermittent saturation of zones of weakness in the limestone below the soil zone and the ponding of runoff. This last set of processes is so complex that it must be viewed simply as a single macroscopically random process.

The two broadscale processes are mutually independent except that it is unlikely that a doline will form in a space comprising a pre-existing collapse sink. Furthermore, the probability of a doline forming in immediate proximity to a collapse sink is small, although a collapse sink may form in a space enclosing a pre-existing doline. The two depression forming processes are spatially intermixed, i.e. the situation is one of inhomogeneity rather than one of areal exclusiveness. With regard to the latter, it has been noted previously that no systematic variation in density of depressions occurs between the Ste Genevieve and St Louis limestones. Furthermore, the ratio of dolines to collapse sinks appears to remain constant throughout.[1]

The statistical analysis

Sampling design

As noted previously, a random sample of 80 square miles in the unglaciated portion of the Mitchell Plain contained 6479 separate identifiable depressions with mean and variance approximately 81 and 2980, respectively.[2] The wide range of x for $Q = 80$ and resulting platykurtic frequency distribution are clearly beyond the scope of a meaningful quadrat analysis. Moreover, the likelihood of a cell as large as one square mile (2·64 in $\times$ 2·64 in) containing more than one clump is high (assuming that the Negative Binomial is a relevant model). Thus, it was deemed necessary to employ an alternative area sampling scheme which would (1) render the frequency distribution tractable in the context of a quadrat analysis (wherein the probability of occurrence of a depression should be small relative to the size of quadrat), (2) provide some assurance that an individual quadrat contains at a maximum the number of points attributable to a single clump, and (3) maintain the inferential basis of the analysis.

---

[1] Collapse sinks are much larger in areal extent than dolines, such that the probability per unit area is large compared to dolines. However, the number of sinks per unit area is smaller.

None of the hypotheses considers the role of vertical shafts or 'cutters' as noted by Howard (1963). Although depressions forming atop vertical shafts might conceivably impart an evenness bias to the distribution, it is likely that such depressions would be too small and too shallow to have topographic expression on the 7·5 minute series quadrangles.

This was accomplished by 'throwing' a disc with 0·0247 mi[1] area in each square mile quadrat about a point located by the stratified systematic unaligned sampling method described by Berry (1962).[2]

Sampling was restricted to those conterminous square miles in which the Ste Genevieve and St Louis limestones comprise at least 75% of the bedrock surface. This reduced the study region to 304 square miles.

The frequency distribution of number of depressions per disc is given in table 4.1.[3] It is again indicative of a more clustered than random spatial distribution with variance–mean ratio equal to 2·246.

Test of the hypothesis of a single random process

As anticipated, the Poisson probability law for $\hat{\mu} = 1·87$ provides an extremely poor fit of the cell counts (table 4.2). Not only is the null hypothesis rejected for $\alpha = 0·20$; it is rejected at even the 0·01 level.[4] It is concluded that the hypothesis that the spatial distribution of karst depressions is the topographic expression of a single random process is statistically as well as logically untenable.

---

[1] A single closed hachured contour somewhat circular in plan or set of concentric hachured contours is defined as a single karst depression. In the event that a single closed hachured contour or set of concentric hachured contours contains several closed hachured contours, each individual closed contour or concentric set thereof is regarded as an individual depression. Any circular water body above the water table representing a sinkhole pond or karst lake is also regarded as a karst depression if it does not meet the above requirements. The depressions are abstracted to points for purposes of allocation to quadrats by locating a point in the centre of each feature.

The more clustered than random nature of the distribution is exemplified by the fact that estimates of the Negative Binomial parameter k remain essentially constant as the area is reduced by a factor of four. Where the area is one square mile, $\hat{k} = 2·26$, for one-fourth square mile, $\hat{k} = 2·04$, and for one-sixteenth square mile, $\hat{k} = 2·00$.

[2] The first two criteria pertain roughly to shape and size of cell. With respect to shape, the circle or disc is the preferable sampling unit, since it is the most geometrically compact in that it minimises the ratio of length of edge to area. As regards size, Curtis and McIntosh (1950) have stated that the optimal quadrat size may be estimated by $2A/n$ where A/n is the probable area surrounding any point in an array. Here, $2A/n = 160/6479 = 0·0247$ square miles, or 0·1721 square inches at a scale of 1 : 24,000. The map diameter of a disc is 0·4682 in.

[3] The mean number of points per disc (1·87) compares favourably with that which might have been expected (2·00) from the random sample of 80 square miles; (0·0247) (81) $\simeq$ 2·00.

[4] The computational form for $K_{0·20}$ is (Massey 1951)

$$K_{0·20} = \frac{1·07}{Q^{0·5}} Q$$

Critical values for K for $\alpha = 0·20$, 0·10, 0.05 and 0·01 are 18·66, 21·27, 23·71 and 28·42, respectively, where $Q = 304$.

Table 4.1   *Frequency of depressions per sampling disc*

x	$O(Fx)$
0	112
1	53
2	35
3	43
4	22
5	23
6	6
7	5
8	3
9	1
10	1

$$\bar{X} = 1{\cdot}87 \qquad V = 4{\cdot}20 \qquad D = 2{\cdot}246$$

Table 4.2   *Cell counts fit with Poisson distribution*
$$\hat{\mu} = 1{\cdot}87$$

x	$O(Fx)$	$E(Fx)$	$Cum\ O(Fx)$	$Cum\ E(Fx)$	K^*
0	112	48	112	48	64
1	53	88	165	136	29
2	35	82	200	218	18
3	43	50	243	268	25
4	22	23	265	291	26
5	23	9	288	300	12
6	6	3	294	303	9
7	5	1	299	304	5
8	3	0	302	304	2
9	1	0	303	304	1
10	1	0	304	304	0

$$*K_0 = 64$$

Test of the hypothesis of a contagious process

Although the distribution is more clustered than random, the hypothesis that it is attributable to contagion (whereby the presence of a
randomly distributed depression increases the likelihood of another) is
almost as untenable statistically as the hypothesis of a single random

process.[1] Rejection of the null hypothesis for the Negative Binomial model (table 4.3) is indicated for both $\alpha = 0.20$ and 0.10 where $\hat{p} = 0.445$ and $\hat{k} = 1.501$. Acceptance is indicated for the 0.05 level; however, this level of significance is regarded as too small for meaningful acceptance of the null hypothesis in a Type II error situation.

Test of the hypothesis of multiple random processes

The Mixed Poisson model (see Table 4.4 for parameters and k's) provides an adequate fit of the cell counts at the 0.20 level of significance. Although the fit is less than perfect, the hypothesis that the spatial

Table 4.3   *Cell counts fit with negative binomial distribution*
$$\hat{p} = 0.4452, \ \hat{k} = 1.5008$$

x	$O(Fx)$	$E(Fx)$	Cum $O(Fx)$	Cum $E(Fx)$	$K*$
0	112	90	112	90	22
1	53	75	165	165	0
2	35	52	200	217	17
3	43	34	243	251	8
4	22	21	265	272	7
5	23	13	288	285	3
6	6	8	294	293	1
7	5	5	299	298	1
8	3	3	302	301	1
9	1	2	303	303	0
10	1	1	304	304	0

$*K_0 = 22$

distribution of karst depressions is reflective of two mutually independent and non-equally important random processes of cavern roof collapse and corrosion is statistically as well as logically tenable. Moreover, it is certainly more plausible in light of this analysis than either of the two less robust hypotheses.

An attempt at interpretation of the parameters is undertaken at this point. It was noted earlier that most depressions in the unglaciated Mitchell Plain appear to have been formed by near-surface corrosion. However, dolines are small in areal extent compared to collapse sinks.

[1] Our conclusion would have been the same had our hypothesis been that the distribution is the result of a single random process, but that the density and Poisson parameter vary over space according to minor inequalities in likelihood of occurrence, due, say, to differences in thickness of beds, or purity as measured by insoluble residue content. In any event, our conclusions differ markedly from those of LaValle (1967), who conducted his tests at the 0.01 level of significance.

Table 4.4   *Cell counts fit with Mixed Possion distribution*

$$k_1 = 0{\cdot}5481, \quad k_2 = 0{\cdot}4519$$
$$\hat{\gamma}_1 = 0{\cdot}4839, \quad \hat{\gamma}_2 = 3{\cdot}551$$

x	$O(Fx)$	$E(Fx)$	Cum $O(Fx)$	Cum $E(Fx)$	K^*
0	112	107	112	107	1
1	53	64	165	171	6
2	35	37	200	208	8
3	43	29	243	237	6
4	22	26	265	263	2
5	23	19	288	282	6
6	6	11	294	293	1
7	5	6	299	299	0
8	3	2	302	301	1
9	1	1	303	302	1
10	1	1	304	303	1

$$*K_0 = 8$$

Cavern roof collapse is further associated with the spacing of conduits in the joint system. With respect to cell size used, cavern roof collapse is the more probable (due to their large size) but less frequent event and doline formation is the less probable but more frequent event.

Thus, it is reasonable to identify $\hat{\gamma}_2$ as the expectation for a disc affected only by corrosion, k_2 is the probability of encountering a corrosion sink in any disc selected at random, such that $k_2\hat{\gamma}_2 = 1{\cdot}60$ is the probable number of dolines in any disc.

$\hat{\gamma}_1$ is taken to be the collapse sink parameter, k_1 is the probability of encountering cavern collapse in any disc, and $k_1\hat{\gamma}_1 = 0{\cdot}27$ is the probable number of collapse sinks in any disc.

Note that $k_1\hat{\gamma}_1 + k_2\hat{\gamma}_2 = 1{\cdot}87 = \bar{X}$.

It is hoped that this exploratory analysis of the utility of hypotheses of macroscopic randomness as regards spatial variation of surface karst will provide a convenient frame of reference and the stimulation for subsequent analyses. Although the results and conclusions are plausible within the conceptual framework employed, verification rests with a detailed analysis of the set of events operationalised by the Mixed Poisson model.

References

BERRY, B. J. L. (1962) Sampling, coding and storing flood plain data; *Agricultural Handbook, United States Department of Agriculture*, No. 237.

BLISS, C. I. (1953) Fitting the negative binomial distribution; *Biometrics* 9, 176–200.

CHORLEY, R. J. (1962) Geomorphology and general systems theory; *U.S. Geological Survey, Professional Paper* 500-B.

CURL, R. J. (1959) Stochastic models of cavern development (Abst.); *Bulletin of the Geological Society of America* 70, 1802.

CURL, R. J. (1966) Caves as a measure of karst; *Journal of Geology* 74, 789–830.

CURTIS, J. T. and MCINTOSH, R. P. (1950) The interrelations of certain analytic and synthetic phytosociological characters; *Ecology* 31, 434–55.

DACEY, M. F. (1964) Two-dimensional random point patterns: A review and an interpretation; *Papers of the Regional Science Association* 13, 41–55.

DACEY, M. F. (1967) *An empirical study of the areal distribution of houses in Puerto Rico;* Mimeographed paper, Northwestern University, Department of Geography.

DACEY, M. F. (1968A) Stream length and elevation for the model of Leopold and Langbein; *Water Resources Research* 4, 163–6.

DACEY, M. F. (1968B) The profile of a random stream; *Water Resources Research* 4, 651–4.

DICKEN, S. N. (1935) Kentucky karst landscapes; *Journal of Geology* 43, 708–28.

FREY, R. W. and LANE, M. A. (1966) *A Survey of Indiana Geology* (Rho Chapter, Sigma Gamma Epsilon, Indiana University, Department of Geology).

GRAY, H. H. (1970) Personal communication; Department of Geology, Indiana University, Bloomington, Indiana.

GREIG-SMITH, P. (1964) *Quantitative Plant Ecology* (Butterworths, London).

HARVEY, D. W. (1966) Geographical processes and the analysis of point patterns; *Transactions of the Institute of British Geographers* 40, 81–95.

HARVEY, D. W. (1968) Some methodological problems in the use of the Neyman Type A and Negative Binomial probability distributions; *Transactions of the Institute of British Geographers* 43, 85–95.

HOWARD, A. D. (1963) Development of karst features; *National Speleological Society Bulletin* 25, Part 2, 45–65.

HOWARD, A. D. (1968) Stratigraphic and structural controls on landform development in the central Kentucky karst; *National Speleological Society Bulletin* 30, No. 4, 95–114.

HUDSON, J. C. (1969) A location theory for rural settlement; *Annals of the Association of American Geographers* 59, 365–81.

KAYE, C. A. (1957) The effect of solution motion on limestone solution; *Journal of Geology* 65, 35–47.

KING, L. J. (1969) *Statistical Analysis in Geography* (Prentice-Hall Inc., Englewood Cliffs, N.J.).

LAVALLE, P. D. (1967) Geographic processes and the analysis of karst depression distributions within limestone regions (Abst.); *Annals of the Association of American Geographers* 57, 794.

LEOPOLD, L. B. and LANGBEIN, W. B. (1962) The concept of entropy in landscape evolution; *U.S. Geological Survey Professional Paper* 500-A.

MALOTT, C. A., ESAREY, R. E. and BIEBERMAN, D. F. (1948) *Upper and Middle Mississippian Formations* (Second Annual Geological Field Conference, Bloomington, Indiana).

MASSEY, J. F. (1951) The Kolmogorov–Smirnov test for goodness of fit; *Journal of the American Statistical Association* 46, 70 (tables).

MCCONNELL, H. (1967) Quadrat methods in map analysis; *University of Iowa, Department of Geography, Discussion Paper* No. 3.

MCGREGOR, D. J. and RARICK, R. D. (1962) Some Features of Karst Topography in Indiana (*Indiana Geological Survey Guidebook, Indiana Academy of Science*).

MILLER, R. L. and KAHN, J. S. (1962) *Statistical Analysis in the Geological Sciences* (Wiley, New York).

PERRY, T. G., SMITH, N. M. and WAYNE, W. J. (1954) Salem Limestone and Associated Formations in South-Central Indiana; *Indiana Department of Conservation, Geological Survey*, Guidebook No. 7.

POWELL, R. L. (1961) Caves of Indiana, *Indiana Department of Conservation, Geological Survey, Circular* No. 8.

POWELL, R. L. (1966) Caves: speleology and karst hydrology; In 'Natural Features of Indiana', *Indiana Academy of Science, Sesquicentennial Volume* 116–30.

RHOADES, R. and SINACORI, N. (1941) Patterns of ground water flow and solution; *Journal of Geology* 49, 785–94.

ROGERS, A. (1965) A stochastic analysis of the spatial clustering of retail establishments; *Journal of the American Statistical Association* 60, 1094–103.

SCHEIDEGGER, A. E. (1966) Stochastic branching processes and the law of stream orders; *Water Resources Research* 2, No. 3, 199–203.

SCHEIDEGGER, A. E. and LANGBEIN, W. B. (1966) Probability concepts in geomorphology; *U.S. Geological Survey Professional Paper* 500-C.

SCHENCK, H. (1963) Simulation of the evolution of drainage basin networks with a digital computer; *Journal of Geophysical Research* 68, 5739–5745.

SCHILLING, W. (1947) A frequency distribution represented as the sum of two Poisson distributions; *Journal of the American Statistical Association* 42, 407–24.

SHREVE, R. J. (1967) Infinite topologically random channel networks; *Journal of Geology* 75, 178–86.

STRAHLER, A. (1952) Dynamic basis of geomorphology; *Bulletin of the Geological Society of America* 63, 928–38.

SWINNERTON, A. C. (1929) Changes in baselevel indicated by caves in Kentucky and Bermuda (Abst.); *Bulletin of the Geological Society of America* 40, 194.

SWINNERTON, A. C. (1932) Origin of limestone caverns; *Bulletin of the Geological Society of America* 43, 663–93.

THORNBURY, W. D. (1969) *Principles of Geomorphology* (Wiley, New York), 303–44.

WERNER, C. (1970) Horton's law of stream numbers for topologically random channel networks; *Canadian Geographer* 24, 57–65.

WHITE, W. B., WATSON, R. A., POHL, E. R. and BRUCKER, R. (1970) The central Kentucky karst; *Geographical Review* 60, 88–115.

WILLIAMSON, E. and BRETHERTON, M. H. (1963) *Tables of the Negative Binomial Distribution* (Wiley, London).

WOODWARD, H. P. (1961) A stream piracy theory of cave formation; *National Speleological Society Bulletin* 23, 39–57.

5 The analysis of spatial characteristics of karst terrains

PAUL W. WILLIAMS

Department of Geography, University of Auckland

Introduction

One of the basic tenets of climatic geomorphology is that essentially similar landforms will result from comparable morphogenetic conditions. This view is widely held in Europe and particularly so amongst karst geomorphologists, for it is argued that karst processes are especially susceptible to climatic control. Recently, however, doubt and confusion has grown over the strict similarity of karst forms in grossly comparable morphogenetic regions. This has arisen from contrasting descriptions of karst morphology in what were taken to be similar process environments. The uncertainty has been further intensified as a consequence of such research as that by Jennings and Bik (1962) and Verstappen (1964) in New Guinea, where they demonstrated not only that many karst styles can co-exist, but also that many geological influences are just as important determinants of topography as climate.

The universal tenability of morpho-climatological principles in the study of karst, as in other branches of geomorphology (Stoddart 1968), has therefore been challenged. There is thus a need to find an independent way of testing the reality and internal variation of the morpho-climatic assemblages proposed, and to this end morphometric, spatial-analytical techniques seem the most appropriate.

Quantitative methods in karst research

Karst morphologists have for a long time adopted a quantitative approach to process study (e.g. Ewing 1885), although there has been a recent surge of activity, stimulated by the controversial findings of Corbel (1957, 1959). By contrast, the quantitative appraisal of landform has developed slowly, despite the early leadership of Cvijić (1893) in classifying dolines according to their geometry. His tripartite division of these depressions into bowl-, funnel- and well-shaped forms was based on side slope angles and diameter : depth ratios, and his conclusions were founded on hundreds of field measurements. He also made observations on doline densities. Further quantitative work of this type was accomplished by other investigators in the first half of this century, by far the most important single contributor being Cramer (1941). In the section of his classical paper that deals with 'Dolinenstatistik', he summarises the statistical findings of numerous earlier workers, in addition to presenting the results of a map study of closed depressions in diverse regions of the world (table 5.1).

Since Cramer's major contribution, until very recently there has been little quantitative work on karst morphology. From an aerial photograph study of the Shenandoah Valley, Virginia, Hack (1960) concluded doline density to vary with rock type. Ordovician carbonates support 1·64 per km² (4·25 per ml²), whereas Cambrian formations support only 0·58 per km² (1·5 per ml²). In the Mendip Hills of Somerset, England, Ford (1964) found 80% of depressions to be in the floors of dry valleys; depression frequency being inversely related to thalweg gradient. He also showed an earlier attempt at morphometric study of doline evolution in the district by Coleman and Balchin (1960) to be invalid, because some of the areas analysed are simply former mined ground, not karst.

Matschinski (1962A, B, 1964, 1968) has examined the problem of the distribution of features, such as mares and dolines, which at first sight seem haphazard, but which often reveal the influence of dominant trends. His approach may be illustrated with reference to his 1968 paper in which an area to the north-west of Lake Constance is treated. In a district of about 250 sq km (96 sq ml), 178 dolines were analysed for pattern. Base data were obtained from a recent 1/50,000 map; thus only the larger depressions could have been identified.

The problem was tackled by considering the possibility of alignments in the doline fields. A distinction is drawn between 'local' and 'general' alignment, the former referring to a restricted number of features round a single point and the latter applying to an extended region. Matschinski's analytical procedure is first to determine the local alignment common to every three points (e.g. dolines), as illustrated in fig. 5.1, and then to plot the distribution of these local alignment directions on a histogram (fig. 5.2). The histogram is then subjectively interpreted.

In the Lake Constance example, the orientation of the first maximum (NW–SE) corresponds precisely with the structural trend of the region bordering the study area to the south-east; the second peak (W–E) coincides with the strike of numerous almost parallel faults in

Table 5.1 *Doline dimensions and doline densities in karst regions of different Climate* (from Cramer 1941, 318)

Karst region	Area km²	No. of dolines	Total doline area km²	Mean Surface area m²	Biggest Surface area m²	Smallest Surface area m²	Doline density per km²
Solution dolines in covered karst (Schwunddolinen)							
1. South Harz	0·42	34	0·0056	166	1300	12	80·95
2. Blaubeurer Alb	137·67	327	0·0440	134	1850	7	2·38
3. Gräfenberger Alb	41·25	36	0·0400	1112	5000	125	0·87
4. Wiesentalb	10·58	64	0·0123	203	1125	20	6·04
5. Altmühlalb	202·42	165	0·1207	736	7500	72	0·81
6. St Louis, Missouri	104·00	230	0·8488	3690	78000	100	2·21
Solution dolines in naked and covered karst							
7. Mährischer karst	6·35	124	0·0586	472	31500	6	19·50
8. Oberungarischer karst	59·06	102	0·7360	7216	157500	1200	1·72
9. Karst of Interlachen	78·90	165	23·610	144000	4200000	150	2·09
10. Karst of Williston	199·08	140	22·294	159200	3150000	4800	1·42
Solution dolines in naked karst (Lösungsdolinen)							
11. Zahmer Kaiser Plateau	0·93	972	0·017	17	1200	1	2460·00
12. French Jura	6·63	296	0·284	960	5400	80	31·60
13. Velebit Mts	60·81	70	2·764	39400	843700	2800	1·15
14. Generlaski Stol, Croatia	64·38	275	0·7983	2900	37500	150	4·27
15. Kanzian, Istria	2·98	46	0·088	1900	12500	900	15·40
16. Doberdo, Istria	8·35	412	0·408	990	22500	84	49·32
17. Standingstone, Tenn.	233·00	92	5·1033	21880	562500	1200	0·65
18. Bristol, Virginia	16·18	28	1·8025	64387	625000	1024	0·57
19. Otavi Mts SW Africa	21·00	22	2·446	106000	312000	25000	1·04
20. Gunung Sewu, Java	30·80	151	4·225	28200	187500	2500	4·90

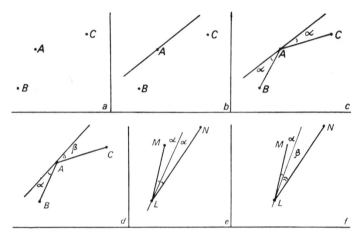

Fig. 5.1 Graphical determination of 'local' alignments, according to Matschinski (1968, fig. 6).
(1a) to determine the alignment at A, find the two nearest points, B and C.
(1b) shows the principal position of the segment describing the local alignment.(1c) illustrates the first approximation to the exact determination of this segment, and (1d) illustrates the final position where the angle β is greater than α in the same proportion as the distance AC is greater than AB.(1e) and (1f) illustrate the procedure when the two nearest points (M and N) are both to one side of the initial point (L).

neighbouring peripheral localities; whilst the third weak maximum may be related to the direction of the principal chain of the Black Forest. Matschinski concludes from this that the local alignments of dolines are influenced essentially by tectonics of the adjoining regions and not at all by the immediately underlying beds. He treats dolines simply as points and does not consider the evidence of their individual morphology, such as the extent to which their elongation reflects local structural conditions. The latter approach is taken up by LaValle (1967, 1968).

LaValle made a detailed study of karst depression morphology in south central Kentucky, basing his findings on fieldwork and measurements from accurate maps at 1/24,000. His data were obtained from a random sample of mile square plots, covering 25% of a karst region of 2072 sq km (800 sq ml). In the 1967 paper he sets out to analyse factors influencing the elongation and orientation of karst depressions by examining interrelationships between these two characteristics and various geological and hydrological parameters. This study was extended in the 1968 paper to consider three other salient attributes of depressions, namely depth, area, and flank slope inclination (fig. 5.3). It was hoped that an insight into the nature of karst would be gained by accounting for the spatial variation of the selected morphological indices by means of multiple regression analysis.

LaValle claims that such analyses relating the geological and hydrological parameters to the morphological ones accounted for 57% of the spatial variation of the distribution of structurally aligned depressions, 70% of the variations of depression areas, 73% of the variations of

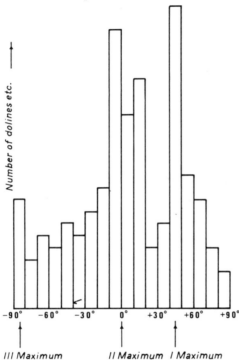

Fig. 5.2 Histogram of the number of dolines corresponding to different geographical directions (from Matschinski, 1968, fig. 2). Numbers indicate the angle between an alignment direction and the west. Positive figures indicate an inclination to the NW, negative to the SW.

flank slopes, 88% of the variations of elongation ratios and 92% of the variations of depression relief. However, despite these impressively high levels of statistical 'explanation', the conclusions are not altogether convincing. This is because some of the geomorphic assumptions relating to the variables are suspect; collinearity probably exists between some of the variables, as he noted; and some of the indirect measures are of doubtful value. Without being able to enter into detail here, the karst relief ratio, for example, seems generally quite inadequate as a measure of the hydraulic gradient, and it can therefore have little connection with discharge rates and karst erosion rates. There is in any case no observational data supporting the assumption that karst erosion rates are directly related to hydraulic gradients.

Nevertheless, while criticisms such as these and those relating to Coleman and Balchin's (1960) work underline the need for a firm geomorphic foundation prior to morphometric studies, exploratory work of this kind is by nature subject to many unknowns and deliberately sets up working hypotheses that may later be disproven. There

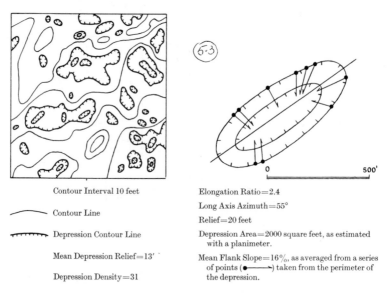

Contour Interval 10 feet

——— Contour Line

⌢⌢⌢⌢⌢ Depression Contour Line

Mean Depression Relief=13′

Depression Density=31

Elongation Ratio=2.4

Long Axis Azimuth=55°

Relief=20 feet

Depression Area=2000 square feet, as estimated with a planimeter.

Mean Flank Slope=16%, as averaged from a series of points (●——➤) taken from the perimeter of the depression.

Fig. 5.3 Measurement of karst features (from LaValle, 1968, fig. 3).

is also much value in the multiple regression analysis technique used by LaValle, and a number of his basic conclusions on dolines in Kentucky will probably be sustained by further work (Table 5.2).

In another study of dolines, but this time in the Craigmore area of South Canterbury, New Zealand, J. N. Jennings (pers. comm.) also adopted a regression model approach. The length, width, depth and long axis azimuth of 77 depressions were measured in the field, and analysis of the data revealed the dolines to tend closely to circularity and thus to display little evidence of preferred orientation. Linear relationships, significant at the 99% confidence level, were found to hold between doline depth and long axis length and between depth and width, indicating there to be little variation in side slope angle as the dolines grow in size. Jennings concludes that this tendency towards a particular side slope angle suggests that surface processes, rather than collapse, play the dominant role in fashioning the dolines.

Table 5.2   (from LaValle, 1967 and 1968)

5.2A   *Mean depression flank slope regression analysis*

Variable	Partial regression coefficient	Standardised partial regression coefficient	Correlation coefficient
X_1 (loglog mean distance from drainage system mouth)	−1·410	−0·490*	−0·52*
X_2 (log % area drained by subterranean drainage systems)	0·279	0·461*	0·79*
X_3 (log karst relief ratio)	0·295	0·334*	0·42*
X_4 (log % structurally aligned depressions)	0·192	0·281*	0·67*
X_5 (limestone density index)	0·149	0·144*	0·19*
X_6 (log mean insoluble residue content)	−0·189	−0·144*	−0·51*

* = significant at 0·05 level; no variables eliminated.

Multiple correlation R = 0·854; multiple coefficient of determination R^2 = 0·729*.

Predictor equation: depression flank slope

$$= 0 \cdot 72 X_1^{-1 \cdot 41} \cdot X_2^{0 \cdot 28} \cdot X_3^{0 \cdot 30} \cdot X_4^{0 \cdot 19} \cdot e^{15 \cdot X_5} \cdot X_6^{-0.19}$$

5.2B   *Mean depression area regression analysis*

Variable	Partial regression coefficient	Standardised partial regression coefficient	Correlation coefficient
X_1 (log % structurally aligned depressions)	0·297	0·519*	0·76*
X_2 (log % unit area drained by subterranean systems)	0·225	0·434*	0·74*
X_3 (log karst relief ratio)	0·129	0·174*	0·48*
X_4 (log mean insoluble residue content)	−0·151	−0·135*	−0·38*
X_5 (limestone density index)	0·090	0·104*	0·11
X_6 (loglog mean distance from drainage system mouth)**	−0·200	−0·080	−0·41*

* = significant at 0·05 level; ** = eliminated from regression model as an insignificant variable.

Multiple correlation R = 0·837; multiple coefficient of determination R^2 = 0·701*.

Predictor equation: depression area = $0 \cdot 33 X_1^{0 \cdot 30} \cdot X_2^{0 \cdot 23} \cdot X_3^{0 \cdot 13} \cdot X_4^{-0 \cdot 15} e^{\cdot 0 \cdot 09 \cdot X_5}$

5.2C *Mean depression relief regression analysis*

Variable	Partial regression coefficient	Standardised partial regression coefficient	Correlation coefficient
X_1 (log % structurally aligned depression)	0·537	0·782*	0·94*
X_2 (log % unit area drained by subterranean systems)	0·138	0·226*	0·77*
X_3 (loglog mean distance from 3 drainage system mouth)	−0·463	−0·160*	−0·39*
X_4 (log karst relief ratio)	0·133	0·151*	0·33*
X_5 (bedding thickness)	0·037	0·061*	0·39*
X_6 (log mean insoluble residue content)**	0·008	0·002	−0·56*
X_7 (limestone density)**	0·030	0·003	0·02

* = significant at 0·05 level; ** = eliminated from regression model as an insignificant variable.

Multiple correlation $R = 0.961$; multiple coefficient of determination $R^2 = 0.923$*.

Predictor equation: depression relief $= X_1^{0.54} \cdot X_2^{0.14} \cdot X_3^{-0.46} \cdot \log X_4^{0.13} \cdot e^{0.04 \cdot X_5}$. 2·13

5.2D *Mean depression elongation ratio multiple regression analysis*

Variable	Simple correlation with elongation ratio	b (partial regression coefficient)	Beta weight (standardised partial regression coefficient)
X_1 (loglog mean distance from drainage system mouth)	−0·21*	−0·249	−0·201*
X_2 (log % structurally aligned depressions)	0·92*	0·281	0·927*
X_3 (log insoluble residue content)	−0·46*	−0·052	−0·090*
X_4 (log karst system relief ratio)	0·23*	0·062	0·163*
X_5 (rock unit bedding thickness)	0·21*	0·025	0·095*
X_6 (rock unit density)	0·09	0·038	0·085*
X_7 (log % area drained by underground conduits)	0·56*	0·003	0·001

* = Significant at 0·05 level.

$R = 0.927$ multiple correlation coefficient
$R^2 = 0.88$ coefficient of determination
s_y 0·1234567 $= 0.079$ standard error of estimate
Regression equation:
$\log (Y_c) = -0.38 - 0.249X_1 + 0.281X_2 - 0.052X_3 + 0.062X_4 + 0.025X_5 + 0.038X_6$

Another feature of the karst landscape to receive some morphometric attention is the stream-link (swallow hole), although investigations have not proceeded beyond reconnaissance level. Williams (1966, 1969) examined swallet relationships in the Ingleborough district of northern England where, with stream-sinks ranked according to the Strahler order (Strahler, 1957) of the stream they absorb (fig. 5.4), various semi-

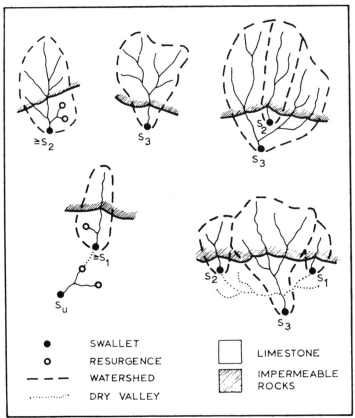

Fig. 5.4 Procedure for ordering swallets and delimiting their catchments, according to Williams (1966, fig. 2).

logarithmic relationships were found to appear to hold between swallet order and other variables (fig. 5.5). It was tentatively concluded that the fluvial morphometry 'laws' of stream numbers and basin areas also seem to apply to swallet catchments and that, in addition, the mean distance between swallets of the same order tends to approximate a direct geometric series. Similar general relationships have since been found to hold for the Gower karst in South Wales (W. J. Chambers, pers. comm.).

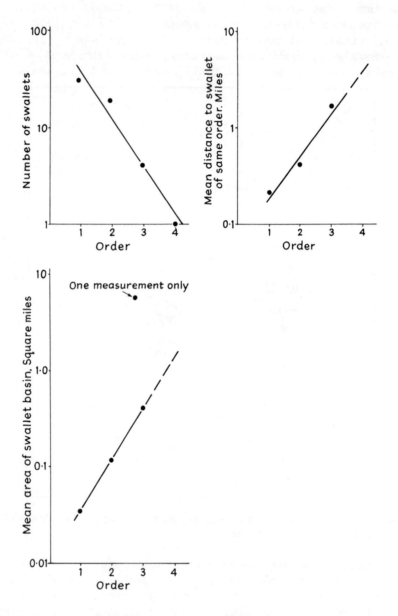

Fig. 5.5 Swallet relationships in the Ingleborough district, Yorkshire, England (from Williams, 1966, fig. 4).

Subterranean aspects of karst have also attracted morphometric interest, from considerations of the changing geometry of cave structures (Lange, 1959, 1968A, B, C) and stochastic models of cave evolution (Curl, 1960, 1966A), to the minute forms of scallops and flutes (Curl, 1966B; Goodchild and Ford, 1971). The interesting problem of meandering cave passages has also been studied, and in particular the question has been asked as to whether or not cavern geometry has parameters similar to those of surface meanders. Ongley (1968) concludes from an analysis of Serpentine Cave, New South Wales, that the passage system winds but does not meander and that it is entirely possible that there may not be an underground geometry comparable with that of surface fluvial forms. From a larger sample of caves, Deike and White (1969) distinguish two types of non-linearity in cave passage traces: (1) an angulate form generated by water flow down a hydraulic gradient diagonal to a rectangular joint set and (2) a curvilinear form with sweeping S-bends apparently similar to meanders in surface streams. The average bend spacing (L) and channel width (W) of the sinuous forms they found related by a power function $L = KW^n$. The coefficient K and the exponent n are respectively 6·8 and 1·05 for the Missouri caves sampled and 8·2 and 0·92 for the other cases. These constants, the authors point out, are similar to those proposed for alluvial rivers (see also Howard, 1971).

The morphometric research described thus far indicates that only the barest beginnings have been made in the study of spatial distributions and associations in karst. This is probably partly because of the mentally deadening effect of the traditional notion that karst landscapes are a random, chaotic jumble of collapse and solution forms. Such ingrained ideas on the lack of order will clearly discourage any thoughts of spatial analysis. Nevertheless, to return to the introductory remarks, the work of climatic geomorphologists (e.g. Lehmann, 1954) has strongly suggested that landscapes largely comparable in gross terms are to be found in diverse regions of similar environment; thus a 'normal' or 'most probable' karst may exist in each morphogenetic region. To test this important concept, it is necessary to employ objective, quantitative methods of terrain analysis and to compare the landscape parameters derived for each region. This is basically a problem of analysing spatial patterns.

Spatial distribution and association analysis

The problems of spatial associations, patterns and diffusion have for long been the concern of ecologists (Greig-Smith, 1964), and for this reason their methods, ideas and experience dominate the field. Pielou (1969) provides the most up-to-date critical assessment of the ecological literature pertaining to these problems, although he overlooks the possible application of important recent developments in statistical geography. The latter are well reviewed by King (1969). Pielou stresses that in a continuum, a spatial pattern has two quite distinct aspects, termed *intensity* and *grain*. The intensity of a pattern is the extent to which density varies from place to place, whereas the grain is independent of intensity and concerns only the spacing and areas of high and low density patches in a dispersion pattern (fig. 5.6).

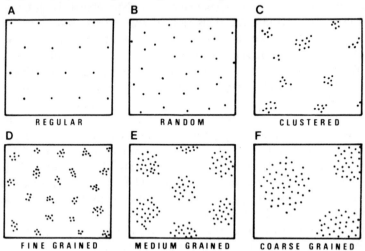

Fig. 5.6 Illustrations of dispersion patterns with different intensities (A, B, C) and grains (D, E, F).

Most methods of pattern analysis measure intensity. The procedure commonly followed is to sample the dispersion by means of randomly placed quadrats. An observed frequency distribution of the number of individuals per quadrat is then obtained, and this is compared to theoretical distributions. However, Pielou states that because of unavoidable assumptions in the methods, the fitting of such theoretical frequency distributions to observational data can never suffice to 'explain' the pattern of a population without additional independent evidence. Nevertheless, measurement or description of a pattern may still be valuable in itself without attempting explanation.

A more serious drawback associated with this technique is that quadrats are arbitrary, unnatural sampling units, and so measures of aggregation based on data derived from them are not as a rule unique; different values being obtained using different quadrat sizes (an attribute employed to advantage in pattern grain analysis). The entire problem may be avoided, however, by using a totally different method of investigating pattern; that of examining spacing of individuals by measuring distance-to-neighbour. Here the data consist of an empirical frequency distribution of a continuous variate, distance.

Pielou examines various indices of aggregation based on nearest neighbour distances and concludes that Clark and Evans's (1954) index is possibly the best if pattern intensity is the main feature of interest. Examples of the use of this index in geography may be found in King (1962), Getis (1964), Smalley and Unwin (1968) and Williams (1972). The application of this index to the analysis of spatial pattern in karst will be illustrated here with reference to the writer's work in New Guinea (Williams, 1971, 1972; in press).

Spatial analysis of karst in New Guinea

Almost nothing is known about the spatial characteristics of karst; so our problem is basically of an exploratory nature. Humid tropical karst is of topical interest, particularly that variety known as cone karst (Ger. *Kegelkarst*). Furthermore, it has a certain regularity of form that suggests it may be susceptible to morphometric analysis. An attempt will therefore be made to define and compare spatial attributes of certain karsts in New Guinea.

Abstraction of data

Conceptual models of karsts in New Guinea were obtained by abstracting from aerial photographs certain information that was considered possibly relevant to the study. The stereoscopic interpretation of the photographs was aided by having first made a field examination of the terrain and the problem. The topographic features mapped were (1) divides, (2) summits, (3) channels and (4) stream-sinks. The depressions were also ranked hierarchically according to the highest Strahler order channel system draining to their stream-sink. Mapping was done directly on to transparent film overlay, that later served as a photographic positive from which working maps were produced at 1/15,000. The data for analysis were then obtained from the morphological maps (fig. 5.7).

The eight localities studied range in topographic style from conventional cone karst to the unusual arête-and-pinnacle karst (Williams, 1971). The limestones in each region were found to be so completely pitted by closed depressions that there is room for no more. The topographic divides of the depressions (fig. 5.8) form polygonal networks, indicating that the terrains are completely partitioned by adjoining basins. Thus from a two-dimensional point of view, all the areas may be considered *polygonal karst*. At ground level, the conical or tower-like hills that form parts of the depression divides dominate the scene, but the basins are the dynamic centres of the landscape. The hills

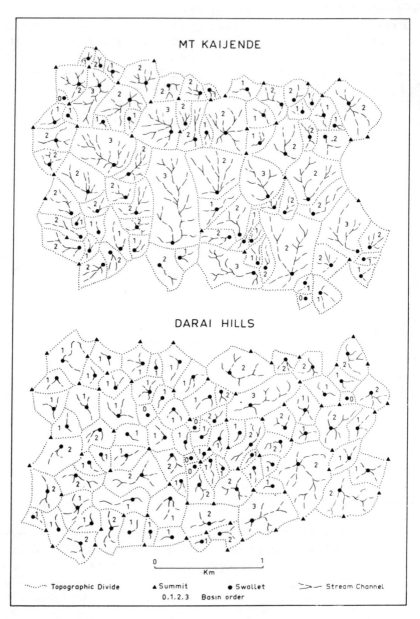

Fig. 5.7 Morphological maps from aerial photographs of parts of two polygonal karsts in eastern New Guinea (from Williams, 1971, fig. 4).

are merely residuals, and while their spatial attributes could be studied, it is more meaningful to consider the depressions (cockpits) at first.

DELIMITATION OF TEMPERATE AND TROPICAL CLOSED DEPRESSIONS

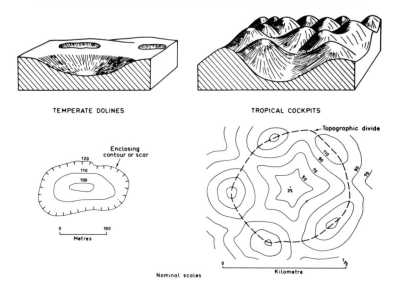

TEMPERATE DOLINES   TROPICAL COCKPITS

Fig. 5.8 Delimitation of closed depressions in karst (from Williams, 1969, fig. 6.II.9).

Analytical procedure

(a) *Pattern.* The adjoining depressions produce a cellular mosaic in plan, but for each area the lattice appears different in detail. In an hexagonal system, the number of lattice nodes or vertices per cell is six, whereas in the New Guinea karsts the average vertices values range from 4·9 to 5·7 per area, with a grand mean of 5·3. Thus the polygonal depressions are found more to resemble pentagons than hexagons.

The location of a closed depression in space may be taken as that of its focus, its stream-sink. Maps of stream-sinks of the different areas showed there to be variations in dispersion pattern (fig. 5.9). Thus it was decided to assess the variation from randomness of the patterns and to compare them, by means of Clark and Evans's index, R. R measures the departure from randomness of a dispersion pattern by comparing the observed mean distance between nearest neighbours with the expected mean distance. Hence to analyse a pattern of stream-sinks, it is necessary first to measure the actual distance (L_a) from each individual stream-sink to its closest neighbour, and then to calculate the mean value ($\bar{L}_a$). Next, the expected mean distance ($\bar{L}_e$) between

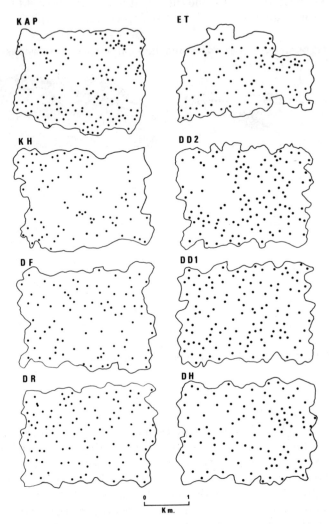

Fig. 5.9 Dispersion patterns of stream-sinks in some New Guinea karsts (from Williams, in press).

nearest neighbours in an infinitely large random population with the same density (D) of stream-sinks as the sample area is computed from the formula:

$$\bar{L}_e = \frac{1}{2 . \sqrt{D}}$$

The index ratio (R) of the observed to expected distance, $\bar{L}_a / \bar{L}_e$, is obtained and this forms the basis of the conclusions. R ranges in value from 0 for a dispersion with maximum aggregation or clustering, through 1 for a random case, to 2·1491 for a regular pattern that is as evenly and widely spaced as possible. The significance of departure from random expectation may be tested by the standard variate of the normal curve while the significance of differences between all values of *R* from various populations may be tested by a one-way analysis of variance.

A summary of the above nearest-neighbour measures for stream-sinks in the New Guinea examples is provided in table 5.3. The results

Table 5.3 *Summary of nearest-neighbour analysis measures for eight New Guinea karsts*

Locality name abbreviation	Number of depressions	D (per km²)	L_a (km)	L_e (km)	R	Nature of pattern
KAP	148	20·8	0·120	0·110	1·091*	near random
KH	172	13·1	0·154	0·138	1·116*	near random
ET	93	17·7	0·134	0·119	1·126*	near random
DF	185	10·5	0·193	0·154	1·253*	approaching uniform
DD1	130	18·5	0·154	0·116	1·328*	approaching uniform
DD2	130	22·1	0·130	0·106	1·226*	approaching uniform
DH	188	13·5	0·191	0·136	1·404*	approaching uniform
DR	182	18·3	0·156	0·117	1·333*	approaching uniform

* Significantly different from random ($R = 1$) at the 0·05 level.

indicate that the probability of a greater difference between L_e and L_a occurring purely by chance is considerably less than 0·01 in all cases except two, where it is less than 0·03 and 0·02 respectively. It may therefore be concluded that within the confines of the selected study areas the stream-sink dispersion patterns (and presumably therefore the depressions) are not random. In three examples (table 5.3), while divergence is on the side of uniformity, the pattern is almost random; but in the other cases a greater tendency towards uniformity of spacing is displayed. Analysis of variance indicates there to be a significant difference at the 0·01 level of confidence between the R values. Thus while all patterns lie on the systematic side of random, some are more regular than others.

Probably the best explanation for this tendency towards uniformity is that it is the result of a rather finely balanced competition for space as neighbouring depressions evolve through time. However, as the karsts are not possessed of a perfectly regular polygonal system, the nature of the variation from regularity also requires investigation, in order to try to cast some light on the underlying geomorphic processes.

Two possible approaches to the assessment of pattern disturbances present themselves: (1) to examine the orientation of individual polygons and (2) to explore the possibility of alignment of neighbouring stream-sinks. In both cases the measurements may be analysed as vectors. The orientation of polygonal depressions was obtained from the bearing of the long axes, and the topographic importance of the trend was estimated from the axis length. Stream-sink alignment may be represented by the orientation and length of vectors joining nearest neighbours. For both sets of measurements, bearings were entered in eighteen 10° classes from 0° to 180°. The accumulated length of axes in each class was then found and expressed as a percentage of the summed lengths of every class. The percentage frequencies were then plotted on rose diagrams for ease of interpretation (fig. 5.10).

Statistical interpretation of circular frequency distributions presents various difficulties, as discussed by Pincus (1953, 1956), Curray (1956) and Batschelet (1965). The present requirement is to identify multiple orientation peaks should they occur and to discriminate between those arising by chance and those that are significant. Krumbein (1939) and Gumbel (1954) discuss the problem of handling polymodality, but a wholly satisfactory procedure has apparently not been developed. The method adopted here is therefore only a heuristic measure. It was decided to test the orientation data against the null hypothesis that they are uniformly distributed, by means of a chi-square (χ^2) test. The alternative hypothesis is that they are not uniformly distributed. A model of uniformity seems appropriate in view of the tendency for the polygons to be regularly dispersed. Using such a model, the statistic χ^2 may be calculated from (Pincus, 1953):

$$\chi^2 = \frac{1}{e} \sum_{1}^{k} (o_i - e)^2$$

where o_i = the observed frequencies in each class;
 e = the expected frequency in a uniform distribution with equal class intervals; and
 k = the number of pairs of frequencies to be compared.

The number of degrees of freedom (d.f.) for entering the tables of the χ^2 function is the number of classes less one. Since all terms of the equation are known except for o_i, it may be solved for the latter at any selected

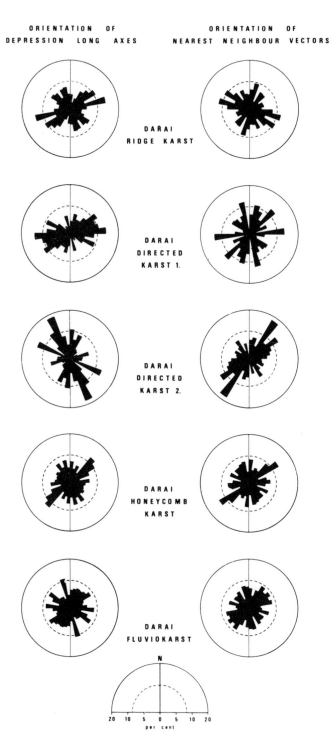

Fig. 5.10 Preferred orientation of depression long axes and stream-sink nearest-neighbour vectors (from Williams, 1972).

probability level. Thus with $e = 5.55$, d.f. $= 17$, and a probability level of 0.05, the calculated value of o_i is 8.47. If this test value is equalled or exceeded in any class interval, then the observed distribution may be considered to depart significantly from the model of uniformity, and the alternative hypothesis must thus be accepted. This is a one-tailed test, since the calculated o_i value takes only positive discrepancies from e into account.

To facilitate rapid visual examination of the class values, a pecked circle with radius equivalent to 8.47% was drawn on the orientation roses (fig. 5.10). Only in one case can the null hypothesis be accepted. The geomorphic significance of the various deviations from uniformity are explained in detail elsewhere (Williams, 1972), but the main conclusions are as follows.

1. All eight karsts examined are to some extent aligned, even those having no obvious visual 'grain' on aerial photographs.
2. The general land slope is the most consistent influence on both the orientation of depressions and the alignment of their stream-sinks. It is thus the most widespread factor preventing the attainment of the hypothetical regular hexagonal network.
3. Master joints as deduced from photo-lineaments often coincide with and thus probably cause the orientation of many depressions and stream-sinks.
4. In the cases examined, tectonic strike is less important than jointing in determining topographic trends in the karsts.

Thus in the eight karsts and involving measurements of more than 1200 depressions, it is possible to identify (a) a gravity effect and (b) a structural effect. These are the influences most responsible for distorting the hexagonal pattern expected from simple competition alone.

Having learnt something about the general framework of the polygonal karsts, attention may now be focused on the individual cells and their spatial associations.

(b) *Spatial associations.* Simple measurements may be made on depressions in order to estimate their two-dimensional geometry (fig. 5.11). Thus for each depression data were obtained on length, width and area characteristics and Strahler order. The association of these variables was determined using correlation and regression analysis (fig. 5.12) and the following conclusions were drawn.

1. There are relatively few 0 and 3 order basins and none of 4 order or above. The modal order is 1 or 2, being higher in those areas that have been subjected to erosion for a longer period of time.
2. The relations of length to width and length/width ratio to order

indicate the shape of depressions to be largely unaltered as they grow in size. In only a minority of cases does elongation occur.

3. While depressions do not normally become more elongate with increasing order, their internal asymmetry (as estimated by the

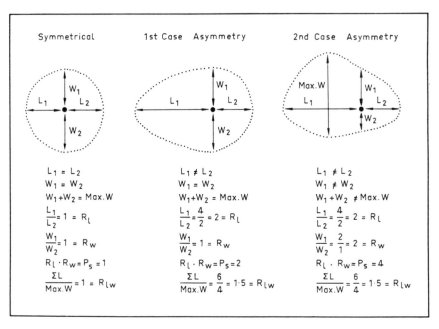

Fig. 5.11 Measurements made to estimate the geometry of depressions (from Williams 1971).

Product of Symmetry, P_s, fig. 5.11) usually increases geometrically (fig. 5.12).

4. There is a geometric growth of depression area with increasing order in all cases (significance level 99·9%). The resemblance of this relationship to the Law of Drainage Areas (Schumm, 1956) in normal river systems is unlikely to be mere coincidence. Thus one is led to conclude that the polygonal depressions are small river basins of a special kind, characterised by intermittent and centripetally directed flow down poorly developed, gulley-like channels. Following from this, they must be considered to be mainly the product of superficial erosion, rather than collapse.

5. Should an increase in depression order (and thus area) be accompanied by no significant change in the dimensionless form parameters R_{1w} and P_s (fig. 5.11), then the growth will have been accomplished by parallel slope retreat, provided the mean width/depth

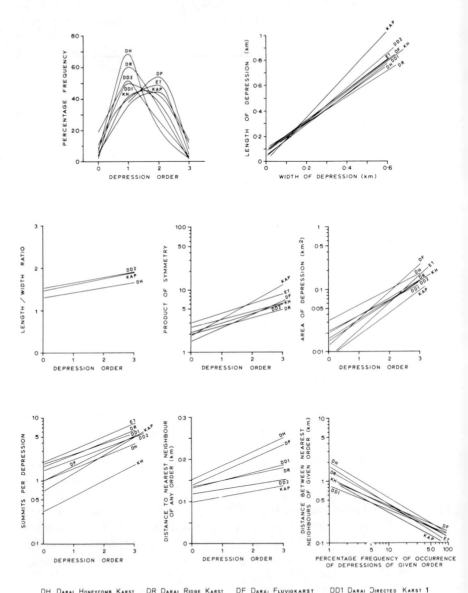

DH Darai Honeycomb Karst   DR Darai Ridge Karst   DF Darai Fluviokarst   DD1 Darai Directed Karst 1

DD2 Darai Directed Karst 2   ET Emia Tower Karst   KH Kaijende Honeycomb Karst   KAP Kaijende Arête and Pinnacle Karst

Fig. 5.12 Graphs illustrating the association of landform attributes in New Guinea karsts (from Williams, 1972). (All regression lines are significant at the 0·05 level.)

Table 5.4 *Homogeneous subsets as identified with Duncan's multiple range test* (from Williams, 1972)

$$\alpha_2 = 0.05$$

Rank ($\bar{x}_1 < \bar{x}_2 \ldots < \bar{x}_8$)

	$\bar{x}_1$	$\bar{x}_2$	$\bar{x}_3$	$\bar{x}_4$	$\bar{x}_5$	$\bar{x}_6$	$\bar{x}_7$	$\bar{x}_8$	Variable
Area identity	DH	DD1	DR	DD2	DF	KH	ET	KAP	$\bar{x}_i = \bar{P}_s$
Sub-sets									

	$\bar{x}_1$	$\bar{x}_2$	$\bar{x}_3$	$\bar{x}_4$	$\bar{x}_5$	$\bar{x}_6$	$\bar{x}_7$	$\bar{x}_8$	Variable
Area identity	DD1	DH	KH	DR	DF	DD2	KAP	ET	$\bar{x}_i = \bar{R}_{1w}$
Sub-sets									

	$\bar{x}_1$	$\bar{x}_2$	$\bar{x}_3$	$\bar{x}_4$	$\bar{x}_5$	$\bar{x}_6$	$\bar{x}_7$	$\bar{x}_8$	Variable
Area identity	KH	DD1	DH	DD2	KAP	DF	DR	ET	$\bar{x}_i = \bar{H}$
Sub-sets									

	$\bar{x}_1$	$\bar{x}_2$	$\bar{x}_3$	$\bar{x}_4$	$\bar{x}_5$	$\bar{x}_6$	$\bar{x}_7$	$\bar{x}_8$	Variable
Area identity	KAP	DD2	ET	KH	DD1	DR	DH	DF	$\bar{x}_i = \bar{L}_a$
Sub-sets									

	$\bar{x}_1$	$\bar{x}_2$	$\bar{x}_3$	$\bar{x}_4$	$\bar{x}_5$	$\bar{x}_6$	$\bar{x}_7$	$\bar{x}_8$	Variable
Area identity	DD2	KAP	DR	ET	DD1	KH	DH	DF	$\bar{x}_i = \bar{L}_0$
Sub-sets									

	$\bar{x}_1$	$\bar{x}_2$	$\bar{x}_3$	$\bar{x}_4$	$\bar{x}_5$	$\bar{x}_6$	$\bar{x}_7$	$\bar{x}_8$	Variable
Area identity	DD2	KAP	DD1	DR	ET	DH	KH	DF	$\bar{x}_i = \bar{A}$
Sub-sets									

[*Key to table appears at foot of page 158*

ratio also remains constant. This may be the case in one area only (DD2), but cannot hold generally.

Thus far the spatial analysis has revealed striking similarities between the areas, demonstrating forcibly that the terrains possess a characteristic style of organisation, not the chaos that has too often been ascribed to karst. Nevertheless, fieldwork and air photo examination showed the areas to have contrasting styles, in detail if not in general. So an attempt should be made to define these perceived differences.

(c) *Terrain differentiation.* Analysis of variance of six landform variables in turn over the eight areas showed there always to be a significant difference between the areas. However, the technique does not isolate the set or sets of data responsible for the significant variations. Hence a follow-up procedure is required that will discriminate between sample data, for example means, at a known level of significance. The oldest method is the least significant difference test, but there are other more recent superior techniques, amongst which Duncan's (1955) multiple range test seems to be one of the most useful. To perform the test, the parameter means (k) for each area are arranged in increasing order of magnitude and range tests are applied to all possible combinations of the k means taken p at a time ($p = 2, \ldots, k$). If any combination of small p means has a non-significant range, then the decision is made that these p means are homogeneous. For any combination of means that is not homogeneous, the highest mean is significantly larger than the lowest.

Duncan's test was performed on the data at a 0·05 level of significance. The results are presented in table 5.4, where underscoring is used to emphasise the statistically indistinguishable (homogeneous) subsets. The results show exactly where the similarities and divergences in the data appear, and indicate that some variables are more discriminating than others. It is also apparent from the results that some areas quite frequently have several geomorphic attributes in common, although there is clearly a range of similarity amongst the karsts. However, if we consider groups of areas in which each member has at least three (out of six) parameters in common with every other member, then groups of

---

$\bar{P}_s$ = mean product of symmetry of depressions
$\bar{R}_{1w}$ = mean length/width ratio of depressions
$\bar{H}$ = mean no. of residual hills per depression
$\bar{L}_a$ = mean distance from a stream-sink to its nearest neighbour of any order
$\bar{L}_0$ = mean distance from a stream-sink to its nearest neighbour of the same order
$\bar{A}$ = mean area of depressions

karsts can be defined in which there is a relatively large degree of morphological homogeneity (table 5.5). Since all eight localities are from polygonal karst landscapes, the collections of karsts defined can be considered to be subsets within the set of polygonal karst. The idea is represented diagrammatically in fig. 5.13.

	KH	DH	DF	DD1	DR	DD2	ET	KAP
KAP	1	1	2	2	2	5	4	6
ET	2	2	1	2	3	4	6	
DD2	2	2	3	3	3	6		
DR	3	4	3	5	6			
DD1	3	5	3	6				
DF	3	4	6					
DH	3	6						
KH	6							

Pinnacle sub-set

Intersecting linear sub-sets

Conic sub-set

Table 5.5   *Association matrix developed from table 5.4, showing the number of parameters in common (out of six) between all pairs of eight karsts in New Guinea* (from Williams, 1972)

The range tests thus permit the identification of three types of polygonal karst that were named (Williams, 1972) the Pinnacle, Conic and Linear styles. The disjoint Pinnacle and Conic subsets are intersected by Linear subsets, which thus possess intermediate landscape types. The Conic and Linear styles correspond roughly to 'cone karst' and 'directed karst' in conventional terms, although the Pinnacle style is quite different from normally accepted 'tower karst'.

Conclusions on the spatial organisation of the karsts

The karsts display a remarkable similarity in spatial organisation, differences being of detail not of kind. The landscapes are characterised by a cellular mosaic of closed depressions, the divides of which form a polygonal network. The cells approximate in shape rather more closely to pentagons than to hexagons and their dispersion pattern tends towards uniformity. The attainment of a perfectly regular hexagonal network is prevented by two main factors: (1) a gravity effect which is transmitted to the landscape via the work of running water and operates parallel to the general slope of the region and (2) a structural effect

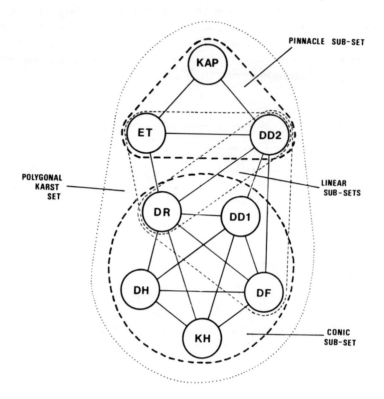

A link indicates there to be no significant difference
in at least three (of six) morphological features examined.

Defines a karstic sub-set in which at least three
elements have mutual links.

Indicates contained within the set of polygonal karst.

Fig. 5.13 Association structure of New Guinea karsts
(from Williams, in press).

which is transferred to the landscape through the bedrock fissure system;
the fractures providing zones of weakness that are preferentially
exploited by erosion.

The polygonal cells in which these oriented erosive forces operate are
small river basins of a special kind, typified by centripetal drainage
down poorly developed gulley-like channels. The mutual competition
of neighbouring polygonal basins probably stabilises their divides, and
as the limestone mass is reduced, the basic geometry of the karst is
substantially unchanged. A breakdown of this steady state should only

occur when a major environmental factor alters, as for example when the limestones are no longer free draining but saturated at the level of the watertable.

References

BATSCHELET, E. (1965) Statistical methods for the analysis of problems in animal orientation and certain biological rhythms; *American Institute of Biological Sciences* (Monograph), Washington, 57p.

CLARK, P. J. and EVANS, F. C. (1954) Distance to nearest neighbour as a measure of spatial relationships in populations; *Ecology* 35, 445–53.

COLEMAN, A. M. and BALCHIN, W. G. W. (1960) The origin and development of surface depressions in the Mendip Hills; *Proceedings of the Geologists' Association* 32, 291–309.

CORBEL, J. (1957) *Les karsts du Nord-Ouest de l'Europe;* Mémoires et Documents no. 12, Institut des Études Rhodaniennes de l'Université de Lyon, 544p.

CORBEL, J. (1959) Erosion en terrain calcaire – vitesse d'érosion et morphologie; *Annales de Géographie* 68, 97–120.

CRAMER, H. (1941) Die Systematik der Karstdolinen; *Neues Jahrbuch für Mineralogie, Geologie, und Paleontologie*, Beilage-Band, Abt. B, 85, 293–382.

CURL, R. L. (1960) Stochastic models of cavern development; *Bulletin of the National Speleological Society* (U.S.A.) 22, 66–74.

CURL, R. L. (1966A) Caves as a measure of karst; *Journal of Geology* 74, 798–830.

CURL, R. L. (1966B) Scallops and flutes; *Transactions of the Cave Research Group of Great Britain* 7, 121–60.

CURRAY, J. R. (1956) The analysis of two-dimensional orientation data; *Journal of Geology* 64, 117–31.

CVIJIĆ, J. (1893) Das Karstphänomen; *Geographische Abhandlungen herausgegeben von A. Penck* 5, 217–330.

DEIKE, G. H. and WHITE, W. B. (1969) Sinuosity in limestone solution conduits; *American Journal of Science* 267, 230–41.

DUNCAN, D. B. (1955) Multiple range and multiple F tests; *Biometrics* 11, 1–42.

EWING, A. L. (1885) The amount and rate of chemical erosion in the limestone of Centre County, Pennsylvania; *American Journal of Science* 29, 29–31.

FORD, D. C. (1964) Origin of closed depressions in the central Mendip Hills; *20th International Geographical Congress, London, Abstracts of Papers*, 105–6.

GETIS, A. (1964) Temporal land-use patterns analysis with the use of nearest-neighbor and quadrat methods; *Annals of the Association of American Geographers* 54, 391–9.

GOODCHILD, M. F. and FORD, D. C. (1971) Analysis of scallop patterns by simulation under controlled conditions; *Journal of Geology*, 79, 52–62.

GREIG-SMITH, P. (1964) *Quantitative Plant Ecology*; 2nd Edn (Butterworths, London), 256p.

GUMBEL, E. J. (1954) Applications of the circular normal distribution; *Journal of the American Statistical Association* 49, 267–97.

HACK, J. T. (1960) Relation of solution features to chemical character of water in the Shenandoah valley, Virginia; *U.S. Geological Survey Professional Paper* 400-B, 387–90.

HOWARD, A. D. (1971) Quantitative measures of cave patterns; *Caves and Karst* 13, 1–7.

JENNINGS, J. N. and BIK, M. J. (1962) Karst morphology in Australian New Guinea; *Nature* 194, 1036–8.

KING, L. J. (1962) A quantitative expression of the pattern of urban settlements in selected areas of the United States; *Tijdschrift voor Economische en Sociale Geografie* 53, 1–7.

KING, L. J. (1969) *Statistical Analysis in Geography* (Prentice-Hall Inc., Englewood Cliffs, New Jersey), 288p.

KRUMBEIN, W. C. (1939) Preferred orientation of pebbles in sedimentary deposits; *Journal of Geology* 47, 673–706.

LANGE, A. L. (1959) Introductory notes on the changing geometry of cave structures; *Cave Studies* 11, 69–90.

LANGE, A. L. (1968A) The changing geometry of cave structures, Part 1: the constant solution gradient; *Caves and Karst* 10(1), 1–10.

LANGE, A. L. (1968B) The changing geometry of cave structures, Part 2: the exponential solution gradient; *Caves and Karst* 10(2), 13–19.

LANGE, A. L. (1968C) The changing geometry of cave structures, Part 3: summary of solution processes; *Caves and Karst* 10(3), 29–32.

LAVALLE, P. (1967) Some aspects of linear karst depression development in south central Kentucky; *Annals of the Association of American Geographers* 57, 49–71.

LAVALLE, P. (1968) Karst depression morphology in south central Kentucky; *Geografiska Annaler* 50A, 94–108.

LEHMANN, H. (Ed.) (1954) Das Karstphänomen in den verschiedenen Klimazonen; *Erdkunde* 8, 112–39.

MATSCHINSKI, M. (1962A) Sur l'alignment des mares de Sucy-en-Brie (Seine et Oise); *Compte Rendu des Séances de la Société Géologique de France;* fasc. 7, 173–4.

MATSCHINSKI, M. (1962B) Sur le problème 'd'alignment' de données apparement dispersées; *Compte Rendu des Séances de l'Académie des Sciences*, Paris, 254, 806–9.

MATSCHINSKI, M. (1964) Le problème statistique et le problème d'alignment des phénomènes de karst; *20th International Geographical Congress, London, Abstracts of Papers*, 107.

MATSCHINSKI, M. (1968) Alignment of dolines north-west of Lake Constance, Germany; *Geological Magazine* 105, 56–61.

ONGLEY, E. D. (1968) An analysis of the meandering tendency of Serpentine Cave, N.S.W.; *Journal of Hydrology* 6, 15–32.

PIELOU, E. C. (1969) *An Introduction to Mathematical Ecology* (Wiley, New York), 286p.

PINCUS, H. J. (1953) The analysis of aggregates of orientation data in the earth sciences; *Journal of Geology* 61, 482–509.

PINCUS, H. J. (1956) Some vector and arithmetic operations on two-dimensional orientation variates, with applications to geological data; *Journal of Geology* 64, 533–57.

SCHUMM, S. A. (1956) Evolution of drainage systems and slopes in badlands at Perth Amboy, N.J.; *Bulletin of the Geological Society of America* 67, 597–646.

SMALLEY, I. J. and UNWIN, D. J. (1968) The formation and shape of drumlins and their distribution and orientation in drumlin fields; *Journal of Glaciology* 7, 377–90.

STODDART, D. R. (1968) Climatic geomorphology: review and reassessment: *Progress in Geography* 1, 159–222.

STRAHLER, A. N. (1957) Quantitative analysis of watershed geomorphology; *Transactions of the American Geophysical Union* 38, 913–20.

VERSTAPPEN, H. TH. (1964) Karst morphology of the Star Mountains (central New Guinea) and its relation to lithology and climate; *Zeitschrift für Geomorphologie* n.f., 8, 40–9.

WILLIAMS, P. W. (1966) Morphometric analysis of temperate karst landforms; *Irish Speleology* 1, 23–31.

WILLIAMS, P. W. (1969) The geomorphic effects of ground water; In Chorley, R. J. (Ed.), *Water, Earth and Man* (Methuen, London), 269–84.

WILLIAMS, P. W. (1971) Illustrating morphometric analysis of karst with examples from New Guinea; *Zeitschrift für Geomorphologie* n.f., 15, 40–61.

WILLIAMS, P. W. (1972) Morphometric analysis of polygonal karst in New Guinea; *Bulletin of the Geological Society of America*, 83, 761–96.

WILLIAMS, P. W. (in press) Cave and karst areas in east New Guinea; *Acts of the 5th International Congress of Speleology*, Stuttgart, 1969.

PART III
Networks

6 The topology of stream networks

A. WERRITTY

Department of Geography, University of Cambridge

One of the results of the increasing use of mathematics in geomorphology has been a resurgence of interest in the morphology of the physical landscape. This has become most pronounced in fluvial geomorphology where a succession of workers, ultimately deriving their inspiration from Horton's classic paper of 1945, have produced a sequence of models of varying mathematical complexity and sophistication (Schumm 1956; Melton 1957; Strahler 1958; and Chorley and Morgan 1962). At first these models were predominantly deterministic in approach, for although Horton's Laws had the appearance of being statistical laws, the parameters of the distributions implicit in their formulation were never made explicit (Shreve 1966, 17). Certainly in terms of the empirical testing procedures used, they were clearly viewed as being deterministic in nature.

A persistent criticism of Horton's work is that the ordering scheme is very insensitive to variations in structure and lithology. Bifurcation ratios were found to be remarkably stable from one area to another, and generally cluster in the range 3·5 to 4·0. In an attempt to generate a more sensitive ordering scheme and a model devoid of Horton's inconsistencies (Smart 1967), Shreve (1966) has proposed a random topology model based solely upon combinatorial properties. From this initial formulation he and Smart have proceeded to derive laws of stream lengths and areas based largely upon the postulates of the random topology model (Shreve 1969; Smart 1968).

Excellent reviews of the earlier work during the 50's and 60's may be found in Chorley and Haggett (1969) and Strahler (in Chow 1964). This review will concentrate on the more recent work.

Topologically random drainage networks

Shreve's random topology model is based upon a little-known combinatorial expression first derived by Cayley in the mid-nineteenth century (Cayley 1859). Rearranging the formulae and introducing a different terminology we proceed by defining the following terms. The extreme tips of the network are termed sources with the exception of the outlet, the farthest point downstream. Points of confluence of two tributaries are called junctions; triple junctions not being permitted. A link is a section of the channel extending either from a source or a junction downstream to a second junction or the outlet with no intervening junctions. Exterior links are synonymous with Strahler first-order streams and extend from a source to the first junction downstream. Any segment which is bounded by two successive junctions, or the outlet and the immediate junction upstream, is termed an interior link (Shreve 1966, 20; Shreve 1967, 178).

By virtue of these definitions a drainage network possessing n sources will have n exterior links, $n-1$ interior links and $n-1$ junctions. The network can also be ordered by a far more sensitive and consistent procedure than that adopted hitherto (fig. 6.1) – each link in the network is coded according to its magnitude, the magnitude of a link being equal to the number of sources upstream (Shreve 1967, 179).

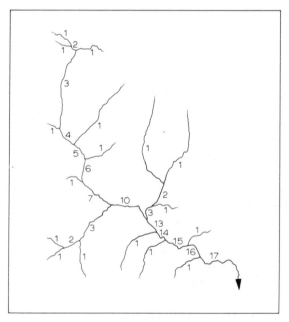

Fig. 6.1 A drainage network with 17 sources, links ordered by magnitude.

Having established the primitive terms of the random topology model, we may now examine the idea of topologically distinct networks. This is based upon the fact that networks having the same number of sources, whilst of similar topologic complexity (i.e. possessing the same number of links and junctions) are nevertheless topologically distinguishable. This is illustrated in fig. 6.2 which shows the 14 possible

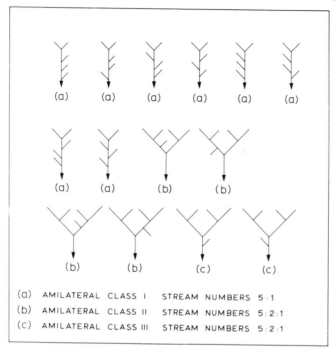

(a) AMILATERAL CLASS I   STREAM NUMBERS 5:1
(b) AMILATERAL CLASS II   STREAM NUMBERS 5:2:1
(c) AMILATERAL CLASS III STREAM NUMBERS 5:2:1

Fig. 6.2 The 14 TDCN and three ambilateral classes for magnitude-5 networks.

topologically distinct channel networks (TDCN) for magnitude-5 networks. By visual inspection one can readily determine that the number of TDCN for networks with one to six sources are 1, 1, 2, 5, 14, 42 respectively. But for networks with more than six sources the number increases very rapidly. The number of TDCN for a given number of sources, $N(n)$, is given by Cayley's expression:

$$N(n) = \frac{1}{2n-1}\binom{2n-1}{n} \tag{1}$$

where n equals the number of sources, i.e. first-order streams.

For $n > 5$ the value of $N(n)$ increases so rapidly that a regrouping of TDCN becomes necessary for the purposes of testing the model (table 6.1). For example, $N(6) = 42$, which would necessitate generating

Table 6.1   *Topologically distinct networks $N(n)$*

Number of sources n	$N(n)$	Number of sources n	$N(n)$
1	1	10	4862
2	1	20	$1 \cdot 767 \times 10^9$
3	2	30	$1 \cdot 002 \times 10^{15}$
4	5	40	$6 \cdot 804 \times 10^{20}$
5	14	50	$5 \cdot 095 \times 10^{26}$
6	42	100	$2 \cdot 275 \times 10^{56}$
7	132	200	$1 \cdot 29 \ \times 10^{116}$
8	429	500	$1 \cdot 35 \ \times 10^{296}$
9	1430	1000	$5 \cdot 12 \ \times 10^{596}$

a sample which would contain several hundred magnitude-6 networks
if the minimum cell requirements of goodness-of-fit tests are to be met
(Cochran 1952). One solution is to group the TDCN according to their
order. This involves solving for the awkward recursion formula (Shreve
1966, 29):

$$N(n; \Omega) = \sum_{i=1}^{n-1} [N(i; \Omega - 1) \times N(n - i; \Omega - 1) + 2N(i; \Omega)$$

$$\times \sum_{\omega=1}^{\Omega-1} N(n - i; \omega)] \tag{2}$$

$$N(1; 1) = 1, \qquad N(n; 1) = 0$$
$$N(1; \Omega) = 0, \qquad n = 2, 3, \ldots,$$
$$\Omega = 2, 3, \ldots$$

Representative values of this expression for networks of up to nine
sources are given in table 6.2. One could use these values and test for

Table 6.2   *Topologically distinct networks grouped according to order $N(n; \Omega)$*

	$\Omega = 1$	$\Omega = 2$	$\Omega = 3$	$\Omega = 4$	$N(n) = \sum_{i=1}^{\Omega-1} N(n; \Omega_i)$
$n = 1$	1	0	0	0	1
$n = 2$	0	1	0	0	1
$n = 3$	0	2	0	0	2
$n = 4$	0	4	1	0	5
$n = 5$	0	8	6	0	14
$n = 6$	0	16	26	0	42
$n = 7$	0	32	100	0	132
$n = 8$	0	64	364	1	429
$n = 9$	0	128	1288	14	1430

randomness, but it is much more convenient if one further regroups the TDCN with respect to specific sets of stream numbers. Thus for a given magnitude, different TDCN may possess the same Strahler stream numbers. For example, magnitude-5 networks have two possible sets of stream numbers – 8 TDCN are second-order networks with stream numbers (5, 1) and 6 TDCN are third-order networks with stream numbers (5, 2, 1) (cf. fig. 6.2). The general expression for $N(n_1, n_2, \ldots n_{\Omega-1}, 1)$ denoting the number of TDCN of order Ω having $n_1, n_2, \ldots, n_{\Omega-1}, 1$ streams of order $1, 2, \ldots, \Omega - 1, \Omega$, respectively is given by (Shreve 1966, 29):

$$N[n_1, n_2, \ldots, n_{\Omega-1}, 1] = \prod_{\omega=1}^{\Omega-1} 2^{(n_\omega - 2n_{\omega+1})} \binom{n_\omega - 2}{n_\omega - 2n_{\omega+1}} \qquad (3)$$

The general term of the product in equation (3) is the number of topologically distinct ways that the n_ω streams of order ω may be attached to the higher-order streams in the network. Since at least two tributaries of order ω are necessary to increase the order of the main stream to $\omega + 1$, this requires that $2n_{\omega+1}$ of the n_ω streams join in pairs to form the $n_{\omega+1}$ streams of order $\omega + 1$. The remainder, $r_\omega = n_\omega - 2n_{\omega+1}$, can be attached in various ways to the $l_{\omega+1} = 2n_{\omega+1} - 1$ links of order $\omega + 1$ or greater. The binomial coefficient gives the number of ways the r_ω streams may join these $l_{\omega+1}$ links. The factor 2^{r_ω} occurs because each tributary may enter from either the right or left.

Values for equation (3) are listed in table 6.3. For magnitude-9

Table 6.3 *TDCN grouped according to Strahler stream numbers with nine sources*

Stream numbers	Number of TDCN		Probability for each of the stream number sets
$N(9, 4, 2, 1)$	14		$p_1 = 0 \cdot 00979$
$N(9, 4, 1)$	56		$p_2 = 0 \cdot 03916$
$N(9, 3, 1)$	560	1288	$p_3 = 0 \cdot 39161$
$N(9, 2, 1)$	672		$p_4 = 0 \cdot 46993$
$N(9, 1)$	128		$p_5 = 0 \cdot 08951$

Total = 1430

$N(9) = 1430$
$N(9; 3) = 1288$

networks there are five possible sets of stream numbers which by inspection are seen to be exhaustive. That equation (3) is a special case of equation (2) can be seen by summing the number of TDCN for stream sets (9, 4, 1), (9, 3, 1) and (9, 2, 1). This is found to be identical to the value of $N(9; 3)$ in table 6.2.

One of the interesting results of Shreve's work is his observation that the most probable stream sets derived from equation (3) always approximate very closely Horton's Law of Stream Numbers. More specifically, for networks with a given number of sources n, the most probable network of order Ω is that which makes the geometric bifurcation ratio

$$B = n^{1/(\Omega-1)}$$

closest to 4. If both n and Ω are specified the most probable networks have a bifurcation ratio $B_2 = n_1/n_2$ which is always close to 4; their Horton diagrams are characterised by concave upward, straight, or concave downward curves relative to the value of B being less than, equal to, or greater than 4 respectively (Shreve 1966, 31). If Shreve's assertion that the random topology model is substantially correct is well proven, the above argument provides a theoretical explanation for the clustering of bifurcation ratios in the range 3·5 to 4·0 noted earlier.

Confirmation of this result may be observed in table 6.4 listing the

Table 6.4   *Magnitude-50   networks,*
ten most probable stream
number sets

Stream numbers				Probability
n_1	n_2	n_3	n_4	
50	12	3	1	0·09695
50	13	3	1	0·08970
50	11	3	1	0·07110
50	13	4	1	0·06279
50	14	3	I	0·05713
50	14	4	1	0·05332
50	12	4	1	0·04847
50	12	2	1	0·04155
50	11	2	1	0·04063
50	10	3	1	0·03450
				$\Sigma = 0·59614$

ten most probable sets of stream numbers for magnitude-50 networks. They are all of fourth order and cumulatively account for 59·61% of all possible TDCN. Their bifurcation ratios, $B_\omega = n_\omega/n_{\omega+1}$, vary between 5 and 2. As predicted they all conform to Horton's Law very closely.

Less probable networks of magnitude-50 and orders 2, 3, 5 and 6 deviate markedly from Horton's Law (fig. 6.3).

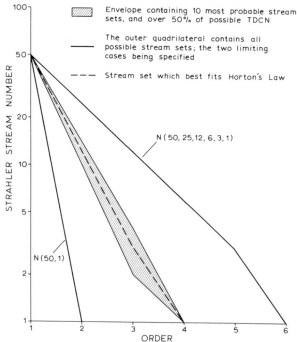

Fig. 6.3. Horton diagram for all TDCN of magnitude-50.

Infinite topologically random drainage networks

A natural extension to the random topology model is from topologically finite to topologically infinite networks. The results are both mathematically more elegant and simpler to interpret. The overall network is termed an infinite topologically random network if the population of sub-networks within it are topologically random. For such infinite networks Shreve has derived the following probability distributions (Shreve 1967, 179–82).

The probability $p(\mu, \omega)$ that a link drawn at random from an infinite topologically random network will have magnitude μ and order ω is given by the recursive relationship:

$$p(\mu, \omega) = \frac{1}{2} \sum_{\alpha=1}^{\mu-1} [p(\alpha, \omega - 1) \times p(\mu - \alpha, \omega - 1) + 2p(\alpha, \omega)$$

$$\times \sum_{\beta=1}^{\omega-1} p(\mu - \alpha, \beta)], \qquad (4)$$

$p(1, 1) = \frac{1}{2}, \qquad p(\mu, 1) = 0,$

$p(1, \omega) = 0, \qquad \mu, \omega = 2, 3, \ldots$

Comparing this relationship with that for $N(n; \Omega)$ it may be observed that both possess the same underlying structure. Values for $p(\mu, \omega)$ are given in table 6.5.

Table 6.5 $p(\mu, \omega)$: *Infinite topologically random networks*

	$\omega = 2$	$\omega = 3$	$\omega = 4$	$\omega = 5$	$\cdots$	$v(\mu)$
1						0·50000
2	0·12500					·12500
3	·06250					·06250
4	·03125	0·00781				·03906
μ 5	·01562	·01172				·02734
6	·00781	·01270				·02051
7	·00391	·01221				·01611
8	·00195	·01111	0·00003			·01309
9	·00098	·00983	·00011			·01091
10	·00049	·00856	·00022		$\cdots$	·00927
.	.	.	.	.		
.	.	.	.	.		
.	.	.	.	.		
$u(\omega)$	0·25000	0·12500	0·06250	0·03125		

$$u(\omega) = \sum_{\mu=1}^{\infty} p(\mu, \omega)$$

$$v(\mu) = \sum_{\omega=1}^{\infty} p(\mu, \omega)$$

Recursive relations for the marginal probabilities can be derived in a similar manner to equation (4) or by summation (see above) or by using the two closed expressions

$$u(\omega) = 1/2^{\omega}, \qquad\qquad \omega = 1, 2, \ldots \qquad (5)$$

$$v(\mu) = \frac{2^{-(2\mu-1)}}{2\mu - 1}\binom{2\mu - 1}{\mu}, \qquad \mu = 1, 2, \ldots \qquad (6)$$

For a given magnitude the distribution of probabilities with respect to order has a relatively pronounced peak, whereas for any given order the distribution with respect to magnitude has a comparatively flat peak (table 6.5). For these reasons the magnitude of a network is a far more sensitive measure of network characteristics than is order.

Further probability distributions which are useful for characterising the properties of these idealised networks have been derived by Shreve (1969, 398–400)

(a) The probability of drawing a Strahler stream of order ω at random from the streams (not links) of an infinite topologically random network is

$$s(\omega) = 3/4^\omega, \qquad \omega = 1, 2, \ldots \tag{7}$$

(b) The expected magnitude of a randomly drawn link of order ω is

$$E_\omega(\mu) = (2^{2\omega-1} + 1)/3, \qquad \omega = 1, 2, \ldots \tag{8}$$

(c) The expected number of links in streams of order ω is

$$E_\omega(\lambda) = 2^{\omega-1}, \qquad \omega = 1, 2, \ldots \tag{9}$$

Using these expressions it is now possible to re-write Horton's Law of Stream Numbers in the following terms. A random sample of Strahler streams drawn from a finite topologically random network will yield an expectation value of the ratio of streams of successive orders which converges on 4 as the size of the sample tends to infinity.

$$E(B_\omega) \lim_{n_\omega \to \infty} n_\omega/u_{\omega+1} = 4 \tag{10}$$

n_ω = number of streams of order ω.

Statistical tests on the random topology model

Only where small-magnitude networks are involved is it possible to test Shreve's model directly (Ranalli and Scheidegger 1968; Smart 1969A; Krumbein and Shreve 1970). Having set up the null hypothesis that in a topologically random population all TDCN will be of equal frequency, one proceeds to group the networks of a given magnitude into their respective classes of TDCN. For networks with more than five sources the value of $N(n)$ increases so astronomically that a regrouping of TDCN becomes necessary for meeting the requirements of goodness-of-fit tests. For networks of magnitudes up to 8 Smart has suggested an ambilateral classification in which 'two TDCN belong to the same class if one can be converted into the other by reversal of the right/left order at one or more junctions' (Smart 1969A, 1761). For magnitude-5 networks there are three ambilateral classes corresponding to the first 8, the next 6, and the last 2 TDCN respectively in fig. 6.2.

Further aggregation becomes necessary for networks with more than 8 sources. Grouping according to Strahler stream numbers, equation (3) allows one to test up to magnitude-20 networks providing a large enough sample is involved. For very high magnitude networks a crude but useful test is to plot the ratio p/p_{max}, where p is the probability of a given set of stream numbers occurring, and p_{max} is the probability of the most probable stream number set for a given magnitude. If the network being considered is topologically random one would expect this ratio to fluctuate around a mean value of 1·00 for most of the main channel length. If the confluence of successive sub-networks does not restore a diminishing value of p/p_{max} to a value of approximately 1·00, one may identify the master network as being topologically non-random. An example of such a non-random network is the headwater section of the River Barle in west Somerset (fig. 6.4).

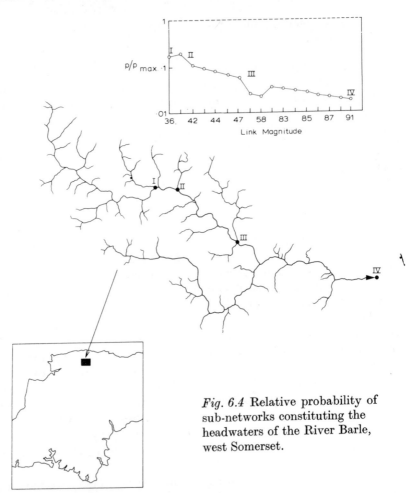

Fig. 6.4 Relative probability of sub-networks constituting the headwaters of the River Barle, west Somerset.

If Shreve's model is to be of use to the geomorphologist and capable of further refinement, one would hope that large-scale lithologic, structural and climatic controls would give rise to distinctive deviations from random expectations. With this in mind, the author is currently testing the model over two major physiographic regions in the south-west peninsula of England. The two areas selected for analysis are (i) the Devonian upland of Exmoor and (ii) the Culm Measures of central and north Devon (fig. 6.5). Whereas the Devonian formations are predominantly slates, shales and grits (Ussher 1881; Evans 1922; Webby 1965), and possess no large-scale lithologic units, the Culm is divisible into two distinct formations – the Crackington Formation (shales and thin turbite sandstones) and the Bude Formation (massive sandstones

Fig. 6.5 Generalised Geology of Devon and West Somerset.

with shales) (Edmonds *et. al.* 1968, 1969; Ussher 1901). The drainage networks have been derived from the blue lines of the Second Edition of the 1/25,000 series, the stream channels of the new edition of the 1/10,560 series and those of the Second Edition of the old 1/2500 County Series. This last source has been used where coverage by neither of the two other series is complete. But since the Second Edition of the 1/25,000 series and the new edition of the 1/10,560 series have stream channels based upon those of the old County Series, the operational definition is a consistent one ultimately based upon that survey of

Table 6.6 *Significance levels for goodness-of-fit tests for TDCN of magnitudes 3 to 5*

	Exmoor North	Exmoor South	R. Taw	R. Torridge	R. Tamar
$n_1 = 3$ (2 TDCN)	* $P = ·0332$	NS $P = ·7338$	NS $P = ·4654$	* $P = ·0358$	NS $P = ·0524$
$n_1 = 4$ (5 TDCN)	NS $·10 > P > ·05$	NS $·20 > P > ·10$	NS $·80 > P > ·70$	NS $·98 > P > ·95$	NS $·80 > P > ·70$
$n_1 = 5$ (14 TDCN)		$·70 > P > ·50$	NS $·80 > P > ·70$	NS $·70 > P > ·50$	NS $·50 > P > ·30$
$n_1 = 5$ (3 ambila-teral classes)	NS $·95 > P > ·90$	NS $·10 > P > ·05$	NS $·70 > P > ·50$	NS $·95 > P > ·90$	NS $·50 > P > ·30$

NS Not significant   * Significant at the 0·05 level

Table 6.7 *Significance levels for goodness-of-fit tests for stream sets of magnitudes 4 to 10*

Stream Numbers	Exmoor North	Exmoor South	R. Taw	R. Torridge	R. Tamar
4 : 1 4 : 2 : 1	NS $P = ·4771$	NS $P = ·0668$	NS $P = ·4681$	NS $P = ·4364$	NS $P = ·3669$
5 : 1 5 : 2 : 1	NS $P = ·4681$	NS $P = ·4443$	NS $P = ·3085$	NS $P = ·3783$	NS $P = ·4522$

6:1					
6:2:1	NS ·20 > P > ·10	NS ·70 > P > ·50	NS ·50 > P > ·30	NS ·50 > P > ·30	NS ·50 > P > ·30
6:3:1					
7:1					
7:2:1	NS ·90 > P > ·70	NS ·10 > P > ·05	NS ·80 > P > ·70	** ·01 > P > ·001	NS ·80 > P > ·70
7:3:1					
8:1					
8:2:1					
8:3:1	NS ·70 > P > ·50	* ·05 > P > ·02	NS ·50 > P > ·30	NS ·50 > P > ·30	NS ·95 > P > ·90
8:4:1					
8:4:2:1					
9:1					
9:2:1					
9:3:1	NS ·40 > P > ·30	NS ·90 > P > ·80	NS ·20 > P > ·10	NS ·70 > P > ·50	NS ·90 > P > ·80
9:4:1					
9:4:2:1					
10:1					
10:2:1					
10:3:1					
10:4:1	NS ·50 > P > ·30	NS ·70 > P > ·50	NS ·90 > P > ·50	NS ·90 > P > ·80	NS ·70 > P > ·50
10:4:2:1					
10:5:1					
10:5:2:1					

NS Not significant   * Significant at the 0.05 level   ** Significant at the 0.01 level

c. 1907. Contour crenulation methods have been avoided because of problems of interpolation in the periglacially-enlarged valley heads, and the presence of relict valley forms possibly formed during the Pleistocene (Gregory in Barlow 1969, 38). Field checking has confirmed that the operational definition of the drainage networks given above is both consistent and reliable within small confidence limits.

The results for networks of magnitudes 3 to 5, where individual TDCN can be treated directly, are given in table 6.6. Using both the binomial and the likelihood ratio goodness-of-fit tests (Hoel 1962, 220–9), in only two of the nineteen categories is the null hypothesis rejected. Both of these are for very low magnitude networks. Otherwise the model is well substantiated. For higher magnitude networks where sets of stream numbers have been tested (table 6.7), the null hypothesis is accepted in all but two cases. The two deviant cases are magnitude-8 networks in the southern part of Exmoor and magnitude-7 networks in the catchment of the Torridge. There is no discernible difference between low magnitude networks in the Bude and Crackington Formations in the Taw valley. The strong lithologic contrast here is completely masked.

Other large sample studies are in general agreement with the above findings (Krumbein and Shreve 1970; Smart 1969A). Smart was only able to find significant discrepancies between observed and predicted bifurcation ratios of low order streams for networks with more than sixty sources in the western U.S.A. Krumbein and Shreve noted that their sample of TDCN drawn from homogeneous sandstones in eastern Kentucky is 'approximately topologically random with weak, but apparently systematic deviations' which are possibly related to systematic differences in outlet orientation (Krumbein and Shreve 1970, 1.)

Other small-scale topologic properties

An attempt at specifying small-scale topologic features of drainage networks has been described by Krumbein and James (1969). They classify the interior links along the main channel and tributaries according to the pattern of junctions. If the tributaries at each end of the link enter from the same side, the link is termed a cis link. If the tributaries are from opposite sides, a trans link is obtained (fig. 6.6). Sequences of cis and trans links give rise to cis and trans chains. These four types of links should possess certain properties according to the random topology model. One such property is that cis and trans links should occur with equal frequency and their link lengths should possess the same distribution function. In Krumbein and James's study neither of these predictions was found to be true. Based upon a sample of 485 interior links from a network in eastern Kentucky, they observed 293

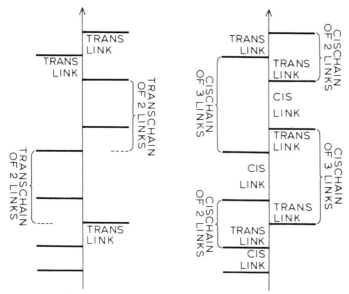

Fig. 6.6 Definition of cis and trans links.

trans links and 192 cis links; a result differing significantly from random expectations. Likewise the link length distributions were found to be different. Of the 192 cis links only 32 fell into the shortest class (0–200 ft), whereas the expected number of shortest cis links was 44·35. In an attempt to explain these disparities Krumbein and James proposed a model which implied that tributaries are generated independently of each other on opposite sides of the channel. Cis chains on each side of the channel were assumed to be independent random samples from a gamma distribution with a shape factor (r) equal to 2.

$$f_{cc}(X) = \beta^2 X e^{-\beta X} \tag{11}$$

$\beta = 0·098$ by maximum likelihood methods for $r = 2$.

Not only was this model found to predict cis chain lengths very accurately, but it also predicted the occurrence of cis and trans links in the ratio 0·375 to 0·625 respectively. These compare very favourably with the observed frequencies of 0·396 and 0·604.

A possible explanation for the use of a gamma density with $r = 2$ is offered by Krumbein and James (1969, 554). If one assumes that the distances of adjacent tributaries from their mutual divide are independent random variables from an exponential distribution, their sum will give rise to a gamma distribution with a shape factor of two. This is an important paper in that it focuses on small-scale topologic effects for the first time, and contradicts some of the basic findings of the random topology model. It also approaches topologic and link length problems simultaneously, an area in which future research might well be profitable.

Stream length and link length properties

The distribution functions of link lengths and stream lengths have been the subject of much theoretical work in the last few years. Using a random-walk simulation model based on that by Leopold and Langbein (1962), Smart *et al.* (1967) noted that exterior link lengths possessed a geometric distribution. Later work also confirmed this result for interior links. Moving from the random-walk to the real world one might therefore anticipate that link lengths of natural networks would be exponentially distributed (Smart 1968, 1011). Although goodness-of-fit tests confirmed this model for two small catchments in Missouri, further investigation has revealed a dearth of very short links. Kirkby has noted field evidence in favour of the exponential model (in Shreve 1969, 400), although Shreve himself, in the same paper, has proposed the gamma density as a more appropriate model based upon 30 networks measured in flat-lying homogeneous sandstone in eastern Kentucky (see also Krumbein and Shreve 1970). More recently Krumbein and James (1969) have analysed small-scale topologic properties of the same networks and have derived a model of link length distribution which combines three gamma densities, each weighted by its respective probability

$$f_{ct}(X) = \tfrac{1}{4}(2\beta e^{-2\beta X}) + \tfrac{1}{2}[(2\beta)^2 X e^{-2\beta X}] + \tfrac{1}{4}\left[\frac{(2\beta)^3}{2}X^2 e^{-2\beta X}\right]$$
$$= \beta(\beta^2 X^2 + 2\beta X + \tfrac{1}{2})e^{-2\beta X} \tag{12}$$

where $\beta = 1/\bar{X}$ as an estimator for β, which equals the estimator for the exponential distribution.

In this instance the fit is not very satisfactory due to the lack of very short links. One possible reason is the minimum practical length which can be measured, which equals 200 ft on the ground. Another possible reason is that opposite-side tributary interactions were excluded.

With the exception of the above example, most workers assume that interior link lengths are independent random variables drawn from a common density. Combining this assumption with that of topological randomness, Smart has derived an expression which predicts $\bar{L}_\omega$ (mean Strahler stream length of order ω) better than Horton's Law of Stream Lengths, yet involves no adjustable parameters. This expression is derived in the following manner (this account draws heavily on Smart (1968, 1004–9)).

Since Strahler streams of order $\omega \geqslant 2$ are composed of stream links, the stream length distribution function for any given order will vary according to the distribution function of individual links and the manner in which these links may be combined to form a stream network. The probability that streams of order ω in a network with stream numbers N_ω will have a total length $\mathcal{L}$ is

$$p(\mathcal{L}; \omega, N_\Omega) \tag{13}$$

This can be regarded as a compound distribution which is the product of the probability that streams of order ω have ν links, and the probability that ν links have a length $\mathcal{L}$. Symbolically this may be written as

$$p(\mathcal{L}; \omega, N_\Omega) = \sum_\nu f(\nu; \omega, N_\Omega) g(\mathcal{L}; \nu, \omega, N_\Omega) \tag{14}$$

where $f(\nu; \omega, N_\Omega)$ is the probability that streams of order ω in a network with stream numbers N_Ω have a total of exactly ν links, and $g(\mathcal{L}; \nu, \omega, N_\Omega)$ is the probability density function for the length $\mathcal{L}$ for a set of ν links of order ω in the network N_Ω.

The properties of these two probabilities may be explicitly determined by the addition of two further assumptions:

1. That all TDCN for a given number of sources are equally probable
2. That the lengths of interior links are independent random variables drawn from a common population

The second assumption means that g is independent of ω, i.e. links of differing orders have the same length distribution. Since the individual link lengths are independent random variables, the average length of ν links equals ν times the average lengths of individual links

$$E(\mathcal{L}; \omega, N_\Omega) = \bar{l}_i \sum_\nu \nu f(\nu; \omega, N_\Omega)$$

$$= \bar{\nu}\bar{l}_i \tag{15}$$

where $\bar{l}_i$ equals the mean length of interior links in the whole network.

The first assumption completely determines the nature of $f(\nu; \omega, N_\Omega)$ since the topology of the network is a function of the number of tributaries possessed by the network. This implies that for a given network the set N_Ω should be related to the set of stream lengths (Smart 1968, 1002). In order to illustrate the relationship between topology and $f(\nu; \omega, N_\Omega)$ consider the following fourth-order stream set: $n_1 = 25$, $n_2 = 10$, $n_3 = 3$, $n_4 = 1$. For each order we have to distribute $n_\omega = n_\omega - 2n_{\omega+1}$ excess tributaries of order ω to the $2n_{\omega+1} - 1$ links of higher order. In combinatorial terms this is equivalent to placing p indistinguishable objects into m cells, the number of different ways in which this may be done being:

$$\binom{m + p - 1}{p} \tag{16}$$

Thus two of the third-order streams must join to create the one fourth-order stream, and the remaining third-order stream must join the fourth-order stream at some point such that the highest-order stream is composed of at least two links. Six of the second-order streams are required to generate the three third-order streams, leaving four to be distributed amongst the five links of third and fourth order. This may be done in 70 different ways. Twenty first-order streams are needed to form the ten second-order streams, which leaves five to be attached to the 19 links of higher order. This may be done in 33,649 different ways.

The total number of possibilities for a network with stream numbers N_Ω is

$$F(N_\Omega) = \prod_{\omega=1}^{\Omega-1} \binom{n_\omega - 2}{n_\omega - 2n_{\omega+1}} \qquad (17)$$

which differs from equation (3) by factors of 2^{r_ω}. These factors allow for the possibility of a junction being from either right or left which is an important topological property, but not important· from the point of view of link length distribution. But in view of Krumbein and James's results perhaps this ought to be included in the derivation of the model, despite Smart's assurance to the contrary.

In order to evaluate $f(v; \omega, N_\Omega)$ we assume that the probability of obtaining a particular link distribution is proportional to the number of configurations in which it can occur. This is best illustrated by means of an actual example – the stream set $n_1 = 10$, $n_2 = 3$, $n_3 = 1$. In this instance there are $r_1 = n_1 - 2n_2 = 4$ first-order streams to be attached to the $2n_2 - 1 = 5$ links of second and third order. Inspection of the possible combinations of three second-order streams with one third-order stream will reveal that three of these five links must be second-order whilst the remaining two are third order (fig. 6.7). If we let k

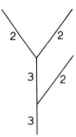

Fig. 6.7 Arrangement of second- and third-order links.

equal the number of tributaries attached to the second-order links, by definition only $r_1 - k = 4 - k$ may be attached to the third-order stream. Since the addition of each first-order tributary adds a new link to the network, the final network possesses

$$v_2 = n_2 + k \qquad \text{links of second order}$$
$$\text{and } v_3 = n_2 - 1 + r_1 - k \qquad \text{links of third order}$$

Using the combinatorial expression (16), the number of different ways in which this assignment may be made is:

$$\binom{n_2 + k - 1}{k}\binom{n_2 - 1 + r_1 - k - 1}{r_1 - k} \tag{18}$$

since k tributaries may be attached to n_2 second-order links in

$$\binom{n_2 + k - 1}{k}$$

different ways; and the remaining $r_1 - k$ tributaries may be attached to the $n_2 - 1$ third-order links in

$$\binom{n_2 - 1 + r_1 - k - 1}{r_1 - k}$$

different ways. Because

$$n_2 - 1 + r_1 + k = n_2 - 1 + n_1 - 2n_2 - k$$

equation (18) can be simplified as

$$\binom{n_2 + k - 1}{k}\binom{n_1 - n_2 - 2 - k}{n_1 - 2n_2 - k} \tag{19}$$

Then

$$f(n_2 + k; 2, N_\Omega) = f(n_1 - n_2 - 1 - k; 3, N_\Omega)$$
$$= \binom{n_2 + k - 1}{k}\binom{n_1 - n_2 - 2 - k}{n_1 - 2n_2 - k} \bigg/ \binom{n_1 - 2}{n_1 - 2n_2}$$
$$k = 0, 1, 2, \ldots, r_1 \tag{20}$$

since from (17) we have

$$F(n_1, n_2, 1) = \binom{n_1 - 2}{n_1 - 2n_2}.$$

The final binomial coefficient of (20) normalises the probability f, as is shown by the following example, which evaluates $f(\nu; \omega, N_\Omega)$ for $\nu = 3, 4, 5, 6, 7, \omega = 2, n_1 = 10, n_2 = 3, n_3 = 1$. Since $r_1 = n_1 - 2n_2 = 4$, k varies from 0 to 4.

$i = 1$	$\nu_2 = 3$	$\nu_3 = 6$	$k = 0$	$f(3; 2, N_\Omega) = \binom{2}{0}\binom{5}{4} \big/ \binom{8}{4} = 5/70$
$i = 2$	$\nu_2 = 4$	$\nu_3 = 5$	$k = 1$	$f(4; 2, N_\Omega) = \binom{3}{1}\binom{4}{3} \big/ \binom{8}{4} = 12/70$
$i = 3$	$\nu_2 = 5$	$\nu_3 = 4$	$k = 2$	$f(5; 2, N_\Omega) = \binom{3}{2}\binom{4}{2} \big/ \binom{8}{4} = 18/70$
$i = 4$	$\nu_2 = 6$	$\nu_3 = 3$	$k = 3$	$f(6; 2, N_\Omega) = \binom{2}{1}\binom{5}{3} \big/ \binom{8}{4} = 20/70$
$i = 5$	$\nu_2 = 7$	$\nu_3 = 2$	$k = 4$	$f(7; 2, N_\Omega) = \binom{1}{0}\binom{6}{4} \big/ \binom{8}{4} = 15/70$

$$\sum_{i=1}^{5} f_i = 1\cdot00$$

Mean values for k, v_2 and v_3 are easily obtained by weighting them according to their appropriate probability f_i.

$$k = \sum_{i=1}^{5} \frac{k_i f_i}{70} = 2\cdot4 \tag{21}$$

$$\bar{v}_2 = \sum_{i=1}^{5} \frac{v_{2i} f_i}{70} = 5\cdot4 \tag{22}$$

Because $f(v; 2, N_\Omega) = f(v; 3, N_\Omega)$ in this example

$$\bar{v}_3 = \sum_{i=1}^{5} \frac{v_{3i} f_i}{70} = 3\cdot6 \tag{23}$$

Manipulation of equation (20) allows us to replace (21), (22) and (23) by

$$k = \frac{n_2(n_1 - 2n_2)}{2n_2 - 1} \tag{24}$$

$$E(v; 2, N_\Omega) = \frac{n_2 (n_1 - 1)}{2n_2 - 1} \tag{25}$$

$$E(v; 3, N_\Omega) = \frac{n_3(n_2 - 1)(n_1 - 1)}{2n_2 - 1} \tag{26}$$

Since from (15) we have

$$E(\mathcal{L}, \omega, N_\Omega) = E(v; \omega, N_\Omega).\bar{l}_i$$

the following may be obtained

$$E(\mathcal{L}; 2, N_\Omega) = \frac{n_2(n_1 - 1)}{2n_2 - 1}.\bar{l}_i \tag{27}$$

$$E(\mathcal{L}; 3, N_\Omega) = \frac{n_3(n_2 - 1)(n_1 - 1)}{2n_2 - 1}.\bar{l}_i \tag{28}$$

By induction equations (25) and (26) can be generalised for all values of $\omega \geqslant 2$

$$E(v; \omega, N_\Omega) = n_\omega \prod_{\alpha=2}^{\omega} (n_{\alpha-1} - 1)/(2n_\alpha - 1) \qquad \omega \geqslant 2 \tag{29}$$

Let

$$E(\mathcal{L}; \omega, N_\Omega)/n_\omega = \bar{L}_\omega$$

where $\bar{L}_\omega$ is the mean value of $\bar{L}_\omega$ for a large set of networks each with the same stream numbers N_Ω, and

$$\bar{L}_\omega = \mathcal{L}/n_\omega$$

In order to generalise equations (27) and (28), we observe that the expressions for $E(\mathscr{L}; 2, N_\Omega)$ and $\bar{L}_2$ are correct for all values of Ω. The fact that $n_2 - 1$ links may be added to links of order $\omega > 2$ has no effect on $E(\mathscr{L}; 2, N_\Omega)$ and $\bar{L}_2$. Therefore

$$\bar{L}_\omega = L_\omega/n_\omega = E(\mathscr{L}; \omega, N_\Omega)/n_\omega \tag{30}$$

From equations (15), (29) and (30) we have

$$\bar{L}_\omega = E(\mathscr{L}; \omega, N_\Omega)/n_\omega = \left[n_\omega \prod_{\alpha=2}^{\omega} (n_{\alpha-1} - 1)/(2n_\alpha - 1).l_i \right] \Big/ n_\omega \quad \omega \geqslant 2$$

$$\bar{L}_\omega/l_i = \prod_{\alpha=2}^{\omega} (n_{\alpha-1} - 1)/(2n_\alpha - 1) \qquad \omega > 2 \tag{31}$$

Using the approximation

$$L_\omega \approx \bar{L}_\omega$$

equation (31) represents a replacement of Horton's Law of Stream Lengths based upon the assumptions of the random topology model. Values comparing the two relationships are given in table 6.8. Whereas

Table 6.8 *Mean stream length ratios for fourth-order basins*

ω	Horton's Law	Smart's expression
2	$\dfrac{L_2}{L_1} \approx R_L$	$\dfrac{L_2}{l_i} \approx \dfrac{n_1 - 1}{2n_2 - 1}$
3	$\dfrac{L_3}{L_1} \approx R_L{}^2$	$\dfrac{L_3}{l_i} \approx \dfrac{(n_1 - 1)(n_2 - 1)}{(2n_2 - 1)(2n_3 - 1)}$
4	$\dfrac{L_4}{L_1} \approx R_L{}^3$	$\dfrac{L_4}{l_i} \approx \dfrac{(n_1 - 1)(n_2 - 1)(n_3 - 1)}{(2n_2 - 1)(2n_3 - 1)(2n_4 - 1)}$

Horton's model gives the ratio of L_ω to L_1 (the mean length of exterior links), Smart's model gives the ratio of L_ω to l_i (the mean length of interior links) whose rate of increase varies according to ω. The latter is an improvement on the former in that it is dependent upon no adjustable parameters. It should be noted that equation (31) converges on Horton's predicted relationship with a length ratio equal to 2 as $n_\omega \to \infty$ and $n_\omega/n_{\omega+1} \to 4$.

Testing the random link length model

At the time of writing, Smart's model has yet to be tested on a large body of data; nevertheless indirect tests upon Melton's data (1957)

suggest a relatively good fit (Smart 1968, 1007–8). A comparison of observed and predicted values of $L_2/\bar{l}_i$ and $L_3/\bar{l}_i$ for Melton's 81 basins gives coefficients of determination greater than 0·60 (fig. 6.8). Similar

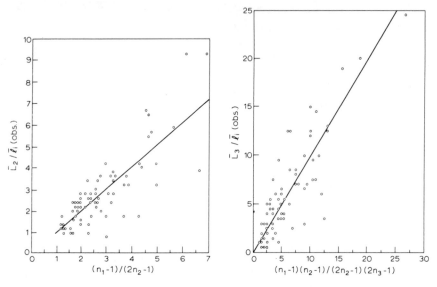

Fig. 6.8 Observed versus predicted values (equation 31) of $L_2/\bar{l}_i$ and $L_3/\bar{l}_i$, for 81 third-order basins (Melton 1957).

results were obtained from Coates' (1958) data. The unexplained variance of $\bar{L}_\omega/\bar{l}_i$ is partly an effect of variation in the number of links per order, partly variation in link length, and partly inadequacies in the model. The first of these can be removed by introducing an additional set of 'tributary numbers'. If $n_{\alpha\beta}$ represents the number of excess tributaries of order α which are attached to streams of order $\beta > \alpha$, in a network N_Ω, there are $\frac{1}{2}\Omega(\Omega - 1)$ such tributary numbers, of which only $\frac{1}{2}(\Omega - 1)(\Omega - 2)$ are independent since

$$\sum_{\beta=\alpha+1}^{\Omega} n_{\alpha\beta} = r_\alpha \qquad \alpha = 1, 2, \ldots, \Omega - 1 \tag{32}$$

must be met. When the tributary numbers $n_{\alpha\beta}$ are known, the number of links per order ω is

$$v_\omega = n_\omega + \sum_{\alpha=1}^{\omega-1} n_{\alpha\omega} \tag{33}$$

Using the same procedures as outlined in the previous section we have

$$\bar{L}_\omega/\bar{l}_i \approx 1 + \sum_{\alpha=1}^{\omega-1} n_{\alpha\omega}/n_\omega \tag{34}$$

Illustrating this by our stream set, $n_1 = 10$, $n_2 = 3$, $n_3 = 1$, and using the TDCN of fig. 6.9, we have

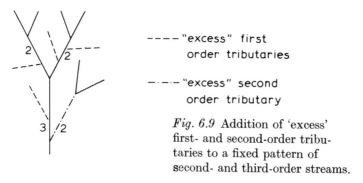

$----$ "excess" first
order tributaries

$-\cdot-\cdot-$ "excess" second
order tributary

Fig. 6.9 Addition of 'excess' first- and second-order tributaries to a fixed pattern of second- and third-order streams.

$$\tfrac{1}{2}\Omega(\Omega - 1) = 1 \cdot 5(2) = 3 \quad \text{sets of tributary numbers}$$

of which

$$\tfrac{1}{2}(\Omega - 1)(\Omega - 2) = 1 \quad \text{is independent.}$$

The three tributary numbers are by inspection $n_{1,2}$, $n_{1,3}$, and $n_{2,3}$ of which $n_{2,3}$ is independent in that one second-order tributary must be attached to the third-order stream in order for the stream set $N_\Omega = 10;\ 3;\ 1$ to exist. Evaluating

$$\sum_{\beta=\alpha+1}^{\Omega} n_{\alpha\beta} = r_\alpha$$

for all possible values of α and β we have

$\alpha = 1,\ \beta = 2$ $n_{1,2} = 3$ (number of first-order tributaries attached to second-order streams)

$\alpha = 1,\ \beta = 3$ $n_{1,3} = 1$ (number of first-order tributaries attached to third-order stream)

$\alpha = 2,\ \beta = 2$ $n_{2,3} = 1$ (number of second-order tributaries attached to third-order stream)

$$\sum_{\beta=\alpha+1}^{3} n_{\alpha\beta} = 5$$

Therefore,

$$v_2 = n_2 + \sum_{\alpha=1} n_{1,2} = 6 \quad \text{(second-order links)}$$

$$v_3 = n_3 + \sum_{\alpha=1}^{2} n_{\alpha,3} = 3 \quad \text{(third-order links)}$$

Comparing equations (31) and (34)

$$\begin{cases} L_2/l_i = 1 + n_{1,2}/n_2 = 2 \\ L_2/l_i = \dfrac{n_1 - 1}{2n_2 - 1} = 9/5 = 1\cdot8 \end{cases}$$

$$\begin{cases} L_3/l_i = 1 + n_{1,3}/n_3 + n_{2,3}/n_3 = 3 \\ \bar{L}_3/l_i = \dfrac{(n_1 - 1)(n_2 - 1)}{(2n_2 - 1)(2n_3 - 1)} = 18/5 = 3\cdot6 \end{cases}$$

gives a significant, although not large, discrepancy between the two equations. Using equation (34) instead of (31) Smart (1968, 1009) was able to reduce the amount of unexplained variance in his test on Melton's data to about 15%.

One of the fundamental assumptions of the random link length model is that interior link lengths are independent random samples from a common population. However, Scheidegger (1968) and Scheidegger and Ghosh (1970) have claimed that link lengths increase with order, which, if correct, would nullify the above assumption. Scheidegger and Ghosh's procedure was based on a model which assumed random topology and links of a unit length. Expected values of length ratios $\bar{L}_\omega/\bar{L}_{\omega+1}$ were ascertained by monte carlo methods for networks of a fixed magnitude. The observed length ratios, based on Morisawa's (1962) data, were found to be greater than the simulated ones. Likewise, link length ratios for 20 basins were found to indicate an increase of link length with order (Scheidegger and Ghosh 1970, 338–9). Smart (1969C) has criticised certain of Scheidegger's procedures in the earlier paper of 1968, and has demonstrated that in one instance (that of Schumm's Perth Amboy data) direct measuring of link lengths results in a decrease of mean link length by a factor of 2·1 (Smart 1970). Ghosh (1970) in reply notes that the method outlined in the earlier paper results in an increase of mean link length of 1·5; see also Scheidegger (1970) for reply. To some extent the dispute rests upon differences in operational definitions and terminology. It can best be resolved by direct measurement. The one attempt, thus far, of testing the assumption by direct methods (Smart 1969B) has produced equivocal results. Using non-parametric methods, four of the ten networks examined showed no significant relation between link length and magnitude; three showed an apparent increase and three an apparent decrease of link length with magnitude. Comparing the results of the four tests, Smart concludes (1969B, 1340) 'that some of the ten networks do show significant changes in link length with magnitude'. Whether or not there is a dependence between link length and magnitude has yet to be established.

In addition to deriving a random link length model Smart has also attempted to model drainage basin areas using the random topology model. The drainage area model is based upon the two assumptions:

1. that drainage density is uniform within the basin
2. that lengths of exterior and interior links are independent random samples from the same population.

Admitting that the second of these assumptions is rather dubious, Smart (1968) has derived expressions for $\bar{A}_\omega/\bar{A}_1$ for $\omega = 2, 3$. An absence of data on basin areas has precluded much testing or refinement of the model. Shreve, working from a different approach, has obtained equations for basin area distribution functions for $\Omega = 2, 3$ based on the premise that a_f (first-order areas) and a_i (areas draining directly into individual interior links) behave as gamma densities with differing expectations. Comparison with Schumm's Perth Amboy data suggests a good measure of agreement with the model's predictions (Shreve 1969, 410–12).

Other approaches to stream network properties

For reasons of brevity this review has made little reference to many other models of river networks which have been proposed in recent years. Many of these were inspired by Leopold and Langbein's (1962) random-walk model (Smart *et al.* 1967; Seginer 1969; Smart and Moruzzi 1970; Howard 1971). Although in many instances a good correspondence between simulated and observed results has been noted, there remains the vexed question of the validity of any given set of input parameters. As Howard (1971, 49) succinctly puts it, 'if, as seems likely, any number of generating processes are able to do as well in simulation, then the argument from simulation process to natural process may be dangerous'.

Of a very different nature is Woldenberg's hexagonal model for basin area (1969) based upon Christaller's hierarchy of service centres with area ratios of 3, 4 and 7. The resulting mixed hierarchy of drainage basins exist in an equilibrium state of 'least work in overland movement of material to the channel and maximum economies of scale at the channel' (Woldenberg 1969, 103). Woldenberg's analogy between central places and river networks is intriguing, but unacceptable. Dimensionally the systems are not isomorphic; one involving movement from an area to a point, the other movement from an area to a line. There is no theoretical justification for the use of the convergent mean, other than that it seems to work; and the arbitrary manner in which the powers of 3, 4 and 7 are grouped seems to be justified solely with reference to the data set in hand. Although Woldenberg has made a spirited defence of his model (1969, 104) against the stochastic formulations of Shreve and Smart, there seems little likelihood of real advances being made from the mixed hierarchy model.

Comments on the random topology and link length models

The results presented in a previous section suggest that the random topology model correctly describes drainage networks in a wide variety of lithologic, structural and climatic regions. Thus far no study has demonstrated regional differences in topology which can be traced to lithologic or structural controls, Only a possible climatic effect may be observed in Smart's findings for high-magnitude networks in the western U.S.A. Such a slavish adherence to the model across such a wide variety of physiographic regions is disquieting for the geomorphologist, for it implies that drainage networks are topologically random irrespective of major geomorphic controls. Furthermore, the randomness implies that the geomorphologist has nothing of importance to say on this topic, a galling conclusion to say the least.

But if one pursues this question further, several disturbing features of the model become apparent. One of the difficulties inherent in the model is the fact that in order to test for non-trivial cases of n, one needs such a large body of data that any small-scale deviations possibly ascribable to local geologic controls tend to be cancelled out by the high degree of aggregation necessary for testing purposes. In reply to this one might claim that these effects are negligible and are correctly described as random fluctuations. But the scale at which such effects make themselves felt has yet to be demonstrated. They may well be more important than has been thought hitherto, and if so, the model is in danger of throwing the baby out with the proverbial bathwater.

A second, and more important, problem is that the testing procedures are measuring different types of randomness as the aggregation of TDCN increases for higher magnitude networks. The topologic properties of a drainage network are completely specified by the following three properties (Smart 1969A):

1. The Strahler stream numbers or bifurcation ratios
2. The location of the $n_\omega - 2n_{\omega+1}$ 'excess' tributaries of order ω
3. The right/left order at each junction in the network.

Only in the testing procedure in which each TDCN is treated as the basic unit are all three topologic properties specified. If one discards 3 then one has in effect combined TDCN into the ambilateral classes. If one discards both 2 and 3 one is reduced to a consideration of stream sets only. At each successive stage a loss of information is involved.

This can easily and efficiently be measured by use of Shannon's classic formula (Quastler 1958, 21):

$$H(x) = -\sum_{i=1}^{n} p(i) \log_2 p(i) \qquad (35)$$

where $H(x)$ equals the information content of x; x being a classification with n categories each with an associated probability of $p(i)$.

Values of $H(x)$ for probabilities derived from equations (1) and (3) are compared for networks with 3 to 40 sources in table 6.9. The

Table 6.9 $H(x)$: *For ungrouped TDCN and TDCN grouped according to stream nunbers*

Magnitude	Maximum entropy for $N(n)$	Entropy for stream sets $N(n_1, n_2- \ldots, n_{\Omega-1}, 1)$	Ratio of columns 2 & 3
3	1·0000	0·0000	—
4	2·3219	0·7219	0·3109
5	3·8074	0·9852	0·2588
6	5·3923	1·2009	0·2227
7	7·0444	1·3460	0·1911
8	8·7448	1·4756	0·1687
9	10·4818	1·6017	0·1528
10	12·2473	1·7389	0·1420
15	21·3508	2·5231	0·1182
20	30·7189	3·1488	0·1025
25	40·2304	3·5774	0·0889
30	49·8322	3·9055	0·0784
35	59·4959	4·1925	0·0705
40	69·2050	4·4675	0·0646

maximum possible value for $H(x)$ is listed in column 2 and is based upon an equal probability for each TDCN. Column 3, on the other hand, gives the value of $H(x)$ when TDCN have been aggregated according to their stream set numbers. For networks of magnitude-20 or greater, the loss of information involved in this grouping procedure is greater than 90%. Since purely topologic measures of network structure have already involved the discarding of a great deal of information of a metric nature, it is not perhaps surprising that less than 10% of the remainder can be described in terms of a random combinatorial process.

In view of these theoretical problems, plus the results presented by various workers, it seems unlikely that the random topology model *per se* will be of much assistance to the geomorphologist. The mathematics has proved both elegant and exciting, but from the geomorphic view-point, exploration of the purely topologic properties of stream networks has proved unrewarding. Nevertheless the random topology model is not to be entirely discarded for it forms an essential component of Smart's random link length model. Here in the world of a tangible metric, the geomorphologist may well feel more at ease; not because the relationships become any simpler, but because it seems intuitively more plausible that the geomorphic components of stream networks are to be identified at this level. Whether this hope is to be substantiated depends upon extensive testing of the various stream and link length models which have been proposed. Certainly one essential prerequisite is establishing the nature of the relationship between link length and order, preferably using data which has been checked in the field.

One might also enter a strong plea for standardisation of procedures, particularly in the problematic area of channel interpolation. Map accuracy standards vary, as do operators' individual criteria of what constitutes the last significant contour crenulation in following a first-order stream channel to its source. With a few praiseworthy exceptions (e.g. Melton 1957; Smart 1968; Gregory and Walling 1968), the operational definition of what constitutes a stream is rarely specified in morphometric work from maps; and much remains to be done in revealing the nature and extent of this operator bias.

The ultimate goal is a far more detailed understanding of the purely morphologic components of drainage networks. Such an understanding will not only be of intrinsic interest for its own sake, but will also be of assistance to the flood hydrologist, and students of process who share our common interest in drainage basins, albeit from different viewpoints.

References

CAYLEY, A. (1859) On the analytical forms called trees; *Philosophical Magazine* 18, 374–8.

COATES, D. R. (1958) Quantitative geomorphology of small drainage basins in Southern Indiana; *Office of Naval Research, Geography Branch, ONR Task No. 389–042, Technical Report No. 10, Department of Geology, Columbia University, New York*.

COCHRAN, W. G. (1952) The chi-square test of goodness of fit; *Annals of Mathematical Statistics* 23, 315–45.

CHORLEY, R. J. and HAGGETT, P. (1969) *Network Analysis in Geography* (Edward Arnold, London).

CHORLEY, R. J. and MORGAN, M. A. (1962) Comparison of morphometric features, Unaka Mountains, Tennessee and North Carolina, and Dartmoor, England; *Bulletin of the Geological Society of America* 73, 17–34.

EDMONDS, E. A. *et al.* (1968) *Geology of the Country around Okehampton*;

Memoir of the Geological Survey, Natural Environmental Research Council (H.M.S.O., London).

EDMONDS, E. A. *et al.* (1969) *South-West England*; British Regional Geology, Natural Environmental Research Council (H.M.S.O., London).

EVANS, J. W. (1922) The geological structure of the country around Combe Martin; *Proceedings of the Geologists' Association* 33, 201–28.

GHOSH, A. K. (1970) Reply: Comment on 'Dependence of stream link lengths and drainage area on order' by A. K. Ghosh and A. E. Scheidegger; *Water Resources Research* 6, 1425–6.

GREGORY, K. J. (1969) Geomorphology; In Barlow, F. (Ed.), *Exeter and its Region* (University of Exeter), 27–42.

GREGORY, K. J. and WALLING, D. E. (1968) The variation of drainage density within a catchment; *Bulletin of the International Association of Scientific Hydrology* Year 13, 61–8.

HOEL, P. G. (1962) *Introduction to Mathematical Statistics* (Wiley, New York).

HORTON, R. E. (1945) Erosional development of streams and their drainage basins: Hydrophysical approach to quantitative morphology; *Bulletin of the Geological Society of America* 56, 275–370.

HOWARD, A. D. (1971) Simulation of stream networks by headward growth and branching; *Geographical Analysis* 3, 29–50.

KRUMBEIN, W. C. and JAMES, W. R. (1969) Frequency distributions of stream link lengths; *Journal of Geology* 77, 544–65.

KRUMBEIN, W. C. and SHREVE, R. L. (1970) Some statistical properties of dendritic channel networks; *ONR Task No. 389-150, Technical Report No. 13, Department of Geological Sciences, Northwestern University, and Special Project Report, NSF Grant GA-1137, Dept. of Geology, U.C.L.A.*

LEOPOLD, L. B. and LANGBEIN, W. B. (1962) The concept of entropy in landscape evolution; *U.S. Geological Survey Professional Paper 500-A.*

MELTON, M. A. (1957) An analysis of the relations among elements of climate, surface properties, and geomorphology; *Office of Naval Research, Geography Branch, ONR Task No. 389-042, Technical Report No. 11, Department of Geology, Columbia University, New York.*

MORISAWA, M. E. (1962) Quantitative geomorphology of some watersheds in the Appalachian Plateau; *Bulletin of the Geological Society of America* 73, 1025–46.

QUASTLER, H. (1958) A primer on information theory; In *Information Theory in Biology*, In Hockley, H.P., Platzman, R. L. and Quastler, H. (Eds.), *Symposium on Information Theory in Biology*, Galinburg, Tennessee.

RANALLI, G. and SCHEIDEGGER, A. E. (1968) A test of topological structure of river nets; *Bulletin of the International Association of Scientific Hydrology* Year 13, 142–53.

SCHEIDEGGER, A. E. (1968) Horton's law of stream lengths and drainage areas; *Water Resources Research* 4, 1015–21.

SCHEIDEGGER, A. E. (1970) Comment on 'Comparison of Smart and Scheidegger stream length models' by J. S. Smart; *Water Resources Research* 6, 996–7.

SCHEIDEGGER, A. E. and GHOSH, A. K. (1970) Dependence of stream link lengths and drainage areas on stream order; *Water Resources Research* 6, 336–40.

SCHUMM, S. A. (1956) Evolution of drainage systems and slopes in badlands at Perth Amboy, New Jersey; *Bulletin of the Geological Society of America* 67, 597–646.

SEGINER, I. (1969) Random walk and random roughness models of drainage networks; *Water Resources Research* 5, 591–607.

SHREVE, R. L. (1966) Statistical law of stream numbers; *Journal of Geology* 74, 17–37.

SHREVE, R. L. (1967) Infinite topologically random channel networks; *Journal of Geology* 75, 178–86.

SHREVE, R. L. (1969) Stream lengths and basin areas in topologically random channel networks; *Journal of Geology* 77, 397–414.

SMART, J. S. (1967) A comment on Horton's law of stream numbers; *Water Resources Research* 3, 773–6.

SMART, J. S. (1968) Statistical properties of stream lengths; *Water Resources Research* 4, 1001–14.

SMART, J. S. (1969A) Topological properties of channel networks; *Bulletin of the Geological Society of America* 80, 1757–74.

SMART, J. S. (1969B) Distribution of interior link lengths in natural channel networks; *Water Resources Research* 5, 1337–42.

SMART, J. S. (1969C) Comparison of Smart and Scheidegger stream length models; *Water Resources Research* 5, 1383–7.

SMART, J. S. (1970) Comment on 'Dependence of stream link length and drainage area on order' by A. K. Ghosh and A. E. Scheidegger; *Water Resources Research* 6, 1424.

SMART, J. S. and MORUZZI, V. L. (1970) Random-walk model of stream network development; *I.B.M. RC 3108* (No. 14216).

SMART, J. S., SURKAN, A. J. and CONSIDINE, J. P. (1967) Digital simulation of channel networks; *Symposium on river morphology, International Association of Scientific Hydrology, Publication No. 75*, 87–98.

STRAHLER, A. N. (1958) Dimensional analysis applied to fluvially eroded landforms; *Bulletin of the Geological Society of America* 69, 279–300.

STRAHLER, A. N. (1964) Quantitative geomorphology of drainage basins and channel networks; In Chow, V. T. (Ed.), *Handbook of Applied Hydrology* (McGraw-Hill, New York), p. 4/40–4/74.

USSHER, W. A. E. (1881) Palaeozoic rocks of Devon and Somerset; *Geological Magazine* 8, 441–8.

USSHER, W. A. E. (1901) The Culm Measure types of Great Britain; *Transactions of the Institute of Mining Engineers* 20, 360–91.

WEBBY, B. D. (1965) The stratigraphy and structure of the Devonian rocks in the Brendon Hills, West Somerset; *Proceedings of the Geologists' Association* 76, 39–60.

WOLDENBERG, M. J. (1969) Spatial order in fluvial systems: Horton's Laws derived from mixed hexagonal hierarchies of drainage basin areas; *Bulletin of the Geological Society of America* 80, 97–112.

7 Modelling hillslope and channel flows

ANN CALVER, M. J. KIRKBY and D. R. WEYMAN
Department of Geography, University of Bristol

Hydrograph models are usually evaluated on the basis of their predictive success, but it is desirable that they should be applicable not only to large basins but also to small catchments and even to sections of a hillslope if they are to have any explanatory power. To fit observations on all scales, a flow model must take account of all the processes involved, so that field measurements of hillslope and channel characteristics may be related to model parameters. This paper reviews some of the models which have attempted to link geomorphic and hydrologic processes more closely, and proposes some alternative models which take account of hillslope flow measurements and can be applied to small catchments.

Rain falling on the land produces a stream hydrograph by first flowing over and through the soil, and secondly along a network of discrete channelways. These two phases of flow are very different and are therefore considered separately.

Hillslope flow models

Although a large number of models to predict runoff from rainfall have been proposed (see Linsley 1967), many of those of practical use are essentially black-box systems in which the operations which convert rainfall to runoff have no counterpart in the physical processes involved. The commonest type of predictive model assumes that rain will infiltrate the soil until the infiltration capacity is exceeded, at which point additional rain is distributed as overland flow (Horton 1933). The original concept of an exponential decay of the infiltration capacity over time was modified to vary the initial capacity according to antecedent moisture conditions. Difficulties in equating measured infiltration rates with observed runoff (Cook 1946) led to the development of the basin infiltration index derived empirically from observed runoff, i.e. a black-box in which only input was variable. The need for greater flexibility resulted in the addition of further parameters to describe the nature of storm input. Those additional parameters, in the absence of an objective reality connected with the processes operating, could only be obtained by optimisation, using either graphical correlation (Kohler and Linsley 1951; Sittner, Schauss and Monro 1969) or multiple regression (e.g. Schreiber and Kincaid 1967). It now seems clear, with the hindsight of recent process experiments, that these models owed much of their success to the accidental intercorrelation of many components within most hydrologic systems. For example, a parameter expressing antecedent soil moisture, normally calculated from antecedent precipitation (Saxton and Lenz 1967), is used and optimised during numerical analysis of observed runoff. Consequently, antecedent soil moisture may generously add to the explanation in a black-box model with an infiltration basis, although in reality it may refer to the operation of some other process, such as rapid subsurface flow.

Despite the predictive success of these black-box models, there remain non-linearities in the relationship of rainfall to runoff which cannot be overcome simply by alteration of the input variables. Recent work has examined three major sources of non-linearity.

1. Apart from the general observation that infiltration-excess overland flow may be rarer in humid areas than was formerly thought, investigation of throughflow discharge from slopes (Whipkey 1965; Dunne and Black 1970A; Weyman 1970) has revealed that runoff processes may vary their location within the soil profile. In particular, it appears that in a multi-layered soil, where hydraulic conductivity decreases with depth below ground level, usually in a discontinuous fashion, saturated conditions may build up at several horizons under suitable antecedent moisture and storm rainfall conditions. The rate of runoff (saturated throughflow) will increase in each saturated layer as it becomes thicker. In particular, where saturated conditions develop in the 'A' horizon the build-up of saturation to the surface leads to overland flow. Even if overland flow is seen as flow through a very porous medium (the vegetation mat) it is clear that the initiation of overland flow will lead to a greatly increased rate of runoff. The rainfall–runoff relationship at a point can only be linearised by taking the depth, and therefore rate of runoff processes into account. In other words linearity implies that the rating curve, $q \propto h^n$, relating discharge q to depth of flow h, has an exponent n equal to 1·0. This may be reasonable for saturated throughflow within a single layer of uniform proper-

ties, but it is certainly not so in a soil for which the properties vary with depth, and it is not so for overland flow with a free water surface ($n = 1\cdot5\text{--}2\cdot0$).

2. Flows within the soil also produce non-linearities because there are delays between rainfall and the beginning of effective lateral flow at a saturated horizon. As saturation levels approach the surface, the delay before lateral flow starts is reduced. Thus if two similar storms follow one another, the delays before flow in the second storm are much reduced and its peak flows are earlier with respect to the rainfall. This source of non-linearity is usually evaded by the use of suitable antecedent moisture functions, but is clearly inherent in the hydrologic properties of the soil.

3. Rainfall–runoff non-linearity also arises as a result of the spatial variability of runoff. Hewlett (1961) demonstrated that the lateral movement of water in soil on a hillslope would lead to a gradient of decreasing water content upslope after drainage. Any runoff process dependent upon antecedent moisture is therefore initially more effective at the slope base. A similar conclusion was reached by Betson (1964), who envisaged only part of the total catchment area contributing directly to runoff. Further empirical research has shown that variations in the soil and topographic characteristics within a basin affect the location of near surface and overland flows. Areas of overland flow and other storm runoff processes have been identified with areas of thin 'A' horizon (Betson and Marius 1969), the downslope parts of long slopes (Jamison and Peters 1967), and topographic hollows (Dunne and Black 1970A).

In combination these three sources of rainfall–runoff non-linearity result in hillslope runoff which varies areally and over time, as a response to input and physiographic controls. This situation has been measured by Dunne and Black (1970B) and inferred analytically by Ragan (1968) and Dickinson and Whiteley (1970). At the moment, however, only conceptual models exist for describing this situation – the 'dynamic watershed model' (Tennessee Valley Authority, 1965; Hewlett and Hibbert, 1965). Process-orientated flow models must take all these non-linearities into account, but so far no model has considered all non-linearities at the same time although several papers have discussed models which consider at least one aspect.

The non-linearity of the rating curve has been considered by Wooding (1965–66) who has shown that the kinematic wave solution to the flow equation is applicable to most overland hillslope flows. The hydrograph is a solution to the continuity equation

$$\frac{\partial q}{\partial x} + \frac{\partial h}{\partial t} = i \tag{1}$$

where x is horizontal distance t is time elapsed and i is the rate of rainfall excess, which is assumed spatially uniform; and (2) the rating equation

$$q = \alpha h^n \tag{2}$$

where α is a constant.

The characteristic equations of the problem are

$$\frac{\partial h}{\partial t} + c . \frac{\partial h}{\partial x} = \frac{dh}{dt} = i \tag{3}$$

and

$$\frac{dx}{dt} = c = n . \alpha . h^{n-1} \tag{4}$$

where the velocity $\quad c(h) = \dfrac{dq}{dh} = n . \alpha . h^{n-1}$

The method of characteristics breaks down if either rainfall excess (i), or hillslope roughness and gradient (α), are spatially variable. Numerical solutions to equations (1) and (2) must then be obtained (Brakensiek 1967). These solutions may be applied in two ways. The first is to allow infiltration and slope gradient to vary spatially during flow (Brakensiek and Onstad 1968), and so consider in addition at least a part of the non-linearity associated with spatial variations of flow. These authors do not, however, explicitly consider the effect of contour curvature in producing flow convergence or divergence.

Secondly, finite-difference solutions to the flow equation may also be applied to subsurface flow to include the effect of some of the non-linearities which arise from differences in soil profile. To do this, the terms in equations (1) and (2) must be re-interpreted as follows. Subsurface saturated flows in a medium of downward-decreasing hydraulic conductivity can be represented by a rating function similar to equation (2) if the depth of flow, h, is interpreted as the total water equivalent depth of the flow. However, rainfall takes some time to reach the saturated level and contribute to lateral flows, and this lag time can be considered as a decreasing function of h. In equation (1) therefore, the rainfall excess, i, must be interpreted to be delayed by this lag time, and, since soil saturation levels usually increase downslope and this lag time consequently also decreases downslope, the effective values of i are no longer spatially uniform. This is one type of model which can take into account the non-linearities described, provided that there are not separated saturated layers within the soil, as have been inferred by Whipkey (1965) and Weyman (1970).

An alternative approach to modelling subsurface flows is to work from the equations for flow in porous media. This approach, in its most general form, can cope with non-linearities arising from topography and soil profile differences, but introduces different non-linearities in the flow equation. Solution of the three-dimensional flow equation is very complex, especially for a layered soil, and it appears justifiable to make considerable simplifications. One method is to consider the soil as an inclined slab without vertical differences in soil moisture (Kirkby and Chorley 1967), but this fails to take into account the observed marked vertical differences and layers of saturation. For this case the flow equation (replacing equation (2)) becomes:

$$q = z\left\{K \tan \theta - D.\frac{\partial m}{\partial x}\right\} \tag{5}$$

where z is the soil thickness
K is the hydraulic conductivity
θ is the slope gradient angle,
D is the moisture diffusivity
and m is the volumetric moisture content.

The approach proposed here deals instead with separated phases of vertical and horizontal flow. This appears justifiable because of the great difference in lateral and vertical extents of the soil body. It also appears attractive because of the stratification of soil into horizons more or less parallel to the surface, within each of which soil properties differ less laterally than vertically. Where saturated layers do build up in the soil, then the lateral flow within these layers will be the major part of any throughflow, and each will occur at a relatively constant hydraulic conductivity ($\partial m/\partial x = 0$ in equation (5)), corresponding to saturated conditions.

The simplification adopted therefore consists of considering vertical drainage, with budgeting for the quantities in each saturated layer, and considering drainage between them to provide a delay. Lateral through-flow is considered to take place in each saturated layer, at a fixed rate for that layer, and not elsewhere. This model appears reasonable, but has been simplified still further in the present example for ease of computation. (1) Drainage from each saturated store is assumed to be directly proportional to the quantity in the store, and (2) the delay time in travelling between saturated layers is assumed to be a constant. In the initial infiltration, a build-up of soil moisture occurs near the surface, usually in an unsaturated state (the transmission zone of Bodman and Colman 1943). In the present model the flow in the highest layer is taken to be a linear function of this quantity, again for computational simplicity. Figure 7.1 shows the parameters adopted for several soil layers at the foot of a straight hillslope in the East Twin catchment, Somerset, England.

Despite simplifications, this model illustrates some of the non-

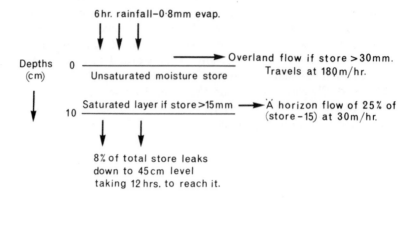

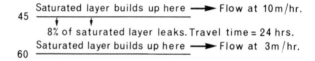

Fig. 7.1 Soil depths and hydraulic parameters used in the example of a multi-layer subsurface flow model.

linearities and interrelationships which exist in hillside flow. For instance, on level ground with no lateral flows, it suggests an infiltration rate which declines with time towards a constant value, although over a much longer period than in normal infiltration tests. It may be inferred that infiltration tests are open to two interpretations: either (as is usual) as a measure of the transmission rate through the surface; or as a measure of the rate at which soil moisture is backed up through saturation of lower layers.

The model also shows how the flow moves into more permeable layers above, or becomes overland flow, as saturation builds up above an interface, so that the flow velocity increases and the lag to peak flow decreases. Thus at higher rainfall intensities, the lag to peak flow becomes less (fig. 7.2). In the model this decrease takes place in definite steps, though in reality the decrease is probably more gradual. The higher flow rates nearer to (or on) the surface also produce much more sharply peaked hydrographs responding to individual short periods of high intensity, so that the peak values predicted for the six-hour periods used are likely to be serious underestimates. This decrease in lag time for the hillslope hydrograph is one major source of non-linearity illustrated by the model.

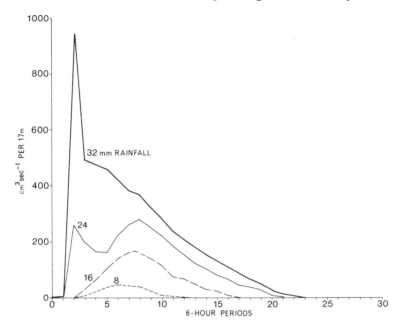

Fig. 7.2 The effect of differences in rainfall on the simulated flow from a multi-layer soil, on a straight slope of uniform gradient. Rainfalls are considered to be of uniform intensity over a 12-hour period.

After selecting a set of values for soil depths and hydraulic parameters, the same values were used to examine the effect of topography in modifying the hydrograph produced by a section of the hillside. The variables examined were first the length of the hillside, second the contour shape and third the gradient. It appears that for points at increasing distances downslope, the initial storage values are greater, although for the stipulated initial conditions, the increase of initial storage in the upper layer is no more than 3 mm of water over 200 m of slope length. However, with even this slight difference, the lower parts of the slope tend to produce saturated overland flow preferentially. In a rainstorm, the main effect of slope length is seen to be a delay and increase of the peak flow. As the intensity of the rainfall is raised, however, there will be situations in which the downslope portions of the slope profile are contributing flow from a nearer-surface soil layer, so that the lag is less from the downslope sections, and the resultant peak flow from all parts of the slope will be much greater. Immediately after a storm, the storage differences between sections of the slope will be at their greatest, so that a second storm following afterwards will show a much greater difference in response between upslope and downslope sections.

The influence of contour shape is to concentrate the flows in hollows, increasing the depth of flow in each layer and so producing overland flow from much lower intensity rains. Figure 7.3 shows the extent of this tendency for a single storm of 32 mm spread over 12 hours, after

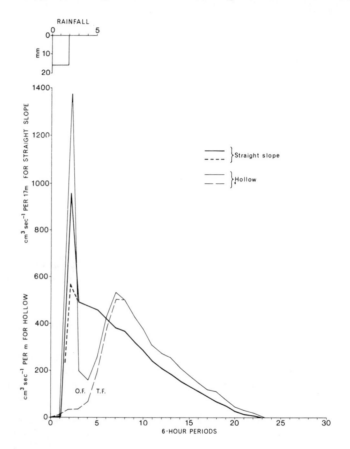

Fig. 7.3 The effect of a hollow in modifying the simulated flow from a multi-layer soil, for slopes of uniform gradient and of equal total drainage area. Broken lines refer to the subsurface flows, solid lines to the total flow, including overland flow. Rainfall in each case is of 32 mm spread uniformly over a 12-hour period.

four days without rain. It is apparent that overland flow is a much greater component of the flow from the hollow, and that in many storms overland flow will only be produced in a hollow. As with the effect of increasing slope length, the increased depths of flow in a hollow during a rainstorm also lead to greater depths of flow after the storm, so that the hollows produce subsurface flow for a longer period after rain and initial storages are appreciably higher at the beginning of the next storm, the extent of this effect being less for longer intervals between storms. It is interesting that the model thereby predicts that the hollow will produce a hydrograph which both peaks higher and sooner and also continues to supply low flows for longer. The hollow thus produces a more markedly right-skewed hydrograph, although of course with a similar total yield per unit area drained. For modelling purposes the hollow was assumed to be of uniform curvature along each contour.

The effect of low gradient is twofold. If the initial storages are the same as for a steeper slope, then the effect of the lower gradient is mainly to delay the flows, and reduce them considerably in soil layers, but only slightly in lower layers. The slower response produces the second effect of gradient, which is to increase initial storage levels of soil moisture. The influence of gradient has only been modelled for slopes of uniform gradient, but it may be seen that the influence of a concave slope profile, for example, would be to localise the high storage values and susceptibility to saturation overland flow in the slope-base area.

If these effects of topography are combined they show marked non-linearities in the hydrograph in addition to those present in the simplest slope flow, although arising because of them. This non-linearity arises because a hillside composed of a series of spurs and hollows does not behave in the same way as a straight hillside of the same drainage area. This means that if an averaged hillside hydrograph is produced from analysis of the stream hydrograph (for example the IUH of Nash 1957), then the hillslope properties which might be derived from it do not correspond to those at an average hillslope point. The lumping of hillslope hydrographs is thus somewhat inadequate even for a simple empirical correlation between basin and hydrograph properties, quite apart from more serious failure of lumping techniques to predict the areal variations in hillslope response.

Since this model has been derived from empirical observations in the East Twin catchment, Somerset, the hydrograph it predicts will be comparable only with that catchment. The more general, and therefore more important, significance of the model lies in how closely it reproduces the hillside hydrologic processes, and in its relative success in overcoming the non-linearities in the rainfall–runoff relationship.

The model slope is based on the side-slopes of the lower part of the East Twin catchment which have straight contours and a broadly convex profile with a steep (25°–30°) basal convexity. A three-layer Brown Earth ('A', 'B', and 'B/C' horizons) is developed on the slopes. Rainfall on the slopes infiltrates unsaturated upper soil layers to feed a saturated zone at the soil (and slope) base. Water is transmitted along this zone as saturated throughflow. The initial extent of the saturated zone, up-profile and upslope, is controlled by antecedent soil moisture conditions. During a storm the zone grows in both directions, and particularly upslope, as long as unsaturated supply exceeds downslope saturated flow. Saturated throughflow discharge at the slope base increases as a power of the vertical and lateral extent of saturation. With dry antecedent moisture and low rainfall, saturation may be confined to the 'B/C' horizon, but will expand into the 'B' horizon with an increase in antecedent moisture or rainfall. Under the observed conditions, saturation has never reached ground level to give rise to overland flow, although 'A' horizon discharge may occur at the slope base.

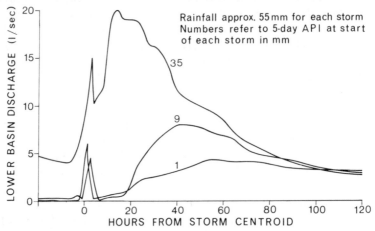

Fig. 7.4 Runoff hydrographs from the lower East Twin basin. The initial peak refers to channel precipitation and A horizon throughflow, and the main peak to B horizon throughflow. As antecedent soil moisture (predicted by an antecedent precipitation index) increases, there is a non-linear increase in runoff volume and peak discharge, and decrease in time lag to peak discharge.

Since they are closely related to the maximum vertical and lateral extent of saturation, total 'quick' runoff volume and peak discharge are direct power functions of total storm rainfall and antecedent soil moisture, while time lag to peak discharge is an inverse function of the same variables. In other words, as storm rainfall or antecedent moisture increases, the basin hydrograph becomes more like a true storm hydrograph resulting from overland flow (fig. 7.4).

The model reproduces most of these characteristics and in particular the non-linear increase of slope runoff with vertical and lateral expansion of saturation. This non-linearity is even more marked in the model, however, since it predicts the growth of saturated conditions in the 'A' horizon and consequent production of saturated overland flow (fig. 7.2.).

In part this is due to the use of rainfall intensities in excess of those actually recorded in the catchment while the tendency of the model to predict small areas of saturated overland flow at lower intensities is probably countered in reality by the basal convexity of the slope which is not reproduced in the model.

The model can therefore predict non-linear increase in runoff as different runoff processes occur within the soil profile and as the area over which those processes operate changes in a spatially uniform situation. There is, however, a good deal of empirical evidence to suggest that spatial variation in soil and topography may lead to increased areal non-uniformity of runoff production. In particular, topographic hollows, stream heads and flat bottom-lands have been observed to give rise to saturated overland flow and thus dominate total catchment runoff (Dunne and Black 1970A). In the East Twin catchment, the headwater area is a shallow topographic bowl with a concave profile (5°–10°) and a dendritic network of minor hollows within the main bowl. Storms which generate only a very delayed throughflow hydrograph in the lower basin give rise to a true storm hydrograph in

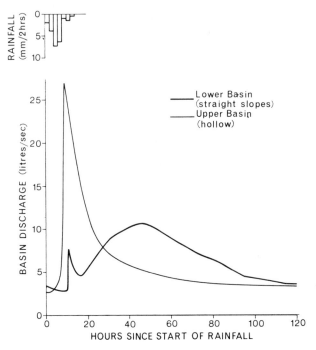

Fig. 7.5 Runoff hydrographs from two areas in the East Twin basin, each of 0·1 km², following a storm. The topography of the upper basin concentrates flow to produce a quick response from saturated overland flow in contrast to the delayed throughflow hydrograph of the lower basin.

the headwater (fig. 7.5). Direct observation suggests that runoff is concentrated by the headwater topography to produce saturated overland flow in the hollows. Overland flow extends rapidly up the hollow network during a storm to produce a variable drainage network similar to that observed by Gregory and Walling (1968) and directly analogous to the dynamic contributing area suggested by Tennessee Valley Authority (1965) and Hewlett and Hibbert (1965).

By adapting the simulation model to a simplified version of the East Twin topography, it has been possible to estimate how far topography alone can account for increased runoff (fig. 7.3). It is clear that the model fails to reproduce the extreme difference between straight slope and hollow which has been observed. Because initial soil storages in the model hollow were altered only for the concentrating effect of concave contours and did not take into account the overall storage increase as a result of lower profile gradient, topography alone may be able to explain a slightly greater proportion of the lower basin/headwater contrast. The increased runoff from the headwater still left unexplained is probably a result of differences between soils in the two parts of the East Twin system. In particular, field measurements suggest that the vertical permeability across the interface at the base of the 'A' horizon is considerably lower in the headwater soils. Overall, however, the model has provided a rational basis for considering concave profile and contour areas within a catchment as the particularly responsive elements of the hydrologic system which are likely to dominate total catchment runoff.

Channel flow models

Causal links between the stream channel network and its associated flows of water and sediment are dependent upon the time and space scales in which the system is considered. For runoff, the relevant time span is that covering the transmission of storm rainfall through small basins as slope and channel flow, and over such time periods the time-distribution of water flow from a catchment is a function of the existing network pattern. Such water flows may, however, also be responsible for local extensions of the network.

The prediction of the form of the channel flood hydrograph from a catchment is commonly approached either by the completely empirical unit hydrograph (Sherman 1932) or by routing procedures involving channel processes (see, for example, Linsley, Kohler and Paulhus 1949, ch. 19). In most of the established routing methods, an upstream flood wave form has been modified during its travel downstream according to the velocity and storage characteristics of its channel. Generation of the flood wave has rarely been considered, channel flow has not been supplemented to allow for inflow from the hillslopes along the banks, and the effects of the geometry of the channel network have been neglected. Routing procedures are thus generally concerned with large areas through which the storm flow from a headwater region is transmitted, and their orientation is towards practical prediction rather than towards an understanding of the geomorphic processes producing the flow.

A hydraulic model of catchment behaviour has been developed by Wooding (1965A and B) which has already been discussed in relation to the modelling of hillslope flows. The same model also presents analytical solutions and examples of stream outflow arising from calculated catchment slope discharge. In terms of the geomorphic properties of the channel, emphasis is placed on the 'equilibrium time' required for the channel flow regime to achieve a steady state, and the relation of this time to a similarly delimited 'catchment equilibrium time' for the hillslopes. By considering infiltration to vary through time, channel head migration is partly brought into the model. The 'catchment-stream' model is fitted to natural situations with a certain degree of success, but with a realisation that the theoretical model could be improved by a better geometrical description of the network than the idealised rectangular V-shaped catchment employed, in which the stream system is generalised to one central channel whose length is linearly dependent upon drainage area (Wooding 1966).

A development of this model deals with a 'subwatershed' divided into a 'predominantly overland flow distributed system' and a 'predominantly channel flow system'. Kinematic routing equations are again used for the channel routing of the flow. The model variables are optimised largely upon the criterion of the equivalence of peak flows with real world conditions, and the optimisation is strongly dependent upon the value of roughness coefficients for slopes and channel (Brakensiek and Onstad 1968; Onstad and Brakensiek 1968).

Surkan (1968) is one of the few writers who has attempted to investigate specifically the effect of network pattern on flow. His simulation study compares observed and predicted flows, and it appears that a good fit can be achieved by using particular storage delay functions, to a large extent independently of the network shape and connectivity, of which the effect *per se* on the channel outflow is thought to be minimal. By transforming naturally occurring networks to a rectangular system, slope and channel flows are simulated using a simple exponential decay function for the channel input hydrograph (Surkan 1968).

In the model presented here the effects of channel network geometry on the streamflow are examined for small catchments, paying particular attention to the effects of drainage area, drainage density and the pattern of network bifurcations. Systematic variations in these factors

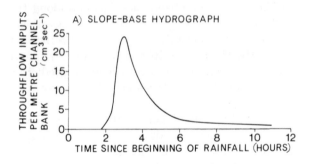

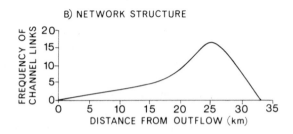

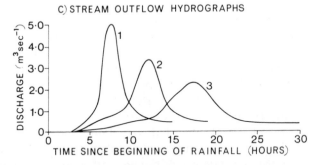

Fig. 7.6 An example of the channel flow model. The slope-base hydrograph (Whipkey 1965) and the drainage network structure are combined to produce stream outflow hydrographs.

Hydrograph	Relative linear scale	Drainage area (km²)	Drainage density (km⁻¹)
1	0·33	29	2·31
2	0·60	94	1·29
3	1·00	262	0·77

are the focus of attention while, to quantify the effect of the channel system, hillslope flows are simplified.

For simplicity, spatially uniform rainstorms and uniform hillslopes are assumed, such that all slopes provide the same water contributions over time. The slope-base hydrographs used here are taken from field observations of Whipkey (1965) and from the East Twin catchment and are associated with storm rainfall producing slope flow which is predominantly within the soil (fig. 7.6A for example). The structure of each drainage network is expressed as a frequency distribution of channel distance from the outflow point against the number of channel sections at that distance. A network characterised, for example, by many headwater streams far distant from the outflow is thus associated with a negative skewness of this distribution (fig. 7.6B). This means of portrayal of network geometry efficiently covers the two-dimensional plan properties of the net, but deals only by implication with properties, such as channel slope, which are a function of elevation.

Hillslope storage is implicit in the measured distribution of slope flow activity over time. For any section of *channel*, the storage equation may be written as

$$\frac{\partial Q}{\partial x} + \frac{\partial A}{\partial t} = q(t) \tag{6}$$

where Q is the stream discharge,
$\quad A$ is the channel cross-section,
and $\quad q(t)$ is the hillslope discharge per unit channel length (both banks together).

If stream velocity is assumed spatially uniform, then

$$A = Q/v(t) \tag{7}$$

where $v(t)$ is the channel flow velocity at time t, and the complete solution, taking account of channel storage, is

$$Q = v(t)\left\{ f\left(x - \int_0^t v.dt\right) + \int_0^t q(t).dt \right\} \tag{8}$$

for an arbitrary function f. Such an assumption of spatially independent velocity appears justifiable within the sensitivity level of this method in the light of the observations of Carlston (1969) and Brush (1961). Velocity is initially considered constant over time, and a value assigned relevant to peak flow for the network, as it is towards discharge peaks that attention is especially directed. A temporally constant stream velocity is employed here for initial simplicity; its use is not a prerequisite of the model.

Comparing these equations with those used by Wooding (1965) for channel flow, which are strictly analogous to equations (1) and (2) above, it may be seen that equation (7) implies an exponent $n = 1.0$

in equation (2). If the solution is examined, it can be seen that values of n can be obtained for both at-a-station and downstream rating curves, and that the latter is relevant in this context. Leopold and Maddock (1953) imply that n should take values of 1·0–1·1 in a downstream direction; instead of the 1·5–1·7 obtained at-a-station, or from Manning's equation and as used by Wooding (1966, 24).

It is appreciated that over small spans of time such as are considered here, length of streamflow, if not of eroded channel, *may* extend up-valley and, in the terms of this model, collect an increased input. Even without extension, the stream head receives water from the bowl-shaped valley-head rather than from two side-slopes. In these zones, too, the assumption of spatially constant velocity is perhaps rarely met. Such local changes together with the slope influences discussed above, are, for the sake of simplicity, excluded from this model of channel flow. Channel flows are perhaps best approached at this stage by attempting to quantify the separate effects of firstly the hillslope contribution and secondly the network geometry.

Hillslope contributions to the channel are routed downstream at the flow velocity, adding lateral channel-bank contributions and summing the elements of streamflow downstream from network junctions. Stream outflow may be expressed as the integral

$$Q = \int_{t=0}^{\infty} N\left(\int_0^t v \,.\, dt\right) . q(t) \,.\, dt \qquad (9)$$

where $N(y)$ is the number of channels at distance y upstream from the outflow point considered.

For simulation purposes a summation over increments of time has been undertaken. Stream outflow at any point is therefore equal to the the sum of the hillslope input to all sections of channel at unit distance from the outflow point one time unit ago, plus the input at time two units ago to all sections twice unit distance from the outflow point, plus the input three time units ago to all sections three times unit distance from the outflow, etc.

This approach was applied to a series of networks whose frequency distributions, $N(y)$, were constructed initially from 1/63,360 maps, and which then underwent a series of scale modifications in order to provide networks of changing values of density and area whilst maintaining a constant particular bifurcation pattern. When the linear scale of a network is varied by a factor (1/r), drainage area is changed by a factor (1/r)² and hence drainage density by a factor r.

An example of the input components of the model and calculated outputs are shown in fig. 7.6. In general the effect of a large number of channel branches is to combine their peak slope flow contributions in the channel flow downstream, thus producing a sharp-peaked hydrograph. In contrast, in sections with few branches, peak inputs to the channel are distributed over a period of time and the resulting stream hydrograph peak is broader and of lower magnitude.

Peak values of the computed storm hydrographs increased almost linearly with drainage area (for drainage areas <150 km²) and according to a 2·6–2·7 power function of drainage density. Network frequency distributions characterised by high kurtosis and low variance values bring about high values of storm peak discharge through the grouping effect of a concentration of channels at a particular distance from the outflow point. Larger drainage areas increase the overall potential outflow given the uniform inputs used, and higher drainage density values more readily transmit this water to the outflow point with minimum lowering of the peak. Drainage area in this context is most usefully interpreted as a variable linearly related to the contributing area for storm runoff, the linear relation being justified in that the nature of hillslope flow remains the same. When non-spatially uniform rainstorms are modelled in this way, the relation between drainage area and storm

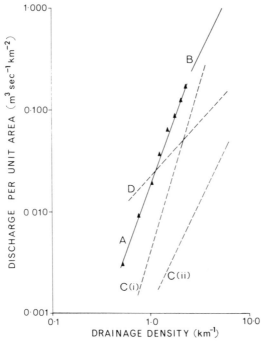

Fig. 7.7 The relationship of stream discharge to drainage density.

A Computed data
B Carlston (1963)
C Gregory and Walling (1968)
D East Twin catchment.

peak discharge need no longer hold, and the spatial characteristics of rainfall inputs become of overriding importance.

Computed observations are shown in fig. 7.7 in conjunction with field data obtained by other workers. Carlston's (1963) discharge data refers to the 2·33-year flood determined from recurrence interval plots for streams of the eastern United States; the drainage density value is that measured from contour crenulations at the 1/24,000 map scale. Gregory and Walling (1968) relate instantaneous discharges to drainage density by field measurement of both variables over time within two small catchments in South Devon, and comparable data are available for the East Twin catchment. The computed data and Carlston's observations thus refer to the relation of peak discharge to a spatial range of drainage density values, while the remaining data relate instantaneous discharges to time-dependent values of drainage density. It is to be borne in mind that the position of the point above which temporally changing values of drainage density are measured will influence the form of the relationship between discharge and density. The relation of discharge to length or drainage density above the limit of minimum flow may be expected to be very different from that below this limit and, if both situations are included in a single plot, further care in interpretation is needed. There seems no necessity that data for spatial and temporal variations should conform to a similar relation except in the most general manner. The correspondence between Carlston's and the computed data does however suggest that we may place a reasonable degree of confidence on the channel flow model for dealing with spatial differences between small basins.

The time interval between the beginning of the storm rainfall and the occurrence of the stream discharge peak is found to increase with the 0·3–0·5 power of drainage area. It is particularly dependent upon the distribution of branches within the network, a marked delaying effect being associated with increasing negative skewness of the network as the maximum concentration of channels becomes further from the outflow point. Drainage area fixes the potential *range* of times of channel travel, while the network pattern provides the *distribution* of actual travel times. Taylor and Schwarz (1952) have investigated empirically the relation of lag time (from the centre of mass of rainfall excess to the observed unit hydrograph peak) to basin characteristics. The relation of lag time to the 'slope of a uniform channel having the same length as the longest watercourse and an equal time of travel' was selected in preference to relationships with the total length of the longest watercourse and to the distance along the main channel from the outflow point to the 'computed centre of gravity of the drainage area'. It is believed that better correlations are obtained if the above variables are computed for 'that sub-area which gives the critical concentration for the basin'. The effect of basin drainage density on the lapse time appears indeterminate; little other information is available on the relationship of these variables, although Hickok, Keppel and Rafferty (1959) relate 'lag time', defined here as the time from the 'centre of mass of a limited block of intense rainfall to the resulting peak of the hydrograph', to the -0.3 power of drainage density measured as the 'total length of visible channels per unit area'. If, as in the cases modelled here, time of hillslope travel is very much smaller than the range of channel time as determined by the maximum network dimension, the timing of the hydrograph is largely controlled by channel travel time and thus changes of drainage density will have little effect on the timing of the stream hydrograph. In the contrasting case of hillslope travel time being very much longer than the channel travel time, it is the hillslope time which controls the time dimension of the hydrograph: through its influence on slope length and gradient, drainage density may be expected to play a part in determining the time to peak discharge, an increase in density more quickly transmitting the storm rainfall to the outflow point. Examples investigated by this model do not, at present, cover the effect of drainage density differences on slope gradients, nor are high enough density values reached for adjacent streams to compete for the flow from areas contributing to storm runoff.

So far, the implications of the interactions of various network structures have been considered for a single type of slope-base hydrograph. The response of a particular network to a series of slope-flow hydrographs of similar magnitude peak but varying time-distribution suggests that the influences of the network are greatest on more rapid inputs. This is a more general manifestation of the comment made above in the context of the influence of drainage density changes, namely, that in considering rapidly responding hillslopes, the channel network properties have a more profound effect on the time distribution of the outflow, whereas in the case of delayed slope inputs, the network effects are secondary to the details of timing of the dominant hillslope flow. These conclusions are compatible with those of Wooding (1965, 1966).

Conclusion

These models all attempt to replace the empirical elements in hydrograph prediction with explanatory subsystems, an intention which is becoming more widespread. Empirical models have relied implicitly on the statistical summation of hydrographs along many flow paths to produce a total stream hydrograph in which the details of the individual path can be ignored. It appears that this assumption may be unjustified and that, instead, the degree of correlation between neighbouring paths on each part of the slope, for example in a hollow, is so high that persistent non-linearities and the cumulative effects of topography are maintained and must be taken into account. Only variations at the micro-topographic scale can cancel one another out statistically.

It is suggested that an accurate and realistic model of hillslope flow need take into account only a single set of hydraulic parameters, describing a typical soil profile for a small basin; but that these must be combined with measures at many points of contour curvature, slope gradients and perhaps thickness of an upper, rapidly conducting soil layer. These values, which are relatively easy to obtain, must be built into an essentially *non-linear* model of flow, to predict quantities and positions of surface and near-surface flows. A very simple basis for such a non-linear model might lie in the physical identification of the 'contributing area' from which overland flow is produced; and the variation of this area over time.

Channel flows have been more fully studied in the past, but few attempts have previously been made to quantify the influence of the channel network and of the channel hydraulic geometry. The effect of the network appears to be particularly important at the medium scale (10–1000 km^2), its pattern influencing both the timing and discharge amount of the hydrograph peak.

In combining the channel and hillslope flow models it is essential to decide at what point in space to make the somewhat arbitrary division between them. It is here suggested that the division should be made at the *low-flow* channel head and that the area of extending stream heads should be included in the hillslope flows as part of the dynamic contributing area. This dividing point maintains the simplicity of a fixed drainage net, and above this point stream velocities are no longer constant, so that an assumption of the channel model is no longer met. Areal differences of slope output, as a result of topography and soil, can be used as a spatially variable input to the fixed drainage net of the channel model to produce a total catchment model.

If the hillslope and channel influences on the hydrograph are compared, the ratio between the flow travel times in the two phases appears to be the most important parameter of the combined flow. If either time is appreciably the longer, then the hydrograph from this phase of the flow is the dominant one. Thus for small streams and at low rainfalls, the hillslope hydrograph tends to be the dominant one, whereas for larger networks and storm rainfalls, the stream response becomes increasingly important. It is therefore natural that the tendency to make more detailed studies in smaller and smaller catchments has highlighted the failures of empirical basin hydrograph models which do not satisfactorily model hillslope flow conditions.

References

BETSON, R. P. (1964) What is watershed runoff?; *Journal of Geophysical Research* 69, 1541–52.

BETSON, R. P. and MARIUS, J. B. (1969) Source areas of storm runoff; *Water Resources Research* 5, 574–82.

BODMAN, G. B. and COLMAN, C. A. (1943) Moisture and energy conditions during downward entry of water into soils; *Proceedings of the Soil Science Society of America* 8, 116–22.

BRAKENSIEK, D. L. (1967) Finite difference methods; *Water Resources Research* 3, 847–60.

BRAKENSIEK, D. L. and ONSTAD, C. A. (1968) The synthesis of distributed inputs for hydrograph predictions; *Water Resources Research* 4, 79–85.

BRUSH, L. M. (1961) Drainage basins, channels, and flow characteristics of selected streams in Central Pennsylvania; *U.S. Geological Survey, Professional Paper* 282-E, 40p.

CARLSTON, C. W. (1963) Drainage density and streamflow; *U.S. Geological Survey, Professional Paper* 422-C, 8p.

CARLSTON, C. W. (1969) Downstream variations in the hydraulic geometry of streams: special emphasis in mean velocity; *American Journal of Science* 267, 499–509.

COOK, H. L. (1946) The infiltration approach to the calculation of surface runoff; *Transactions of the American Geophysical Union* 27, 726–47.

DICKINSON, W. P. and WHITELEY, H. (1970) Watershed areas contributing to runoff; *Proceedings of the General Assembly, International Association of Scientific Hydrology, New Zealand*, 12–26.

DUNNE, T. and BLACK, R. D. (1970A) An experimental investigation of runoff production in permeable soils; *Water Resources Research* 6, 478–90.

DUNNE, T. and BLACK, R. D. (1970B) Partial area contributions to storm runoff in a small New England watershed; *Water Resources Research* 6, 12 96–311.

GREGORY, K. J. and WALLING, D. E. (1968) The variation of drainage density within a catchment; *Bulletin of the International Association of Scientific Hydrology* 13, No. 2, 61–8.

HEWLETT, J. D. (1961) Soil moisture as a source of base flow from steep mountain watersheds; *Southeastern Forest Experiment Station, Asheville, North Carolina, U.S. Department of Agriculture – Forest Service, Station Paper* No. 132, 11p.

HEWLETT, J. D. and HIBBERT, A. R. (1965) Factors affecting the response of small watersheds to precipitation in humid areas; *International Symposium on Forest Hydrology, Pennsylvania State University* (Pergamon), 275–90.

HICKOK, R. B., KEPPEL, R. V. and RAFFERTY, B. R. (1959) Hydrograph synthesis for small arid-land watersheds; *Agricultural Engineering* 40, 608–11 and 615.

HORTON, R. E. (1933) The role of infiltration in the hydrological cycle; *Transactions of the American Geophysical Union* 14, 446–60.

JAMISON, V. C. and PETERS, D. B. (1967) Slope length of claypan soil affects runoff; *Water Resources Research* 3, 471–80.

H

KIRKBY, M. J. and CHORLEY, R. J. (1967) Throughflow, overland flow and erosion; *Bulletin of the International Association of Scientific Hydrology* 12, 5–21.

KOHLER, M. A. and LINSLEY, R. K. (1951) Predicting the runoff from storm rainfall; *U.S. Weather Bureau Research Papers* 34.

LEOPOLD, L. B. and MADDOCK, T. (1953) The hydraulic geometry of stream channels and some physiographic implications; *U.S. Geological Survey Professional Paper* 252.

LINSLEY, R. K. (1967) The relation between rainfall and runoff; *Journal of Hydrology*, 5, 297–311.

LINSLEY, R. K., KOHLER, M. A. and PAULHUS, J. L. H. (1949) *Applied Hydrology* (McGraw-Hill, New York), 689p.

NASH, J. E. (1957) The form of the instantaneous unit hydrograph; *Proceedings of the General Assembly, Toronto, International Association of Scientific Hydrology* 3, 144.

ONSTAD, C. A. and BRAKENSIEK, D. L. (1968) Watershed simulation by stream path analogy; *Water Resources Research* 4, 965–71.

RAGAN, R. M. (1968) An experimental investigation of partial area contributions; *Proceedings of the General Assembly, International Association of Scientific Hydrology, Geneva.*

SAXTON, K. E. and LENZ, A. T. (1967) Antecedent retention indexes predict soil moisture; *Proceedings of the American Society of Civil Engineers, Journal of Hydraulics Division* 93, HY3, 223–41.

SCHREIBER, H. A. and KINCAID, D. R. (1967) Regression models for predicting on-site runoff from short-duration convective storms; *Water Resources Research* 3, 389–96.

SHERMAN, L. K. (1932) Streamflow from rainfall by the unit-graph method; *Engineering News-Record* 108, 501–5.

SITTNER, W. T., SCHAUSS, C. F. and MONRO, J. C. (1969) Continuous hydrograph synthesis with an API-type hydrologic model; *Water Resources Research* 5, 1007–22.

SURKAN, A. J. (1968) Synthetic hydrographs: effects of network geometry; *Water Resources Research* 5, 112–28.

TAYLOR, A. B. and SCHWARZ, H. E. (1952) Unit-hydrograph lag and peak flow related to basin characteristics; *Transactions of the American Geophysical Union* 33, 235–46.

TENNESSEE VALLEY AUTHORITY (1965) Area–stream factor correlation; *Bulletin of the International Association of Scientific Hydrology* 10(2), 22–37.

WEYMAN, D. R. (1970) Throughflow on hillslopes and its relation to the stream hydrograph; *Bulletin of the International Association of Scientific Hydrology* 15(3), 25–33.

WHIPKEY, R. Z. (1965) Subsurface stormflow on forested slopes; *Bulletin of the International Association of Scientific Hydrology* 10(2), 74–85.

WOODING, R. A. (1965–66) A hydraulic model for the catchment-stream problem; *Journal of Hydrology* 3, 254–67, I Kinematic wave theory; 3, 268–82. II Numerical solutions; 4, 21–37, III Comparison with runoff observations.

PART IV
Continuous distributions

—

8 Surface roughness in topography: quantitative approach

R. D. HOBSON

Department of Geology, Emory University, Atlanta

Introduction

Geologists and geomorphologists are beginning to require certain kinds of information which heretofore have been of little importance to the field investigator. Part of these needs for new information stems from the desire to relate topographic characteristics of areas to their underlying geology and to the effects of active geologic processes. Earth sensing and 'ground truth' studies using various airborne and field-stationed sensing apparatus have also demanded different sorts of geologic data. For example, a map showing some characteristic (perhaps vegetation density or moisture content) or loose overburden covering an area may be of more importance to a sensing experiment than a classical geologic map showing distribution of rock types beneath that overburden. Experimentation with sensing apparatus indicates that certain factors, such as moisture content and specific gravity, strongly affect the sensed image of any particular area. Another of these factors is surface roughness of topography (Dellwig and Moore 1966).

A single concise definition of surface roughness is probably impossible. The only usable definitions are incomplete because they describe only a few of the physical or mathematical properties of a surface. There may be as many of these definitions as there are roughness studies themselves.

In the realm of earth sensing in which some type of electromagnetic wave form is directed upon an area, the surface roughness may be defined as a value on a scale ranging between equal ('smooth' surface) and random ('rough' surface) reflectance of those waves. On the other hand, terrain-analysis investigations may require roughness parameters describing larger-scale irregularities of the surface than those affecting most remote sensing instruments. It quickly becomes apparent that different types of investigations require particular sets of roughness parameters. For example, in Beckmann and Spizzichino's (1963) study of wave reflection, mathematical expressions of the surface are required. Kirchhoff approximations of the Rayleigh equations are used to generate these surfaces and then theoretical wave reflection patterns can be studied. It would be difficult to determine exact equations describing natural surfaces and thus other methods are necessary.

In the description of natural terrains, roughness parameters should be established that can be used to describe surface irregularities ranging from a few tenths of an inch to several tens of feet. For roughness parameters to be useful within this realm of terrain analysis, they must fulfil several basic requirements.

First, the parameters should be conceptually descriptive so that a value for any particular test area gives the investigator a mental image of the physical character of that area. The parameters should be easily measurable in the field so that large test sites can be quickly sampled. If possible, roughness parameters should be selected that require similar types of field measurements with a minimal amount of equipment. Parameters should be chosen which can be measured and compared at several different sampling scales and finally, they should be in a digital form suitable for numerical analysis.

The three terrain roughness parameters described below are: (1) comparison of estimated actual surface area with the corresponding planar area; (2) estimate of 'bump' or elevation frequency distribution; and (3) comparison of the distribution and orientation of approximated planar surfaces within sampling domains. Three CDC 3400 Fortran IV computer programs for converting field measurements of these 'mega-roughness' features to standardised surface-roughness attributes are presented by the writer in an earlier paper (1967).

Surface area
Description
This parameter is designed to determine the amount of similarity between the test area surface and a planar surface. It is hypothesised that the surface area increases with surface irregularity. Because there is a definite interplay between the number and magnitude of terrain irregularities such that similar surface area estimates could arise from different manipulations of these two variables, the surface area para-

meter is most effective when accompanied by other roughness parameters (e.g. bump frequency distribution and distribution of planes)
describing the irregularities.

The basic field data for this parameter are a series of orthogonal
traverse measurements ('l' and 'w' of fig. 8.1A), which are used to
estimate the surface area by: (1) subdividing the traverses into segments, (2) forming rectangles from adjacent segments, and (3) summing the area contained within the rectangles (fig. 8.1B). The area
estimate (A') of the test site is then compared to the area of a plane (A)
whose outer dimensions are the same as those of the site. The ratio

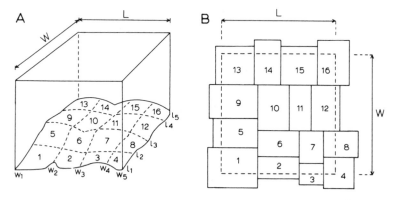

Fig. 8.1 A. Series of orthogonal traverse measurements from field data
to estimate surface area.

B. Rectangles formed from adjacent segments used to sum areas.

(A'/A) shows a curvilinear relationship which asymptotically approaches infinity with increases in A'.

The number of traverses necessary to obtain a reliable estimate of A'
is usually determined during field inspection of the types and areal
distribution of irregularities at a site and this number may be different
at different scales of sampling. Several methods for measuring traverses,
which have been tested at Pisgah Crater, California, are: (1) trigonometric calculation based on data obtained from topographic maps;
(2) measuring wheel (of about 14 in diameter) for sample areas with
outside dimensions between ten and a few hundred feet; and (3) flexible
cables and map-measuring devices for areas less than 10 ft on a side.

Description of program calculations

The theoretical dimensions of the sample area (ZLSA and WSA), the
number of length and width traverses (NL and NW), and the actual
lengths of these traverses (ZL(I) and W(I)) are the basic data. Average
length and width segments (ZLS(I) and WD(I)) are formed for each

traverse (1, 2). Average widths (AWS(I)) and lengths (ALS(I)) are computed (3, 4) and used to estimate sub-areas (AREA(I,J)) which are themselves summed to give an estimate (TOTA) of the actual surface area of the sample location (5, 6). The theoretical sample area (TRUA) is computed (7) as well as ratios of the two area estimates (8).

Basic Equations

$$ZLS(I) = ZL(I)/(NW - 1) \tag{1}$$
$$\text{for } I = 1, NL; J = 1, NW \tag{2}$$
$$WS(J) = W(J)/(NL - 1)$$
$$ALS(I) = (ZLS(I) + ZLS(I + 1))/2 \tag{3}$$
$$AWS(J) = (WA(J) + WS(J + 1))/2 \tag{4}$$
$$\text{for } I = 1, (NW - 1); J = 1, (NL - 1)$$
$$AREA (I,J) = (ALS(I)) (AWS(J)) \tag{5}$$
$$TOTA = TOTA + AREA(I,J) \tag{6}$$
$$TRUA = (ZLSA) (WSA) \tag{7}$$
$$TRTOTA = TRUA/TOTA \tag{8}$$

These equations can be solved using the standard electric calculator. If a large number of areas are to be compared, however, use of a high-speed computer is preferred. Program Comparea is a Fortran IV program described by Hobson (1967) that makes these calculations for any number of sample areas with as many as twenty length and twenty width measurements per area.

Bump frequency distribution

Description

The bump frequency parameters are mean and variance statistics describing size distributions of surface irregularities. They are designed to characterise the magnitude and variation of topographic elevation readings. As presented, these statistics are insensitive to the actual spatial distribution of surface irregularities. Although the spatial distribution could be described by suitable autocorrelation methods (Rice 1951), these techniques are not employed because natural topographic surfaces may be characterised by nearly identical monotonic autocovariance functions (Horton *et al.* 1962).

The bump frequency parameters are originally described (Hobson 1967) to compute elevation deviations (E_n) with respect to three possible orientations of remote sensing apparatus. The orientations shown in fig. 8.2A are: (1) vertically from a horizontal datum surface for the elevations (plane A, fig. 8.2B), (2) in a direction normal to the best-fit planar surface for the elevations (plane B, fig. 8.2B), and (3) vertically

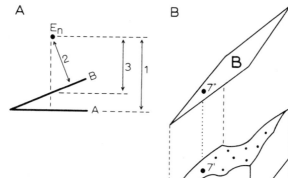

Fig. 8.2 A. Three possible orientations of sensing
apparatus.
B. Array of elevation
readings and their
geographic co-ordinates.

from the best-fit planar surface (the best-fit planar surface is found
using the least-squares method). Actually, orientations '1' and '2' are
similar in being directed normal to planar surfaces, but physically '1'
might be analogous to the case of an airborne instrument directed
normal to the horizon, whereas '2' could describe the case of a stationary
instrument set up normal to a hillside.

Field data required for bump frequency statistics consist of an array
of elevation readings and their geographic co-ordinates (e.g. U and V
for reading no. 7, fig. 8.2B). The three types of bump frequency values
are determined for each reading in the array and are used to calculate
mean and variance statistics.

The number of observations required depends upon the actual irregularities of the terrain. For example, results from the writer's surface
roughness investigations at the Pisgah Crater NASA test site in
California (unpublished) indicate that 16 elevation readings are sufficient to describe major terrain fluctuations of square sampled areas of
1 sq ft and of 10,000 sq ft, whereas fewer readings may be sufficient
for 100 sq ft areas.

Calculations

The basic data required to calculate the bump frequency parameters
are all (NE) sampled elevation readings (E(I)) in an area and their
respective geographic co-ordinates (U(I) and V(I)). These data are first
used to compute the best-fit linear surface for the area.

In matrix form, the coefficients of this surface [B(I)] are found by
solving for the product of the elevation matrix [E(I)] and inverse of

the [U(I)V(I)] matrix (1). Statement 2 defines the contents of each matrix. Computed elevation values (COMPE(I) are then formed for each location in the area (3). Sets of elevation deviations (DEV1(I), DEV2(I) and DEV3(I)) are computed vertically from both a horizontal plane (4a) and the best-fit plane (4b) and the normal from the best-fit plane (4c). Finally, the mean (5), variance (6) and standard deviation (7) are determined for each of the three sets of elevations.

$$[B(I)] = [E(I)] \cdot [U(I)V(I)]^{-1} \tag{1}$$

$$\begin{bmatrix} N & \Sigma U(I) & \Sigma V(I) \\ \Sigma U(I) & \Sigma U(I)^2 & \Sigma U(I)V(I) \\ \Sigma V(I) & \Sigma U(I)V(I) & \Sigma V(I)^2 \end{bmatrix} \cdot \begin{bmatrix} B(1) \\ B(2) \\ B(3) \end{bmatrix} = \begin{bmatrix} \Sigma E(I) \\ \Sigma U(I)E(I) \\ \Sigma V(I)E(I) \end{bmatrix} \tag{2}$$

$$\quad\quad\quad [U(I)V(I)] \quad\quad\quad\quad [B(I)] \quad\quad [E(I)]$$

for $I = 1$, NE

$$COMPE(I) = B(1) + (B(2))(U(I)) + (B(3)V(I)) \tag{3}$$

TRUMN = E(I)/NE

D1(I) = TRUMN − E(I)

QQ = largest negative D2(I)

Q = largest negative D1(I)

D2(1) = E(I) − COMPE(I)

THETA = dip of the best-fit plane

CTHETA = 90·0 − THETA

$$DEV1(I) = E(I) + Q \tag{4a}$$

$$DEV2(I) = D2(I) - QQ \tag{4b}$$

$$DEV3(I) = (DEV2(I))(CTHETA) \tag{4c}$$

$$\bar{X}_1 = \Sigma DEV1(I)/NE \tag{5}$$

$$VARI = (\Sigma(DEV1(I)^2) - ((\Sigma DEV1(I))^2/NE))/(NE - 1) \tag{6}$$

$$STD1 = \sqrt{VARI} \tag{7}$$

Hobson (1967) describes a Fortran IV program (PROGRAM FRE-HUMP) for making these calculations.

Distribution of planes
Description
These parameters are designed to describe the three-dimensional orientation of surfaces within a roughness test site. To do this, the site is simulated by a set of intersecting planar surfaces which are themselves defined by adjacent groups of three elevation readings (e.g. planes 1 and 2, fig. 8.3A). Normals to these planes are represented by unit vectors. Vector mean, vector strength and vector dispersion are computed using methods defined by Fisher (1953) and described by Watson (1957) and Watson and Irving (1957). Vector strength indicates the length of the resultant sum of the unit vectors and is obtained by

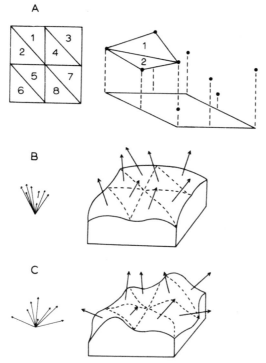

Fig. 8.3 A. Intersecting planar surfaces defined by adjacent groups of three elevation readings.

B. Area with similar elevations producing high vector strength and low vector dispersion.

C. Non-systematic elevation changes yielding low vector strength and high vector dispersion.

using the direction cosine method (Johnson and Kiokemeister 1957). Vector strength, in a standardised form (the square root of the squared sum of the direction cosines divided by the number of unit vectors), ranges in value from zero (no preferred orientation) to one (identical orientation). Dispersion, on the other hand, indicates the variability or spread of the unit vectors in space and is similar, in some respects, to the standard deviation of the normal distribution (Pincus 1953, 1956). In summary, vector strength is usually high and vector dispersion low in areas characterised by similar elevations (fig. 8.3B) or equal rates of elevation change, whereas non-systematic elevation changes yield low vector strength and high vector dispersion (fig. 8.3C).

Figure 8.4 describes a 'typical' area from the Pisgah Crater site (area 1, fig. 8.5B) and shows its topography (A), the groupings of elevation readings defining the planar surface (B) and the topographic simulation (C) created by the intersecting surfaces.

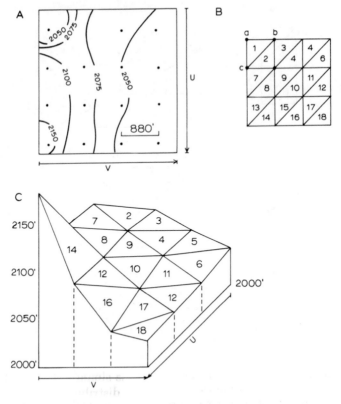

Fig. 8.4 A. Topographic map of area 1, Pisgah Crater site (see fig. 8.5B).
Dots show locations of elevation sampling points.
B. Planar surfaces created by adjacent elevation values.
C. Simulated topography of the area.

The equations presented below and those used in Program Vector (Hobson 1967), a Fortran IV program, calculate vector strength and dispersion using methods described by Watson (1957). Krumbein (1939) and Chayes (1954) describe certain modifications of these methods, entailing doubling of angles for highly variable data. Another addition, when using a computer, might be to include a subroutine that gives an equal-angle Wulff-net stereogram of the distribution of unit vectors as part of the program printout. Loudon (1964) describes such a subroutine.

The basic data is an array of regularly spaced elevation readings. In general, the number of readings necessary for the calculation of bump frequency parameters is adequate for the distribution-of-planes parameters as well.

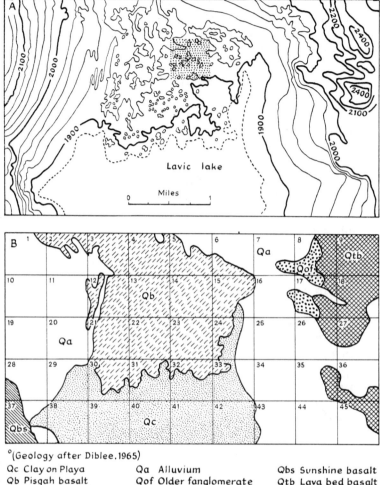

°(Geology after Diblee.1965)

Qc Clay on Playa	Qa Alluvium	Qbs Sunshine basalt
Qb Pisgah basalt	Qof Older fanglomerate	Qtb Lava bed basalt

Fig. 8.5 Comparison of (A) topographic and (B) geologic maps of the Pisgah Crater site, indicating that areas with the same rock type generally have similar topography.

Calculations

The basic input includes the number of rows (M) and columns (N); geographic origin (USTART and VSTART) and spacing (USTEP and VSTEP) of the grid; and the elevation matrix (E(I,J), where $I = 1$, M and $J = 1$, N). The U and V matrices are generated (1a, b) and used with the elevation matrix to compute direction cosines of the normals to the planes defined by successive groups of three elevation readings. (For simplicity, only the direction cosine calculations about the X-axis are included because the cosines relative to the Y- and Z-axes are calculated in a similar manner,) The direction cosines are summed (3) and these sums are used to estimate two forms (4a, b) of the vector strength (R1 and R) and Fisher's dispersion factor (ESTK, 5). The direction cosines (ALPHA, BETA, GAMMA) of the plane normal to the mean vector (R1) are then determined (6a, b, c) and used to calculate the dip (DIP) and strike (STRIKE) of that plane (7 and 8). The final calculations compute the dip distribution of the planes (DPLANE(K)) to each of the individual vectors (9) in terms of their mean (ZMEAN) and standard deviation (ST) statistics (10 and 11). Z-scores of each individual dip (ADIP(K)) are also calculated (12).

$$U(I,J) = USTART + USTEP (I - 1) \qquad (1a)$$
$$V(I,J) = VSTART + USTEP (J - 1) \qquad (1b)$$
$$\text{for } I = 1, M; J = 1, N; K = 1, NK$$
$$II = I + 1$$
$$JJ = J + 1$$
$$NK = (M - 1)(N - 1)(2)$$
$$XDN(K) = (VSTEP(E(II,J))) - (VSTEP(E(I,J)))$$
$$DIV(K) = ((XDN(K))^2 + (YDN(K))^2 + (ZDN(K))^2)^{1/2}$$
$$XNORM(K) = XDN(K)/DIV(K) \qquad (2)$$
$$TOTX = XNORM(K) \qquad (3)$$
$$R1 = ((TOTX^2) + (TOTY^2) + (TOTZ^2))^{1/2} \qquad (4a)$$
$$R = R1/NK - 1 \qquad (4b)$$
$$ESTK = (NK - 1)/(NK - R1) \qquad (5)$$
$$ALPHA = TOTX/R1 \qquad (6a)$$
$$BETA = TOTY/R1 \qquad (6b)$$
$$GAMMA = TOTZ/R1 \qquad (6c)$$
$$ANGG = \arccos (GAMMA)$$
$$AC = \cos (1{\cdot}5708 - ANGG)$$
$$COSTH = BETA/AC$$
$$DIP = (1{\cdot}5708 - ANGG)(57{\cdot}2957) \qquad (7)$$
$$STRIKE = (\arccos (COSTH - 1{\cdot}5708))(57{\cdot}2057) \qquad (8)$$
$$K = 1, NK$$
$$ARCZ(K) = \arccos (ZNORM(K))$$
$$DPLANE(K) = 1{\cdot}5708 - ARCZ(K) \qquad (9)$$
$$ZMEAN = DPLANE(K)/NK \qquad (10)$$
$$SUMDIP = \Sigma DPLANE(K)$$
$$SIMDPS = \Sigma (DPLANE(K)^2)$$
$$ST = ((SUMDPS - (ZMEAN)(SUMDIP))/(NK - 1))^{1/2} \qquad (11)$$
$$ZDIP(K) = (DPLANE(K) - ZMEAN)/ST \qquad (12)$$

Pisgah Crater: an example of map analysis

This section describes a surface-roughness analysis of an area within the NASA test site at Pisgah Crater, California. The basic hypothesis considered is that surface roughness reflects some of the geologic characteristics of the area.

The area, located at the southern margin of the test site, covers approximately 11 sq ml. Six rock types are mapped in the area (Diblee 1966) and comparison of the topographic and geologic maps (fig. 8.5A; 8.5B) shows that those portions of the area containing the same rock type generally have similar topography. For example, the areas of alluvium (Qa) are characterised by gentle, evenly dipping slopes, whereas those areas of Lava Bed Basalt (Qtb) have relatively high, irregular relief.

All data for this analysis were obtained from the Lavic Lake 7·5 minute topographic quadrangle. The basic roughness parameters were computed and these appear as table 8.1. The area was divided into

Table 8.1 *Complete listing of data used for Pisgah Crater test*

	NO U V	1 RATIO AREA	2 MEAN 1 BMP.FREQ.	3 MEAN 2 BMP.FREQ.	4 STD.DEV.1 BMP.FREQ.	5 STD.DEV.2 BMP.FREQ.
SR1	1 1 1	1.0013	1952.7500	74.5750	43.7079	28.6489
SR 1	2 1 2	1.0031	1908.3750	39.7625	41.5583	39.5567
SR 1	3 1 3	1.0020	1925.8750	14.0625	13.4556	6.6779
SR 1	4 1 4	1.0036	1930.0000	11.2000	4.8028	4.7469
SR 1	5 1 5	1.0032	1906.5000	15.0250	8.1199	6.3411
SR 1	6 1 6	1.0014	1905.7500	6.8250	7.7363	4.3386
SR 1	7 1 7	1.0031	1901.1250	13.4375	23.8299	8.1585
SR 1	8 1 8	1.0287	1909.2500	74.9500	91.1496	32.7470
SR 1	9 1 9	1.0234	2138.3750	152.1375	105.1369	70.8290
SR 1	10 2 1	1.0012	1988.1250	86.5125	40.9544	24.9206
SR 1	11 2 2	1.0010	1924.5000	33.0750	19.7400	9.7671
SR 1	12 2 3	1.0045	1915.0000	21.6500	11.3549	9.3904
SR 1	13 2 4	1.0026	1896.6250	17.8125	15.4907	11.3561
SR 1	14 2 5	1.0016	1816.5000	51.9750	48.7777	46.1285
SR 1	15 2 6	1.0035	1819.7500	20.5250	29.8527	25.1793
SR 1	16 2 7	1.0029	1860.8750	15.9625	24.0304	11.2437
SR 1	17 2 8	1.0285	1875.8750	37.6125	59.2143	29.8095
SR 1	18 2 9	1.0808	2002.1250	117.6125	89.9225	79.3735
SR 1	19 3 1	1.0018	1945.7500	7.2000	37.6526	3.8601
SR 1	20 3 2	1.0005	1883.0000	6.3750	20.8423	4.7732
SR 1	21 3 3	1.0068	1873.3750	83.7625	30.3737	25.4883
SR 1	22 3 4	1.0030	1893.5000	14.0000	8.8129	6.5345
SR 1	23 3 5	1.0029	1891.3750	15.3375	8.4358	7.9470
SR 1	24 3 6	1.0011	1871.8750	12.1125	8.7290	8.1740
SR 1	25 3 7	1.0003	1879.6250	3.0875	13.9246	2.1527
SR 1	26 3 8	1.0134	1814.5000	36.4250	53.9178	34.0495
SR, 1	27 3 9	1.0569	1902.3750	152.9375	145.9197	82.7908
SR 1	28 4 1	1.0018	1897.5000	70.2500	29.2449	6.9711
SR 1	29 4 2	1.0001	1877.3750	6.8375	6.2740	4.5359
SR 1	30 4 3	1.0014	1885.7500	7.5250	4.7311	3.4905
SR 1	31 4 4	1.0011	1884.2500	3.6500	6.2490	2.1463
SR 1	32 4 5	1.0005	1880.8750	4.0375	6.4650	2.1574
SR 1	33 4 6	1.0001	1883.7500	3.2750	4.1453	2.1610
SR 1	34 4 7	1.0003	1887.0000	4.4500	14.2642	2.3310
SR 1	35 4 8	1.0049	1899.7500	28.5000	38.4931	17.9901
SR 1	36 4 9	1.0040	1977.3750	31.0625	28.7593	17.5298
SR 1	37 5 1	1.0039	1843.8750	21.8625	37.8329	14.1714
SR 1	38 5 2	1.0000	1886.6250	4.2875	21.5199	2.4264
SR 1	39 5 3	1.0000	1881.6250	4.0625	3.2806	2.3354
SR 1	40 5 4	1.0000	1881.0000	3.3250	2.3664	1.9655
SR 1	41 5 5	1.0000	1883.1250	1.5375	0.7719	0.5605
SR 1	42 5 6	1.0000	1885.1250	0.7375	0.7719	0.3926
SR 1	43 5 7	1.0001	1886.2500	4.4000	7.6409	2.6796
SR 1	44 5 8	1.0002	1904.2500	18.8000	10.0855	5.6939
SR 1	45 5 9	1.0077	1856.5000	31.2000	36.5304	24.9566

Table 8.1 *(continued)*

				6	7	8	9
	NO	U	V	VECTOR STR.	VECTOR DISP.	PLANES DIP. MEAN	PLANES DIP. STD.DEV.
SR 1	1	1	1	17.9779	403.5422	1.5117	2.3858
SR 1	2	1	2	17.8606	121.9512	1.4639	3.7766
SR 1	3	1	3	17.9904	1770.4137	1.5410	0.8176
SR 1	4	1	4	17.9985	11493.2878	1.5592	0.3255
SR 1	5	1	5	17.9966	4958.9267	1.5547	0.6630
SR 1	6	1	6	17.9981	9004.1709	1.5586	0.4750
SR 1	7	1	7	17.9875	1359.5632	1.5345	0.5033
SR 1	8	1	8	17.8145	91.6853	1.4444	4.0379
SR 1	9	1	9	17.6531	49.0000	0.3917	5.1577
SR 1	10	2	1	17.9331	254.2820	0.5015	3.0325
SR 1	11	2	2	17.9890	545.8869	0.5382	0.7425
SR 1	12	2	3	17.9936	2694.7418	0.5483	0.8515
SR 1	13	2	4	17.9888	1519.3473	0.5412	1.1465
SR 1	14	2	5	17.8164	92.5965	0.4543	4.6969
SR 1	15	2	6	17.8944	160.9702	0.4965	4.6623
SR 1	16	2	7	17.9818	932.1909	0.5344	1.5631
SR 1	17	2	8	17.9018	173.1518	0.4897	3.9330
SR 1	18	2	9	17.5917	41.6402	0.3790	5.5832
SR 1	19	3	1	17.9779	770.5749	1.5219	0.4746
SR 1	20	3	2	17.9926	2299.6587	1.5443	0.6526
SR 1	21	3	3	17.9860	1215.8209	1.5338	0.8494
SR 1	22	3	4	17.9964	4774.3272	0.3524	0.4695
SR 1	23	3	5	17.9976	7115.9640	0.5550	0.3364
SR 1	24	3	6	17.9951	3500.0848	0.5504	0.6941
SR 1	25	3	7	17.9967	5089.0615	0.5519	0.2365
SR 1	26	3	8	17.9363	266.9736	0.5123	3.7620
SR 1	27	3	9	17.3781	27.3339	0.3530	8.8899
SR 1	28	4	1	17.9838	1050.1784	0.5313	0.9095
SR 1	29	4	2	17.9991	18996.6854	0.5640	0.4635
SR 1	30	4	3	17.9992	20011.1503	0.5626	0.3209
SR 1	31	4	4	17.9992	20235.2377	0.5621	0.2568
SR 1	32	4	5	17.9991	19050.1651	0.5623	0.3040
SR 1	33	4	6	17.9995	35605.0263	0.5644	0.2118
SR 1	34	4	7	17.9665	4789.8721	0.5511	0.1597
SR 1	35	4	8	17.9604	429.0199	0.5214	2.6182
SR 1	36	4	9	17.9602	552.6186	0.5271	2.2895
SR 1	37	5	1	17.9703	563.3203	1.5199	1.6311
SR 1	38	5	2	17.9989	15465.0466	1.5620	0.3925
SR 1	39	5	3	17.9996	39288.7052	1.5657	0.2782
SR 1	40	5	4	17.9997	59350.7210	1.5665	0.2173
SR 1	41	5	5	18.0000	676598.4398	1.5696	0.0672
SR 1	42	5	6	18.0000	822790.6697	1.5694	0.0385
SR 1	43	5	7	17.9986	12643.4210	1.5595	0.3267
SR 1	44	5	8	17.9948	3248.3833	1.5511	0.8258
SR 1	45	5	9	17.9633	463.8096	1.5293	2.9542

45 sub-areas (fig. 8.5B) and 16 elevation readings and six traverse estimates were gathered from each.

Correlation

Correlation among most variables is generally high (table 8.2). This is not surprising considering the overall correspondence between topography and geology. For example, vector strength would be expected to decrease along with increases in the surface area ratio ($r = -0.86$), mean bump frequency ($r = -0.83$) and standard deviation of bump frequency ($r = -0.92$) if one walked from the smooth even surface of the playa lake on to the irregular surface of a basalt flow.

A few variables, such as vector dispersion (Variable 7, table 8.2), do show surprising correlations suggesting the need for additional analysis. The signs of the coefficients between vector dispersion variable and the others are predictable but their magnitudes are generally lower than would be expected. The map dispersion (not included in this report) shows a pattern almost identical to that of vector strength (fig. 8.8A) despite the low correlation between the two variables ($r = 0.12$). The probable explanation is that these two variables are mathematically related in a curvilinear fashion whereas the correlation coefficient is only a measure of their linear similarity. That is, as vector strength increases from zero to one, vector dispersion changes geometrically from zero to infinity. It is possible that low correlations between dis-

Table 8.2 *Array of correlation coefficients: Pisgah Crater*

Area ratio	Bump Mean 1	Bump Mean 2	Bump S.D. 1	Bump S.D. 2	Vector str.	Vector disp.	Planes mean dip	Planes S.D. dip	
1	2	3	4	5	6	7	8	9	
1·00	0·33	0·72	0·70	0·86	−0·86	−0·12	−0·28	0·74	1
	1·00	0·55	0·37	0·38	−0·36	−0·08	−0·09	0·20	2
		1·00	0·88	0·90	−0·83	−0·20	−0·29	0·80	3
			1·00	0·92	−0·91	−0·23	−0·26	0·91	4
				1·00	−0·92	−0·20	−0·32	0·94	5
					1·00	0·12	0·32	−0·89	6
						1·00	0·26	−0·21	7
							1·00	−0·35	8
								1·00	9

persion and other variables also can be explained in terms of curvilinearity, but the interdependent nature of the variables makes this a multivariate problem which, as yet, has not been resolved.

An inspection of the standardised or z-scores of the data shows that most scores with absolute values greater than three standard deviation units are from sub-areas containing Lava Bed Basalt (Qtb) as the dominant rock type (e.g. sub-areas 9, 18, 27, fig. 8.5B). These 'abnormal' scores are explained, of course, by the high and variable relief that characterises the sub-areas of Qtb, and their deletion from the matrix would probably result in a general increase in the size of most correlation coefficients. A z-score is here defined as the standard deviation of the ith variable divided into the difference of the jth value of that variable and the variable mean. ($z_{ij} = (X_{ij} - \bar{X}_i)/s_i$).

Surface area

Considering the number and density of data control points, the map of the surface area ratio (fig. 8.6A) delineates the major geologic units extremely well. An interesting feature of the map is that although some sub-areas are occupied by two (e.g. 29, 36, fig. 8.5B) or even three (17, 37) rock types, the general shape of the geologic contacts is still maintained. Another interesting relationship is that small

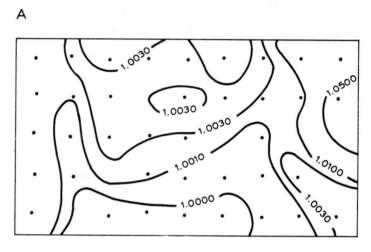

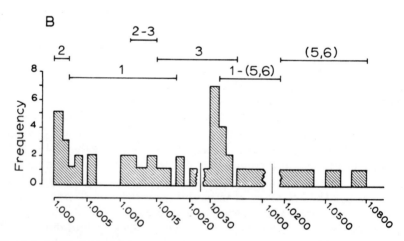

Fig. 8.6 A. Map of surface area ratio (estimated area/theoretical area).
B. Range of surface area ratio values.

1. Alluvium (Qa)   4. Sunshine Basalt (Qbs)
2. Clay on playa (Qc)   5. Lava Bed Basalt (Qbt)
3. Pisgah Basalt (Qb)   6. Older fanglomerate (Qof)

differences in the value of surface area ratios (usually at the third or fourth decimal place) can be used to distinguish between sub-areas containing different rock types. This relationship is illustrated by fig. 8.6B.

In fig. 8.6B the frequency of ratio values and their spread for each rock type is plotted. Overlap is small if ratio values between rock types and some transition sub-areas have narrow, characteristic ranges (e.g. the transition from the alluvium to the Lava Bed Basalt and Older Fanglomerate). Analysis of variance has also been used success-

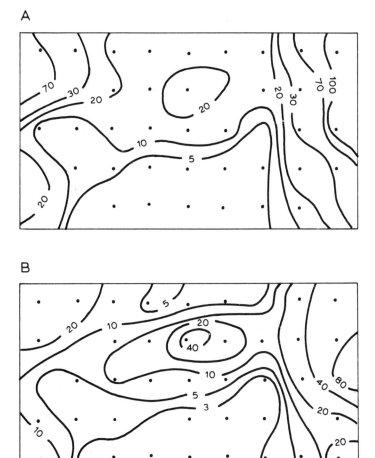

Fig. 8.7 A. Map of bump frequency – mean 2″. Deviations measured (in feet) from the best-fit planar surface.

B. Map of bump frequency – standard deviation 2″ and distribution of planes parameters.

fully to show these differences between the group of ratios associated with each particular rock type.

Bump frequency and distribution of planes

Maps of the bump frequency and distribution of planes parameters (figs. 8.7A, 8.7B, 8.8A and 8.8B), as the surface area ratio, tend to delineate the major geologic patterns in the study area. Again, as shown by vector strength (fig. 8.8A) and by the mean dip of planes (fig. 8B), subtle differences in the value of a variable can be used accurately to predict the rock type characterising the different sub-areas.

One interesting feature common to all parameter maps is a small,

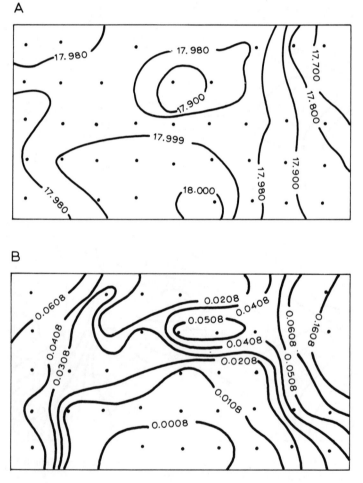

Fig. 8.8 Maps showing (A) vector strength and (B) mean dip of simulated planar surfaces.

closed contour located in the north-central portion of each. The contour always circles sub-area 14 (fig. 8.5B) and inspection of the topographic map shows that this sub-area (shaded area, fig. 8.5A) covers a series of small depressions on the basalt flow. One effect of the locally anomalous nature of the topography within sub-area 14 is that elevation values are more variable than those sampled from other sub-areas on the flow resulting in higher mean bump frequency, bump standard deviation, mean dip of planes, and lower vector strength. Surface area ratio, on the other hand, is surprisingly low. Again, inspection of the sampling procedure provides the explanation: the positioning of the length and width traverses was such that nearly all traverses missed the depressions and thus the estimated surface area was less than that for the sub-areas nearby.

These relationships associated with sub-area 14 are included to emphasise that the adequacy and reliability of any parameter is only as good as the effectiveness of its sampling procedures. In fairly simple analyses such as these, certain sampling inadequacies are apparent. In more rigorous analyses in which there is complicated data interlock, anomalous features of the variables might well go unnoticed and in these situations the original sampling design takes on extreme importance.

Shift of sampling origin

The distribution of planes parameters were recalculated from selected areas in the Pisgah site using different groupings of the three elevation readings used to form each triangular surface. Whereas elevation readings a–b–c define plane one, and b–c–d define plane two (fig. 8.4B) for the calculations of the distribution of planes parameters discussed above, new roughness parameters were calculated using the scheme a–c–d for plane one, a–b–d for plane two, and so forth. The results of analysis of variance of vector strength, vector dispersion and mean dip of planes estimates using the two different groupings of triangular planes in areas 1, 13 and 40 (fig. 8.5A) show no significant difference of the parameters within each area sampled. The highest F-ratio, although still insignificant at the 0·05 level, is in area 13 which is that area with the most irregular topography. One would expect such high topographic variability to be most easily reflected in changes in sampling procedure.

Elevations in areas 1, 13 and 40 were also resampled by shifting the sampling grid (fig. 8.4A) one half sampling unit (440 ft) to the right. The purpose of this resampling was to evaluate the dependence of the roughness parameters on the specific location of the grid. Again, analysis of variance of several of the roughness parameters showed no significant differences between values calculated using either grid location. This analysis has not been extended to include areas that straddle two or more geologic rock types. A shift of the sampling grid in these kinds of areas could change values of specific roughness parameters at positions near geologic contacts but it is difficult to quantitatively evaluate the effects of such changes on the maps of the parameters. A third kind of analysis should be to evaluate the effects on the roughness parameters of changing the number of elevations sampled in each area.

Sonora pass: a field example

Figure 8.9 is the generalised geologic map of an area in the Sierra Nevada Mountains that is located south of Sonora Pass and directly north of the Yosemite National Park boundary, California. This area is part of the one-by-ten miles Sonora Pass test site used in the NASA 'ground truth' sensing program, and was selected because it contains large glaciated areas of clean rock exposure and because there are only a few major rock units that are separated by well-recognised contacts. These characteristics are desirable for testing certain kinds of sensing apparatus over natural terrains. They also qualify it as an excellent place for a field study of surface roughness.

The surface roughness study in the Sonora area was designed to look at two kinds of problems. First, can surface roughness measurements obtained in the field be used to identify different lithologic units? Secondly, at what sampling scale are surface roughness parameters most useful for lithologic identification?

Four rock types were selected for the study (fig. 8.9). Two of these, the Cinko Lake granodiorite (Kcl) and the Dorothy Lake alaskite (Kdl), are medium-grained plutonic rocks of Cretaceous Age that are characterised by large areas of almost completely exposed unweathered rock. The third rock type is a gruss (Qdg) formed from granulation of the Fremont Lake quartz monzonite. Wahrhaftig (1965) postulates that this gruss was formed by excessive weathering of the large euhedral biotites of the monzonite causing hydration expansion and the resulting granulation. This material is coarse-grained, unconsolidated and highly porous, and forms a smooth, nearly plant-free surface (Garrett 1967). The fourth rock type, the Relief Peak Formation (Prp) consists of a series of andesitic flows which, in the areas sampled, are agglomeratic.

A nested sampling scheme was used. Two one-hundred-foot square areas were sampled for each rock type. The areas were selected to contain the maximum exposure of rock surface which, in all cases, exceeded 90%. A random number table was used to select two ten-by-ten-foot areas and two one-by-one-foot areas from each hundred-foot

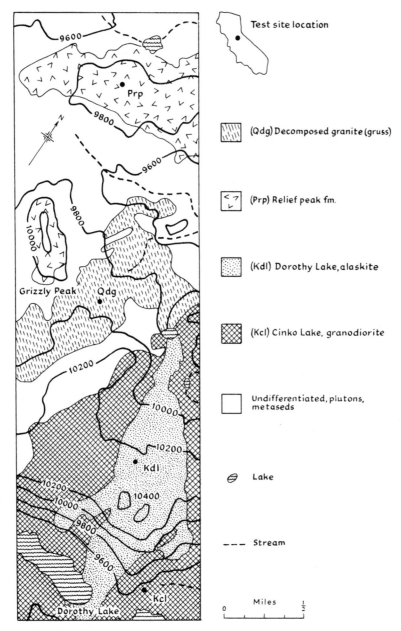

Test site location

(Qdg) Decomposed granite (gruss)

(Prp) Relief peak fm.

(Kdl) Dorothy Lake, alaskite

(Kcl) Cinko Lake, granodiorite

Undifferentiated, plutons, metaseds

Lake

--- Stream

Miles
0   ½

Fig. 8.9 Topographic and geologic map of the Dorothy Lake Area, Sonora Pass Test Site, NASA Remote Sensing Project.

square and ten-foot square respectively. In all, two hundred-foot, four ten-foot and eight one-foot square sample areas were selected for the surface roughness analysis of each of the four rock types.

The field data collected from each sample area consists of sixteen equally spaced elevation readings and four traverse estimates. An alidade and plane table were used to collect the elevation data from the one-hundred- and ten-foot areas whereas elevation data were obtained in the one-foot areas using a metal grid device levelled above each area. Pairs of traverse measurements were taken at right angles using a measuring wheel in the one-hundred- and ten-foot areas and a flexible ruler in the one-foot areas. These field data, and the roughness parameters calculated from them, are available upon request.

Analysis of variance

A one-way analysis of variance model was used to evaluate the source and magnitude of the variability of five surface roughness parameters (table 8.3). The parameters selected were those showing consistently high linear correlation relationships with other roughness parameters at each sampling level. Table 8.3 also shows the degrees of freedom

Table 8.3   *Analysis of variance summary*

| | | Variables[+] | | | | | | |
		1	2	3	6	8	df_B/df_W	Sig.F
Sample square size	100 foot square	0.10	1.09	0.01	1.44	1.79	3/4	6.59
	10 foot square	1.40	1.44	0.01	0.54	0.31	3/12	3.49
	1 foot square	6.00*	3.35	5.59*	3.12	3.93	3/28	2.95

[+] Variables are: (1) area ratio, (2) vector dispersion, (3) mean dip planes, (6) mean 2 bump frequency, and (8) standard deviation 2 bump frequency

associated with the between (df_B) and within (df_W) variance estimates as well as the magnitudes of the F-ratios required to show a significant differences of variability at the 0·05 level. Each F-ratio in the table outlined by heavy lines exceeds the 0·05 null value whereas an asterisk identifies F-ratios significant at the 0·01 level.

Table 8.3 shows that all parameters measured within the one-foot areas vary significantly between rock types sampled. None of the parameters measured in the ten- and one-hundred-foot sample sites shows significant differences between rock types. This second relationship may be due in part to the smaller within group degrees of freedom associated with these two sampling levels, but examination of the table generally shows extremely small F-ratios.

The conclusion drawn from these data is that small-scale sampling is required to distinguish differences in the surface characteristics of the various lithologic units found within the Sonora Pass area. This conclusion can be related in part to several of the geologic and geomorphic aspects of the area's history. In the first place, glaciers have moved across the area and have left most hard rock surfaces fairly smoothed and polished. Although some areas of high relief do exist, such as Grizzley Peak and the steep east–west trending slope in the south (fig. 8.9), the sample sites discussed are all located in glaciated areas of low relief. Therefore, the large surfaces sampled might be expected to reflect glacial processes more than underlying lithology. Secondly, weathering processes at all sampling sites are strongly affected by freeze and thaw conditions. Natural cracks and weaknesses in the rocks would tend to control much of this phase of weathering. Infrared and multiband photography of the area as well as the surface geologic investigation, reveal at least two major joint sets crossing all rock types. The density of these joints would be a major factor affecting weathering and the resulting roughness of the natural topography. In general, the joints are spaced close enough together so that practically all ten-foot areas sampled contained from two to five individual cracks. Joints seldom affected relief within most of the one-foot areas sampled and thus their roughness characteristics seem to reflect more intrinsic responses of individual rock types to weathering rather than being controlled by cross-cutting structural features.

The Qdg (gruss), itself a product of weathering, perhaps requires some special consideration. This coarse material is unconsolidated and quite porous but, like the very resistant rock types, has an extremely smooth surface whose evenness is broken only by occasional clumps of vegetation or by a drainage channel. The drainage network developed on the gruss is dendritic, and has a coarse drainage density. Individual channels are fairly large and incised, but usually have quite smooth walls. The probability is remote of encountering a channel in a one-foot sample area or having a nearby channel affect its surface roughness, whereas channels do affect the roughness in the ten- and one-hundred-foot areas.

Correlation

Table 8.5 shows the correlation coefficients among roughness variables squares (table 8·4) calculated from the data describing the one-foot sample squares. The coefficient arrays for the ten- and one-hundred-foot squares show generally similar patterns of sign and magnitude

Table 8.4   *Roughness parameters; one-foot sample squares Sonora Pass, California*

Area*	Area Ratio	Vect. Disp.	Plane X Dip	Plane S.D.	Bump X 1	Bump X 2	Bump S.D.1	Bump S.D.2
	1	2	3	4	5	6	7	8
DG-111	1.04	832.47	1.53	1.28	9.46	0.11	0.06	0.06
DG-112	1.03	362.72	1.51	1.95	9.52	0.13	0.10	0.08
DG-121	1.06	75.72	1.43	4.61	9.61	0.30	0.35	0.18
DG-122	1.03	100.65	1.44	2.50	9.65	0.17	0.38	0.10
DG-211	1.05	97.02	1.44	3.21	9.81	0.32	0.26	0.19
DG-212	1.08	36.76	1.37	6.09	9.41	0.60	0.48	0.27
DG-221	1.01	115.74	1.45	2.93	9.46	0.20	0.20	0.13
DG-222	1.01	41.64	1.38	5.52	9.07	0.28	0.53	0.16
DA-111	1.49	5.08	1.05	21.24	6.59	1.49	1.77	1.12
DA-112	1.30	4.35	0.91	6.92	6.13	1.97	1.34	1.20
DA-121	1.23	22.34	1.30	12.53	11.09	0.53	0.38	0.30
DA-122	1.15	14.35	1.27	20.39	9.77	1.15	0.57	0.54
DA-211	1.82	3.29	0.84	6.62	6.73	2.09	2.86	1.65
DA-212	1.14	30.68	1.35	7.92	10.59	0.47	0.34	0.26
DA-221	1.12	23.61	1.32	9.83	7.27	0.58	0.54	0.40
DA-222	1.18	7.19	1.07	6.26	8.90	1.18	1.47	0.72
CL-111	1.06	45.50	1.41	7.38	9.10	0.46	0.50	0.31
CL-112	1.11	111.10	1.45	3.50	10.36	1.27	0.22	0.18
CL-121	1.40	7.85	1.13	13.61	8.23	1.21	1.04	0.77
CL-122	1.17	97.37	1.44	3.11	10.00	0.38	0.22	0.18
CL-211	1.59	3.66	0.87	14.60	7.50	2.79	1.43	1.42
CL-212	1.32	4.65	0.97	15.83	2.96	1.51	2.30	0.71
CL-221	1.04	10.04	1.18	12.60	4.70	1.03	0.90	0.76
CL-222	1.27	24.39	1.30	4.81	10.50	0.61	0.53	0.40
RA-111	1.10	57.53	1.40	3.80	10.57	0.29	0.29	0.20
RA-112	1.30	5.20	1.13	27.06	8.94	2.95	2.07	1.33
RA-121	1.07	50.57	1.40	5.66	7.95	0.27	0.39	0.20
RA-122	1.04	16.11	1.26	8.73	8.79	0.53	0.91	0.24
RA-211	1.06	49.58	1.41	6.27	10.70	0.54	0.29	0.27
RA-212	1.12	22.56	1.30	6.37	10.98	0.89	0.53	0.38
RA-221	1.03	67.75	1.41	3.59	11.10	0.35	0.21	0.20
RA-222	1.05	53.73	1.40	3.99	10.65	0.40	0.28	0.22

RA Relief peak formation   DA Dorothy Lake alaskite
CL Cinko Lake granodiorite DG Decomposed granite (gruss)

but are not discussed here because the one-foot square seems to be the optimum sample area for this study.

As noted in the Pisgah Crater example, correlation among the roughness variables is generally high (table 8.5). Variable 8, the standard deviation surface irregularities as measured from a best-fit plane, shows the greatest number of strong correlations with other variables (e.g. those coefficients greater than 0·75). The signs of all coefficients involving variable 8 are what would be expected. These relationships of magnitude and sign among the coefficients show that variable 8 is probably one of the better parameters for describing natural surfaces.

One advantage of it over the bump frequency standard deviation is that it describes surface fluctuations remaining after the general slope of an area has been removed.

Certain relationships of table 8.5 are difficult to explain. Again, as in the Pisgah example, vector dispersion (variable 2) shows low

Table 8.5 *Array of correlation coefficients: one-foot squares*

Area Ratio	Vector disp.	Planes mean dip	Planes S.D. dip	Bump mean1	Bump mean2	Bump S.D.1	Bump S.D.2	
1	2	3	4	5	6	7	8	
1·00	−0·29	−0·84	0·72	−0·14	0·77	0·81	0·87	1
	1·00	0·46	−0·41	0·18	−0·36	−0·37	−0·37	2
		1·00	−0·77	−0·68	−0·84	−0·90	−0·93	3
			1·00	0·56	0·83	0·88	0·86	4
				1·00	−0·48	−0·69	−0·59	5
					1·00	0·81	0·93	6
						1·00	0·89	7
							1·00	

correlation with the other variables although the signs of the coefficients, as expected, are mostly negative. The strong correlations with variable 3, the mean dip of the simulated planar surfaces describing an area, are also confusing. The strong negative values with the area ratio and bump frequency standard deviation parameters seem to indicate that generally smoother areas have greater dip than rougher ones whereas the positive relationship with vector dispersion seems to suggest the opposite concept that more irregular surfaces have larger average dips. Although the mean dip has only a range of 0·84 to 1·53 degrees for these data (Table 8.4), examination of the analysis of variance summary table 8·3 shows it to be one of the stronger variables for distinguishing between the surfaces of different rock types. Further study of this variable will be required to explain its strong correlation and variance characteristics.

Conclusion

In conclusion, these studies provide adequate, if somewhat simplified, examples of the use of the various surface-roughness parameters. The original hypothesis that surface roughness reflects the geologic nature of rocks appears to be correct although it quickly becomes apparent that the scale of sampling required to demonstrate the hypothesis may vary for each area studied. In most studies it is recommended that sampling be conducted at several scales of measurement and that the

244 *Spatial analysis in geomorphology*

variability of each parameter at each scale be compared. As pointed out by the writer in his earlier paper (1967), several multivariate techniques such as factor analysis and discriminant analysis may be helpful in distinguishing surface-roughness differences between various types of geologic areas, whereas trend-surface and Fourier techniques might be helpful in mapping the systematic and residual trends of variables in large areas that have been continuously sampled. Finally, it seems that these techniques of describing natural surfaces may be of use to the geologist and geomorphologist in relating actual topology to underlying geology and to geologic processes.

References

BECKMAN, P. and SPIZZICHINO, A. (1963) The scattering of electromagnetic waves from rough surfaces; *International Series of Monographs on Electromagnetic Waves* 4 (Pergamon, London), 503p.

CHAYES, F. (1954) Discussion: Effect of change of origin on mean and variance of two-dimensional fabrics; *American Journal of Science* 252, 567–70.

DELLWIG, L. F. and MOORE, R. K. (1966) The geological value of simultaneously produced like- and cross-polarized radar imagery; *Journal of Geophysical Research* 71, 3597–601.

DIBLEE, T. W. (1966) Preliminary geologic map—Pisgah Crater and vicinity, California; *NASA Technical Letter No. 4, U.S. Geological Survey.*

FISHER, R. A. (1953) Dispersion on a sphere; *Proceedings of the Royal Society* Ser. A, 217, 295–305.

GARRETT, R. G. (1967) The geochemistry of the Fremont Lake quartz monzonite and associated gruss, NASA Sonora Pass geologic test site, Sierra Nevada, California; *Northwestern University Report No. 14, N.G.R. 14–007–027*, 38p.

HOBSON, R. D. (1967) Fortran IV programs to determine surface roughness in topography for the CED 3400 computer; *Kansas Computer Contribution* No. 14, 28p.

HORTON, C. W., HOFFMAN, A. A. J. and HEMPKINS, W. B. (1962) Mathematical analysis of the micro structure of an area of the bottom of Lake Travis; *Texas Journal of Science* 14, 131–42.

JOHNSON, R. E. and KIOKEMEISTER, F. L. (1957) *Calculus with Analytic Geometry* (Allyn and Bacon, Inc., New York), 709p.

KRUMBEIN, W. C. (1939) Preferred orientation of pebbles in sedimentary deposits; *Journal of Geology* 47, 673–706.

LOUDON, T. V. (1964) Computer analysis of orientation data in structural geology; *Office Naval Research, Geography Branch, ONR Task No. 389–135, Technical Report* No. 13, 129p.

PINCUS, H. J. (1953) The analysis of aggregates of orientation data in the earth sciences; *Journal of Geology* 61, 484.

PINCUS, H. J. (1956) Some vector and arithmetic operations on two-dimensional orientation variates, with applications to geologic data; *Journal of Geology* 64, 533–57.

RICE, S. O. (1951) Reflection of electromagnetic waves from slightly rough surfaces; *Comm. Pure and Applied Mathematics* 4, 351–78.

WAHRHAFTIG, G. (1965) Stepped topography of the southern Sierra Nevada, California; *Bulletin of the Geological Society of America* 76, 1165–90.

WATSON, G. S. (1957) Analysis of dispersion on a sphere; *Royal Astronomical Society Monthly Notices, Geophysics Supplement* 7, 153–9.

WATSON, G.S. and IRVING, E. (1957) Statistical methods in rock magnetism; *Royal Astronomical Society Monthly Notices, Geophysics Supplement* 7, 289–300.

9 Trend-surface analysis of planation surfaces, with an East African case study

J. C. DOORNKAMP

Department of Geography, University of Nottingham

Trend-surface procedures constitute merely a tool; not yet a whole tool, but nothing more than a tool. As such their usefulness and limitations depend as much on the user as on the designer. They are conceptually simple and, like most simple tools, frequently used in situations for which they were not designed. (ROBINSON 1970, 35)

There is small value in highly skilled interpretation of incorrect or unrepresentative data. (GRANT 1954, 24)

Introduction

Trend-surface analysis is designed to detect the general, or regional trend in a variable which can be mapped across, or within, an area. There are, inevitably, local deviations from any regional trend, and these may be a response to an underlying controlling influence or they may be random error values. Within geomorphology trend-surface analysis has been applied most commonly to data concerning the altitude of subaerial planation surfaces and raised beaches. This study is chiefly concerned with the former. Before reviewing the applications of this technique within geomorphology a brief discussion of the technique itself is necessary in order to establish conditions under which it may be validly applied. An examination of the application of trend-surface analysis to planation surfaces indicates that many studies are weakened by the nature of the data analysed rather than by any inherent limitations of the technique. Some attention is devoted, therefore, to the character of the data which may be obtained for planation surfaces. This is illustrated by a case study to show how trend-surface analysis may be used to define the influence of warping on the altitudes of two planation surfaces in Uganda.

The technique

Trend-surface analysis, like any other regression technique, produces a statement of both the general trend of the data and of the individual deviations of each observation (control point) from that trend. The higher the order of the trend-surface the more it can accommodate undulations in the mapped variable. As the order of the surface increases so the deviations tend to decrease. However, it is frequently the case that beyond a fourth-order trend-surface the increase in complexity of the polynomial equation, with successive increases in surface order, is not compensated by a significant improvement in the fit of the higher-order trend-surfaces.

In trend-surface analysis attention may be focused upon either the form of the computed surface or on the deviation values. A few studies have combined these two approaches. The value of either approach depends on the scale of the investigation, and this also has a bearing on the scale at which the raw data is collected. The underlying regional trend should have an amplitude and a wavelength which is greater than that to be found in any one part of the area under analysis. The higher the order of the trend-surface the more it will try to adapt its shape to local undulations in the data. The stage is inevitably reached, therefore, with the higher-order surfaces, when local undulations in the data influence the form of the computed trend-surface. At this stage the computed surface no longer reflects just the regional trend.

The deviations from a linear trend-surface result from two conditions. The first, and usually the larger amount of deviation, is a result of local oscillations in the observed data. The second results from measurement errors; this can never be sensibly accommodated by a trend-surface. The deviations are calculated as the difference between the observed value (Z_{obs}) and the value (Z_{calc}) which Z would have had, at that same locality, if it had been on the computed trend-surface:

$$\text{deviation} = Z_{obs} - Z_{calc}.$$

Since the trend-surface is fitted by least-squares methods, for each polynomial equation representing each surface:

$$\sum (Z_{obs} - Z_{calc})^2 \quad \text{is kept to a minimum.}$$

For this to be possible some of the data values must lie above the computed surface whilst others will lie below it. The variability of the data to which the surface is being fitted may be expressed as:

$$S_{data} = \sum (Z_{obs} - \bar{Z}_{obs})^2$$

where $\bar{Z}_{obs}$ is the mean value of the observed data values.

When trend-surface analysis is carried out the surface will draw on some of this variation and identify it with a regional trend. This variation can be represented by S_{trend}. The rest of the variation will remain within the deviation values (S_{dev}). The extent to which the computed trend-surface 'explains' the variation in the data is usually stated as a percentage term known as the percentage reduction in the total sum of squares accounted for by the fitted surface (RSS), where:

$$\text{RSS} = \left[1 - \frac{\sum (Z_{\text{obs}} - Z_{\text{calc}})^2}{\sum (Z_{\text{obs}} - \bar{Z}_{\text{obs}})^2} \right] 100\%.$$

Assumptions

When the predominant interest in any trend-surface analysis is to be focused upon the computed surface itself then certain conditions regarding both the data and the deviation values need to be fulfilled.

Assumptions concerning the data are:

1. The observed values are not clustered into groups. They either occur on a regular grid across the area or they occur at random (i.e. data points are independent of each other).
2. The observed values are statistically normally distributed.

Assumptions concerning the deviations are:

3. The deviations are statistically normally distributed about the trend-surface.
4. The deviations are uncorrelated with each other, that is to say, they are trend-free across the area, and therefore they do not display autocorrelation.

These assumptions have been discussed and examined by a number of workers. Unwin (1970) shows that clustering does influence the RSS value, especially in the higher-order surfaces, though he concludes that the technique is fairly robust with respect to departures from the ideal distribution of control points. In non-gridded data complete randomness in the distribution of the data points is rare. Methods have been suggested to counteract clustering. These include a system of weighting the data to decrease the influence of cluster groups (Mandelbaum 1963), and of eliminating the clustering by transforming the map co-ordinates, computing the trend-surface, and then transforming the computed contour positions back to a cartesian plane (Norcliffe 1969). Another alternative, which can only be used satisfactorily when there are many control points, is to select a sample at random (Harbaugh and Merriam 1968). The difficulty in this case is that too large a sample may reintroduce clustering and too small a sample involves an excessive loss of information.

Few, if any, workers in geology or geography have demonstrated that the data they have used for trend-surface analysis is statistically normally distributed. A transformation of the data, into log (as done

I

by Andrews 1970, 162), square root or inverse form, as in many linear regression studies, is always available as a means of obtaining a normal distribution before the trend-surface analysis is undertaken. In regression analysis it is usual to assume that not only the dependent, but also the independent values are normally distributed. In this case the latter are the X and Y location co-ordinates. Normal distribution implies a large number of observations in the centre of the range, and fewer observations towards the extremities of the range. This produces a dilemma, for if X and Y are normally distributed then by definition clustering must occur in the centre of the area being studied. Another difficulty arises in that if all of the conditions are insisted upon (especially the third) then 'we must exclude the attempt to separate regional and local effects; for if the latter are meaningful, they will not be normally distributed' (McIntyre 1967). McIntyre adopts the approach of supplementing a random sample of points with values drawn from strategic locations, on the grounds that a bad distribution may seriously affect the results.

An examination of the deviations cannot take place until after the trend-surface has been computed. If normality does not exist within the deviations then the use of the least-squares technique becomes empirical (Grant 1957). For every surface that is used to make an interpretation of the regional trend in the data there should be a test for normality in the deviations. If it does not exist this should be both recognised and stated before continuing with an interpretation of the computed surface.

The fourth assumption, listed above, is important only when it is the trend-surface, as opposed to the deviations, which is of prime interest. The very purpose of the analysis in such a case is that all of the regional trend should be removed from the observations and contained within the computed surface. No trend should linger within the deviations. Whether or not this is the case can be tested from a one-dimensional array of deviations (e.g. perpendicular to the 'strike' direction of the trend-surface). This array can be tested for autocorrelation by testing the null hypothesis in the simple Markov chain:

$$\varepsilon_i = \rho \varepsilon_{i-1} + \mu_i \qquad |\rho| < 1$$

In practice a plot of the deviation value at i (ε_i) against the deviation at $i-1$ (ε_{i-1}) will indicate, by a straight-line plot, if autocorrelation remains among the deviations. Ideally, when this test is made, the observations should be equally spaced. Except in very rare instances, as in some of the examples when the trend-surface accounts for over 90% of the observed variance, spatial autocorrelation remains within the deviations (Robinson 1970). Some of this autocorrelation may be the result of localised variation, and in many instances it is desirable in any case to decrease the influence of this variation on the regional trend. In such circumstances there is every advantage in keeping within the deviation values both the effect of local undulations as well as random deviations produced, for example, by measurement errors. In practice, apportionment in this way may depend on a subjective assessment based on well-founded knowledge of the case study. A partially objective approach to the problem of apportionment is suggested by Miesch and Connor (1968, 2):

> the trend residuals . . . contain estimates of the second and third components of variation – the local deviations and noise . . . where the residuals are autocorrelated, indicating that adjacent values on the map tend to be similar . . . the second component is presumed to be dominant over the third. Where autocorrelation is low, indicating that adjacent trend residuals tend to be unrelated, the third component is presumed to be dominant over the second.

The behaviour of the deviations is important because the calculated trend-surface is in part dependent upon the local variations in these values. In terms of the case study to be presented below, it is necessary to realise that autocorrelation in the deviations may serve a very useful purpose when it is the deviations themselves that are being examined rather than the surfaces. For example, when negative or positive deviations group together on a map they reflect a systematic variation which is not as broad as the map itself, but broader than the distance between the control points.

Even if the conditions for the application of trend-surface analysis are fully met, the resulting trend-surface map will depend on the grid spacing used in collecting the data. In some instances it is possible to control the spacing between observations, by sampling according to the mesh size of a predetermined control grid. At other times there may be no choice over the spacing between observations, for the investigator may not be able to determine the location of the control points. Where choice can be exercised, however, decisions concerning the grid spacing should be based on a consideration of the nature of the data (see case study below). Only broad regional trends will be detected by a wide-spaced grid. Too close a grid spacing will admit local variations which may be demonstrated to be divorced from any regional trend.

Tests for significance

Several methods exist whereby the statistical significance of a trend-surface may be examined. The percentage reduction in the total sum of squares accounted for by the fitted surface (RSS) is one such measure. It is a gross estimate of efficiency based on the overall goodness-of-fit of the trend-surface. The RSS value increases as the order of the surface increases, though additional powers tend to make a progressively smaller difference to the RSS value. Indeed, in the region of the sixth- to the eighth-order surface increasing order may lead to a reduction in the RSS value. Analysis of data generated to be random both in location and magnitude has shown that if the RSS values fall below 6·0, 12·0 and 16·2% for the linear, quadratic and cubic surfaces respectively, then the distribution of the observed values may be random and no underlying trend should be implied (Howarth 1967). Also, the number of data points has an influence on the significance of the RSS value. For example, high-order surfaces (e.g. third-order and above) based on as few as 25 data points need an RSS value of at least 30% before they can be accepted (Unwin 1970). The use of average values as the observed data values will produce a much higher RSS value than actual values will. This is because the variability of the data is reduced by taking averages. Their use is therefore not justified on statistical grounds (Link and Koch 1962), though on geological or even geomorphological grounds the use of averages may be desirable (Whitten 1961A). Another measure of the overall goodness-of-fit is the standard deviation of the residuals (deviations) (Batcha and Reese 1964).

It is possible to generate, with a computer, upper and lower confidence surfaces about a trend-surface (Krumbein 1963). For example, if confidence surfaces corresponding to a significance level of 95% are fitted then the true regression surface will lie between the upper and lower confidence surfaces 95 times out of 100. In most cases the data and deviation values do not satisfy the conditions implicit in the use of this technique. That is to say, since the sample variance of the deviations is used to calculate the confidence surfaces, it has to be assumed that the deviations are both random (i.e. show no autocorrelation) and are normally distributed. As has already been implied, few cases exist when the deviations are uncorrelated. It is important to know the nature of the underlying forces which produce the regional trend when the upper and lower confidence surfaces are far apart, for only then can meaning be discerned in the trend-surface. Where the confidence surfaces are close to the trend-surface then there may be grounds for considering the use of the trend-surface for predicting unknown values (Krumbein 1963).

The separate trend functions S_{data} and S_{trend}, referred to above, can be used as a basis for an analysis of variance test which assesses the statistical significance of a trend component (Harbaugh and Merriam 1968). The source of variation has to be divided into two parts, S_{trend} and $S_{deviations}$ (table 9.1). The number of degrees of freedom of S_{trend} is the number of terms in the trend component (m), and for S_{dev} is the number of data points (n) minus ($m - 1$).

Table 9.1 *Data for analysis of variance*

Source of variation	Df	Sums of squares	Mean square	Ratio of mean squares
Trend function	m	S_{trend} $= (Z_{\text{calc}} - \bar{Z}_{\text{obs}})^2$	$M_1 = \dfrac{S_{\text{trend}}}{m}$	
				$F = \dfrac{M_1}{M_2}$
Deviations	$n - m - 1$	S_{dev} $= (Z_{\text{obs}} - Z_{\text{calc}})^2$	$M_2 = \dfrac{S_{\text{dev}}}{(n - m - 1)}$	
Total	$n - 1$	S_{data} $= (Z_{\text{obs}} - \bar{Z}_{\text{obs}})^2$		

Other significance tests are described by Mandelbaum (1963) and Chayes (1970). Yet another test, and perhaps the least complex, for detecting the surface which best fits the data was suggested by Agterberg (1964). If two equal-sized sub-samples of the observations are used in separate trend-surface analyses, any similarity in the surfaces produced implies that a true trend in the data has been discerned. When the difference between the two sub-sample surfaces at each order comes closest to zero then that is the order of surface which will best fit the data. The advantage in this type of test for the trend-surface which has the best fit is that no assumptions are made which imply statistical independence of the deviations.

The similarity of any two trend-surfaces may also be assessed by reference either to the coefficients of their equations or to the computed Z values. Merriam and Sneath (1966) describe a method of cluster analysis based on the coefficients of the trend-surface equation. Mandelbaum (1966) on the other hand makes the comparison between surfaces by defining a correlation:

$$r = \frac{(x_{ic} - \bar{x}_{ic})(y_{ic} - \bar{y}_{ic})}{n\, Sx_{ic}Sy_{ic}}$$

where x_{ic} and y_{ic} are the computed values at any particular locality on the two different surfaces. A high r value means a good correspondence in the configuration of the two surfaces.

Other methods

Modified forms of polynomial trend-surface analysis have been proposed and used in order to try to overcome some of the difficulties outlined above. For example Kridge (1964; 1966) uses a weighted moving average technique (see also Esselaar 1969). Miesch and Connor (1967; 1968) indicate a stepwise regression method by which the terms used in the regression equation can be 'selected individually according to their effectiveness in reducing the total sum of squares in the dependent variable and dropped from the equation individually if they are not effective or if they are redundant' (1968, 1–2). Equally well the most efficient terms may not be simple polynomials but include square root, exponential, logarithmic or reciprocal terms. The inclusion of non-polynomial terms makes the trend model less sensitive to sampling and analytical errors in the data or to rounding errors in computation. The published program for this technique does, however, require (even in a modified form) a computer with more than 100K store).

More work has been done not so much with different types of polynomial surfaces as with double Fourier series models. These have made it possible to study periodic (or wave) patterns in the data (James 1966A, B; Krumbein 1966A, B, 1967; Harbaugh and Merriam 1968).

This present study is confined, however, to an examination of the polynomial trend-surface model, and in particular its use in the analysis of planation surfaces.

Applications in geomorphology

Although trend-surface analysis had been used for some years in geology, its use in geomorphology did not begin until 1956 when Svensson suggested its application both to the analysis of planation surfaces and to synchronous shorelines of older sea levels (isobases). Since then these are the two fields, within the context of spatial analysis in geomorphology, in which the application of trend-surface analysis has been most widespread. In general its application in terms of isobase studies has met with greater success than is the case with planation surfaces. This stems from several factors chief amongst which are:

1. Isobase data is more accurate than planation surface data. Low-level raised shorelines have a well-defined back which can be accurately located and its height accurately measured in the field. Planation surfaces have a greater local relief, and height values have to be obtained from published maps which, at best, allow height definition to within ± 25 ft.

2. Unlike work on raised shorelines, planation surfaces themselves have nearly always, in the case of trend-surface analysis studies, been defined from map evidence (e.g. spur flats, enclosed summit contours) and not from field mapping.

Both of these differences mean that planation surface data has been more 'noisy' than isobase data, not least because data collection has been much less accurate.

Isobases

A typical example of an isobase study, within the British Isles, is that by Smith, Sissons and Cullingford (1969). The height of the back of the main Perth raised shoreline was established to 0·1 ft (0·305 m), and its location to the nearest 10 m, at 500 data points. This raised beach occurs in a number of localities separated, for example, by wide estuaries. Trend-surface analysis suggested no grounds for separating the measured beaches into two or more different levels. The data were thus used to test the theory that emergence was in the form of an elliptical isostatic dome. The important point here is that the analysis was carried out *to test* two hypotheses initially conceived through a geomorphological analysis of the field data. This is in contrast to some uses of the technique as a hypothesis-generating procedure.

In a full study of isobase data for Arctic Canada Andrews (1970) freely used trend-surface analysis. Again, however, the emphasis was on model testing. In one case a quadratic trend-surface did not conform to a cartographic and geomorphological assessment of field data, despite a high residual sum of squares value and a high F-ratio. 'This example shows how statistical significance must be appraised and the decision of the surface's relevance made on the basis of the field worker's knowledge' (Andrews 1970, 102).

The point is also made (Smith *et al.* 1969; Andrews 1970) that isobase data do not always conform to the spacing conditions required for the valid application of trend-surface analysis. This is unavoidable, however, as data only exist (i.e. along the line of the former shore) where remnants of the raised beach may now be found.

Planation surfaces

Recent attempts at the application of trend-surface analysis to the study of planation surfaces (Thornes and Jones 1969; King 1969B; Beaumont 1970; and Rodda 1970) have been criticised (Tarrant 1970; Unwin and Lewin 1971); so much so that the very exercise has almost fallen into disrepute. The arguments raised against these particular studies are certainly justified, but to completely abandon the technique as a useless one, in the analysis of planation surfaces, is not.

One common failing in these studies arises from the method of data collection. By the end of the period between 1935 and 1960, when the definition of planation surfaces around Britain was in vogue, no one would have committed his work to print if his only source of information was closed summit contours, or spot heights, as defined by Ordnance Survey maps. It was unanimously agreed that the only source of valid evidence was systematic field mapping. A good example of this type of work is presented by Brown (1960), in which he includes a coloured map defining the position and relative ages of the main planation surfaces of Wales. This analysis resulted from intricate geo-morphological arguments calling upon many lines of evidence. The results of this work could not have been obtained by the cruder methods of cartographic analysis. On the contrary, it provides an invaluable set of data to which trend-surface analysis might be applied to test regional trends or in analysing local deviations. It is somewhat surprising, there-fore, that a trend-surface analysis of a part of the area covered by Brown should revert to data in the form of summit heights and estimates of the highest point in all enclosed contours, collected from maps at a scale of 1/63,360 (Rodda 1970). Nor can the actual application of trend-surface analysis, as carried out in this case, be justified for an important misconception is summed up by: 'It [i.e. *trend-surface analysis*] provides a simple quantitative means for testing the validity and relevance of multiple-surface hypotheses and it can of course, reveal surfaces that have hitherto been unrecognised . . . where the planation surfaces that have been recognised are both numerous and disputed trend-surface analysis presents a judicious method of reaching a final solution' (Rodda 1970, 107). This simply is not the case. This misconception arises from the notion that the trend-surface is a slightly generalised replica of the planation surface, which it is not. It is a mathematical statement of a trend in the data values. Rodda also deduces, in this examination of a part of Wales, that separate surfaces can be recognised because distinct belts of positive and negative deviation values exist. In other words autocorrelation remained within the deviations. The case study to be presented below uses the same line of evidence to support a con-cept of the warping of a single surface. Trend-surface analysis *alone* cannot decide between multiple surfaces and a warped single surface.

King (1969B) suggests that trend-surface analysis may be used in one of four ways: (i) description, (ii) reconstruction, (iii) testing a hypothesis, (iv) as a search procedure. Of these, in the case of the analysis of plana-tion surfaces, only the third is entirely appropriate. Its failure as a descriptive technique is mentioned above and criticised by Tarrant (1970). Likewise, because the trend-surface is not a replica of the planation surface it is unlikely to be an accurate method either of surface reconstruction or of searching for (i.e. predicting the height of) other surface remnants. In some ways these statements can be statis-tically qualified, as is the case with all multiple regression equations of which the trend-surface equation is but one kind. However, it may be a good thing to abandon the application of trend-surface analysis to the study of planation surfaces if the object of the study falls under one of these three headings. There remains, then, its use as a hypothesis-

(or model-) testing device. This approach can be justified, but only on the grounds that:

1. The data conforms as closely as possible to the conditions required for the valid application of the technique.
2. The data is based on field mapping, and if possible on height data accurately measured.
3. The model to be tested is based on geomorphological evidence in addition to that supplied by the planation surface remnants themselves.

Uganda case study

The purpose of the case study is to see if trend-surface analysis can assist in analysing the nature and intensity of the warping which has taken place on the eastern flank of the Western Rift Valley in southern Uganda. A brief review of the geomorphology of the area is necessary as it establishes the model to be tested. The nature of the available data, for trend-surface analysis, is also defined. This data is then used to test the model.

Background geomorphology

Two planation surfaces, the Upland and the Lowland Landscapes, exist in southern Uganda, and both of these have been warped by post-oligocene rift tectonics (Doornkamp and Temple 1966). Since the Upland Landscape is more distorted than the Lowland Landscape some warping is assumed to have taken place between the formation of these two planation surfaces.

The two planation surfaces are fully described by Doornkamp (1970), but it is pertinent to reiterate some of their properties. The Upland Landscape (Buganda Surface) has been recognised because of its low relative relief on mountain and hill summits, and because of its blatant disregard for rock structure and lithological boundaries. It has been found from Lake Victoria to the rift edge by mapping across the whole region, and by tracing it from hill crest to hill crest. Similarly the Lowland Landscape displays very shallow local relief, usually less than 200 ft, which may be compared with its vast extent, mapped here for over 5000 square miles. This lower planation surface is also widespread from Lake Victoria to the rift.

Field mapping has shown that not only do these surfaces rise in height westwards, but they also diverge westwards (Doornkamp and Temple 1966). Close to Lake Victoria their vertical separation is only 300 ft, whereas close to the position of the axis of upwarping (see below) their separation exceeds 1000 ft. This warping, which was a

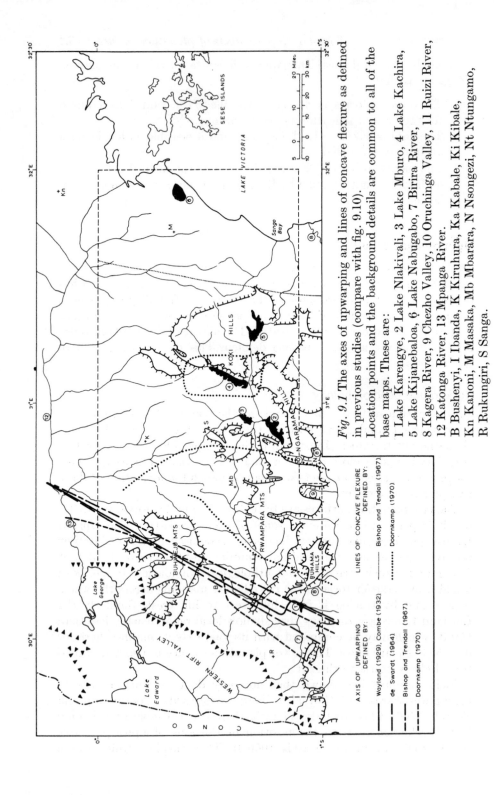

Fig. 9.1 The axes of upwarping and lines of concave flexure as defined in previous studies (compare with fig. 9.10). Location points and the background details are common to all of the base maps. These are:

1 Lake Karengye, 2 Lake Nlakivali, 3 Lake Mburo, 4 Lake Kachira, 5 Lake Kijanebaloa, 6 Lake Nabugabo, 7 Birira River, 8 Kagera River, 9 Chezho Valley, 10 Oruchinga Valley, 11 Ruizi River, 12 Katonga River, 13 Mpanga River.
B Bushenyi, I Ibanda, K Kiruhura, Ka Kabale, Ki Kibale, Kn Kanoni, M Masaka, Mb Mbarara, N Nsongezi, Nt Ntungamo, R Rukungiri, S Sanga.

AXIS OF UPWARPING
DEFINED BY:

—— Wayland (1929), Combe (1932)
– – – de Swardt (1964)
–·–·– Bishop and Trendall (1967)
– – – Doornkamp (1970)

LINES OF CONCAVE FLEXURE
DEFINED BY:

········ Bishop and Tendall (1967)
········ Doornkamp (1970)

part of the tectonic activity associated with the formation of the Western Rift Valley, led to the reversal of the drainage of western Uganda. From an analysis of the reversed drainage systems it has been generally assumed that an axis of upwarping (fig. 9.1) exists to the east of, and running parallel to, the rift edge scarp (Wayland 1929; Combe 1932; de Swardt 1964). It is probable that upwarping was most intense in the extreme south-west (Combe 1932). A variety of positions have been suggested for this main axis. Wayland (1929, 1934A, B) placed it along the line which passes through the swamp-divides of the reversed Kagera and Katonga Rivers. Combe (1932) agrees with this, and stated quite precisely that the country rose along a straight line on a bearing of N 26° 30′ E to S 26° 30′ W and that from this line of upwarp parts of the rivers were reversed. This interpretation has not been questioned, although a map by de Swardt (1964) shows this axis to lie a little west of the Kagera–Birira swamp-divide, rather than passing through it. Bishop and Trendall (1967) place an axis close to that of the earlier workers. An analysis of generalised contours constructed for both the Upland and Lowland Landscapes led Doornkamp (1970) to suspect that the axis of upwarping did not have a simple trend nor that it was quite so conveniently located with respect to the swamp-divides. The work of Bishop and Trendall (1967) and Doornkamp (1970) has also suggested that flexures occur in the back-tilted landscape between the main axis of upwarping and Lake Victoria, to the east (fig. 9.1).

The model

The model suggested by the geomorphological evidence cited above is summarised in fig. 9.2. On to this diagram (fig. 9.2D) has been inserted a hypothetical linear trend-surface to show how it will lie in relation to the supposed form of the warped planation surfaces. A linear trend-surface will, by definition, pass through the data values leaving some points above and others below. The greater the distance between the computed trend surface and the planation surface remnant the greater will be the calculated deviation value. In the present context the highest positive deviations will occur where upwarping has been most intense. Conversely where relative downwarping has occurred the computed deviation will be large and negative. The emphasis in this case study is on an interpretation of these computed deviation values.

The fitting of a linear trend-surface to any planation surface will produce positive and negative deviations. This is the case both because of the definition of a trend-surface and because planation surfaces have some relief, even though it be small. High points on an unwarped planation surface coincide, for example, with the position of divides and areas of more resistant bedrock. In terms of the recognition of

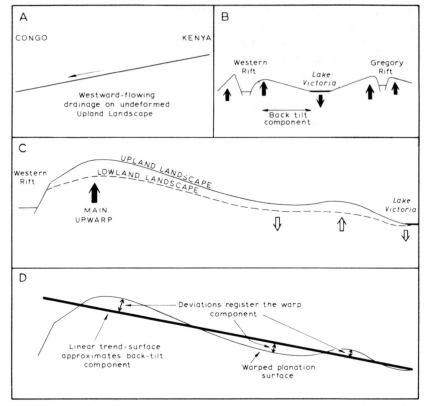

Fig. 9.2 A model for the deformation of the planation surfaces of southern Uganda.

A. Original surface (Upland Landscape) sloping from east to west.

B. Rift valley formation distorts the original surface producing a back-tilt component between an axis of upwarping, close to the rift scarp, and Lake Victoria.

C. Between the Western Rift Valley and Lake Victoria both of the planation surfaces are warped, and have a maximum separation at the main axis of upwarping.

D. Hypothetical relationship between a linear trend-surface and either of the two warped planation surfaces.

warping it is necessary first of all to eliminate features of denudation from the analysis.

Thus, it has to be assumed that upwarping cannot be invoked to explain deviation values where:

1. The computed deviation is less than the normal amplitude of relief;
2. Positive deviations coincide with the position of a divide;

3. Positive deviations coincide only with an area of locally more resistant bedrock.

Again in the present context the scale of warping envisaged is of regional extent. No upwarping will be invoked, therefore, where

4. Positive deviations occur in a local cluster.

Conversely, the possibility of regional warping has to be taken seriously if deviations:

1. Have a value greater than that of the amplitude of the planation surface;
2. Run oblique to the divides;
3. Occur regardless of lithological differences;
4. Have considerable lateral extent.

Collecting the data

The area to be analysed, indicated by the inner broken boundary in fig. 9.1, lies within a larger mapped area (i.e. the whole of the area shown in fig. 9.1). Thus data points occur outside the area of main interest, and this helps to control the performance of the computed trend-surface within the area of particular interest. The whole of the mapped area was covered by a grid whose intersections were spaced at 2-mile (3·2-km) intervals. This spacing 'caught' most of the planation surface remnants within the data, and grid intersections seldom crossed one remnant more than once. The 2-mile (3·2-km) interval is much less than the wavelength of the tectonic deformations being sought. The height of the planation surfaces, as shown on the 1/50,000 topographic base maps, were noted at each point where a grid intersection coincided with the surface. Averages were not taken since the use of averages reduces the local variations from the main trend and gives inflated RSS values (Whitten 1959). Both planation surfaces have some local relief, and this was apparent from the absolute height values recorded against the grid intersections. These local variations, however, are on a much smaller scale (i.e. 200–300 ft (60–90 m)) than those being sought on the regional scale (i.e. 500–1000 ft (150–300 m)). Clear examples of monadnocks, residual hills or hill summits, were omitted in collecting the data. The horizontal amplitude of variations in height are less than the assumed regional amplitude of any periodicity in tectonic deformation zones. The former reflect the relief on the planation surface, the latter are related to gross contortions of the planation surfaces. Indeed, the whole purpose of the use of trend-surface analysis, is to generalise out the local variations in order to be able to establish the main regional variations.

The actual distribution of the available data depends on the present location of preserved remnants of the planation surfaces. The geomorphologist has no control over this. In south-west Uganda this led to a patchy, sometimes clustered, distribution (figs. 9.3 and 9.4). For a part of the analysis therefore, a number of samples of the data were collected on a random basis, so that clustering would have no influence on the results. Points were selected for inclusion by using a table of random numbers to select 10 grid intersects within 20 × 20 miles (32 × 32 km) squares within the superimposed grid.

Data analysis

Mathematical assumptions

Initially the data were analysed to establish how well it fulfilled the mathematical assumptions outlined on page 249. Non-clustering was achieved by sampling, as described above. However, it was found that not only were each of the trend-surfaces obtained from each sample, very similar to each other, they were also very similar to that obtained from using all the data in its somewhat clustered form. This is truer in the case of the Lowland Landscape (fig 9.3, table 9.2A) than in the case of the Upland Landscape (fig. 9.4, table 9.2B). In both cases the trend of the surface between samples is the same but in the case of the Upland Landscape there is a difference in the gradient of the sample surfaces. The surface for all the data in fig. 9.4 adopts

Table 9.2   *A comparison of the equations of the linear trend-surfaces for sample and total data*

Sample	Linear trend-surface
A. Lowland Landscape	
1	$Z = 524 \cdot 8 - 1 \cdot 78X - 0 \cdot 139Y$
2	$Z = 529 \cdot 8 - 1 \cdot 68X - 0 \cdot 317Y$
3	$Z = 526 \cdot 7 - 1 \cdot 71X - 0 \cdot 193Y$
4	$Z = 533 \cdot 7 - 1 \cdot 85X - 0 \cdot 185Y$
5	$Z = 529 \cdot 8 - 1 \cdot 70X - 0 \cdot 302Y$
all data	$Z = 525 \cdot 1 - 1 \cdot 79X - 0 \cdot 059Y$
B. Upland Landscape	
1	$Z = 750 \cdot 86 - 4 \cdot 63X - 0 \cdot 671Y$
2	$Z = 701 \cdot 51 - 3 \cdot 57X - 0 \cdot 934Y$
Random mixture of 1 and 2	$Z = 728 \cdot 30 - 4 \cdot 01X - 0 \cdot 935Y$
all data	$Z = 729 \cdot 52 - 4 \cdot 10X - 0 \cdot 889Y$

The co-ordinate origin for X and Y is in the south-west. Z is in feet.

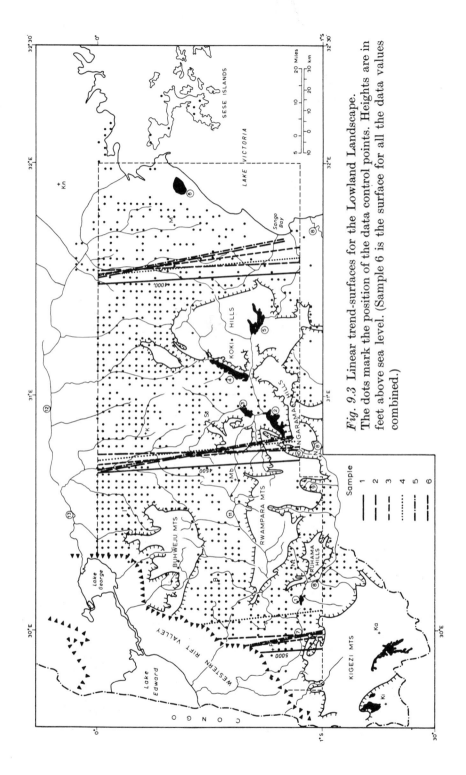

Fig. 9.3 Linear trend-surfaces for the Lowland Landscape. The dots mark the position of the data control points. Heights are in feet above sea level. (Sample 6 is the surface for all the data values combined.)

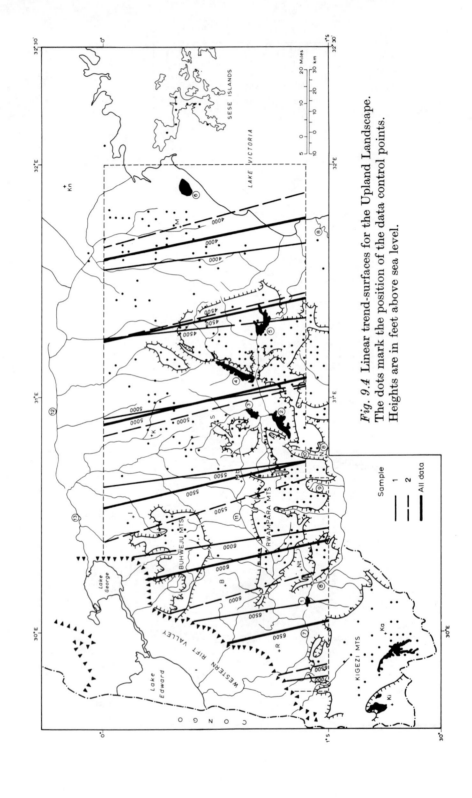

Fig. 9.4 Linear trend-surfaces for the Upland Landscape. The dots mark the position of the data control points. Heights are in feet above sea level.

an intermediate position between the linear trend-surfaces for the two samples. The basic similarity between the sample surfaces (figs. 9.3 and 9.4) implies, for both the Upland and Lowland Landscape, that the computed trend-surfaces obtained are meaningful. It also implies that the analysis of both the trend-surfaces and the computed deviations may sensibly be made, in this case, from the results of running

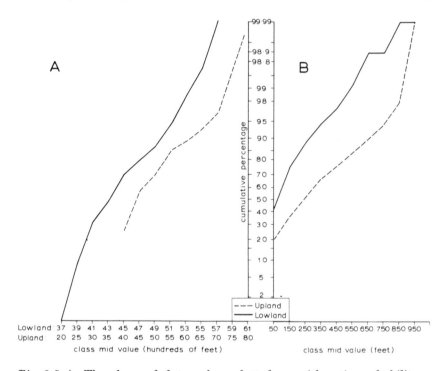

Lowland: 37 39 41 43 45 47 49 51 53 55 57 59 61
Upland: 20 25 30 35 40 45 50 55 60 65 70 75 80
class mid value (hundreds of feet)

50 150 250 350 450 550 650 750 850 950
class mid value (feet)

Fig. 9.5 A. The observed data values plotted on arithmetic probability paper.
B. The deviation values plotted on arithmetic probability paper.

all the data together without considering the separate non-clustered samples.

The small departure from a completely normal distribution of the observed data values, for both planation surfaces, can be shown by plotting them on cumulative probability paper (fig. 9.5A). The closeness to normality in many planation surface studies may not be achieved as it is in this one. In a normal distribution the highest number of values occur in the middle of the total range of height data. The very fact that warping has taken place may make this the case here, whereas an unwarped surface may have a regular number of observations in each height class.

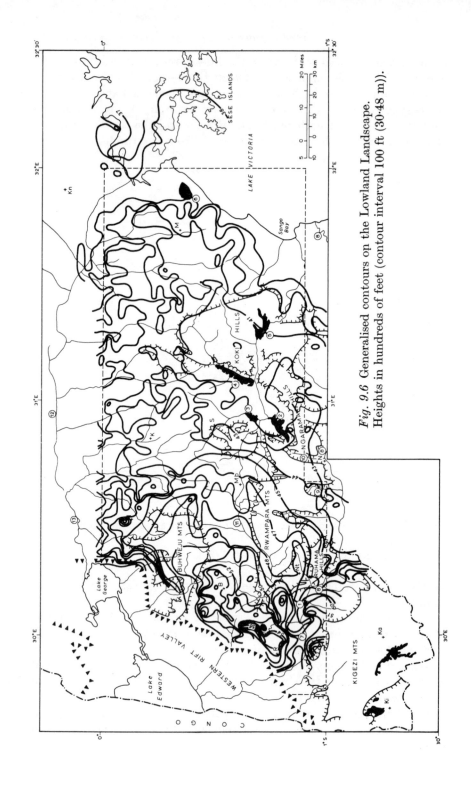

Fig. 9.6 Generalised contours on the Lowland Landscape. Heights in hundreds of feet (contour interval 100 ft (30·48 m)).

Fig. 9.7 Generalised contours on the Upland Landscape.
Heights in hundreds of feet (contour interval 200 ft (60·96 m)).

Generalised contours

The data, in summary form, may be represented by generalised contours (figs. 9.6 and 9.7), based on the same data as that used in the trend-surface analysis (i.e. height at grid intersections). It could be argued that this representation is sufficient in the search for major areas of tectonic deformation. Indeed, this is done in an earlier account (Doornkamp 1970), although in that case the contours were drawn from the data arranged in a slightly different way. Part of the purpose of this account is to see if any more can be learned from a trend-surface analysis. For example, a glance at fig. 9.7 shows that close to the rift valley the Upland Landscape is at its highest and the contours have a strong trend. In the extreme south-west the altitudes reach a peak which is flanked by a sharp decline. These patterns suggest a coincidence with an axis of upwarping in the one case and with a domed uplift in the other. But, what of the country further east? Detailed irregularities in the generalised contours make their interpretation in this area much more difficult.

Trend surfaces

The linear trend-surfaces (figs. 9.3 and 9.4) show that, in conformity with the model outlined above, both the Upland and Lowland Landscapes rise westwards from Lake Victoria towards the Western Rift Valley. The higher surface does this the more steeply of the two, confirming the view that they diverge westwards.

Analysis of deviations

In searching for the main areas of warping the model (fig. 9.2) shows how it is an analysis of deviations that is most useful. The implication is that autocorrelation must exist (i.e. large groups of positive or negative deviations) if tectonic warping is the prime interest. This specifically runs counter to the conditions necessary for the derivation of a significant trend-surface. In this study the trend-surface is only used as an inclined reference plane which has removed the major trend, but which by virtue of the value of the deviations from it, can assist in detecting the warped areas. The deviations are, however, almost normally distributed (fig. 9.5B).

Freehand contouring of the deviations (figs. 9.8 and 9.9), shows that there are some extensive areas of grouped positive and negative deviations. Their significance has to be interpreted in the light of the principles listed on page 249. On this basis shading has been introduced on figs. 9.8 and 9.9 to indicate which areas may be isolated as those most likely to yield evidence concerning warping.

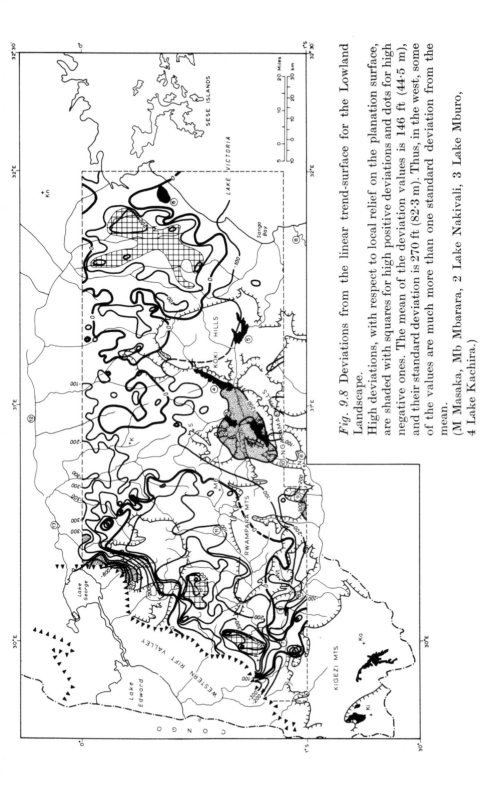

Fig. 9.8 Deviations from the linear trend-surface for the Lowland Landscape.

High deviations, with respect to local relief on the planation surface, are shaded with squares for high positive deviations and dots for high negative ones. The mean of the deviation values is 146 ft (44·5 m), and their standard deviation is 270 ft (82·3 m). Thus, in the west, some of the values are much more than one standard deviation from the mean.

(M Masaka, Mb Mbarara, 2 Lake Nakivali, 3 Lake Mburo, 4 Lake Kachira.)

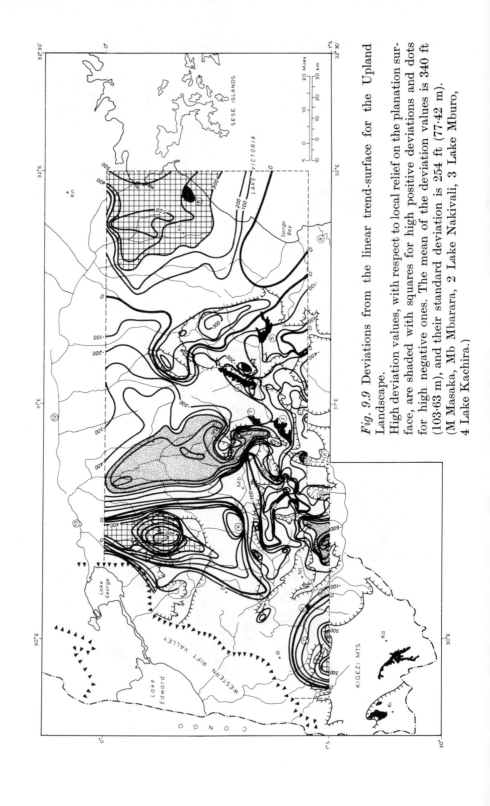

Fig. 9.9 Deviations from the linear trend-surface for the Upland Landscape.

High deviation values, with respect to local relief on the planation surface, are shaded with squares for high positive deviations and dots for high negative ones. The mean of the deviation values is 340 ft (103·63 m), and their standard deviation is 254 ft (77·42 m). (M Masaka, Mb Mbarara, 2 Lake Nakivali, 3 Lake Mburo, 4 Lake Kachira.)

The two sets of deviations (figs. 9.8 and 9.9) are similar in the following respects:

1. High positive deviations occur 10–20 miles east of the rift scarp, and they are set within an elongated belt of positive deviations running parallel to the rift scarp.

2. High negative deviations occur south-east of the Ruhama Hills. In the case of the Upland Landscape this is part of a belt of negative deviations swinging northwards through the Ngarama Hills to the area north of Mbarara. A subsidiary area of high negative deviations occurs at the southern end of Lake Kachira. In the case of the Lowland Landscape the three lakes Kachira, Mburo and Nakivali are seen to coincide with an area of high negative deviations.

3. High positive deviations occur north of Masaka. In the case of the Upland Landscape these are part of a belt trending more or less parallel to the shore of Lake Victoria. The Lowland Landscape indicates high positive deviations are restricted to the north-western corner of Lake Victoria (within the area demarcated by a broken line in fig. 9.8).

The high negative deviations, in the case of the Upland Landscape, north-west of the Buhweju Mts probably represents a rapid downwarp towards the rift scarp. Other deviations appear to be of local significance only, or coincide either with major divides or changes in bedrock lithologies.

Location of axis of warping

Interpretation of the high deviations must be sought in the light of existing knowledge of the geomorphology of the area. An interpretation in terms of tectonic deformation is compatible with the known evolution of the area. The location of the main zones of tectonic deformation, as suggested by this trend-surface analysis, is summarised in fig. 9.10. The strong corroborative evidence, for the existence and position of these zones and axes of warping, lies in their identification during the analysis of each of the two planation surfaces. Conversely this implies that both planation surfaces were deformed. Slight differences in the position of the main axis of unwarping just east of the rift scarp, suggested by the separate analysis of each of the two planation surfaces, may result from a real change in its position between the warping of the upper and lower surfaces. It is perhaps more likely, however, that this change in position reflects the difference in the positions of the data control points.

Intensity of warping

From the cross-sections in figs. 9.11 and 9.12 estimates can be made concerning the intensity of warping of each of the surfaces. The original model (fig. 9.2) showed how the linear trend-surface might be used as

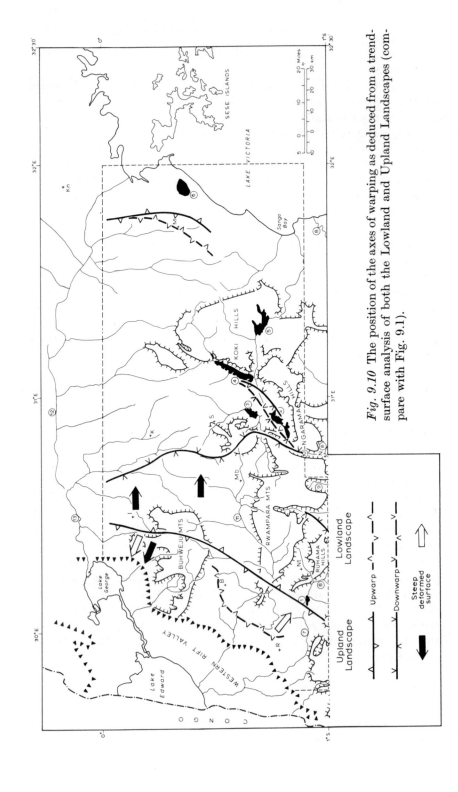

Fig. 9.10 The position of the axes of warping as deduced from a trend-surface analysis of both the Lowland and Upland Landscapes (compare with Fig. 9.1).

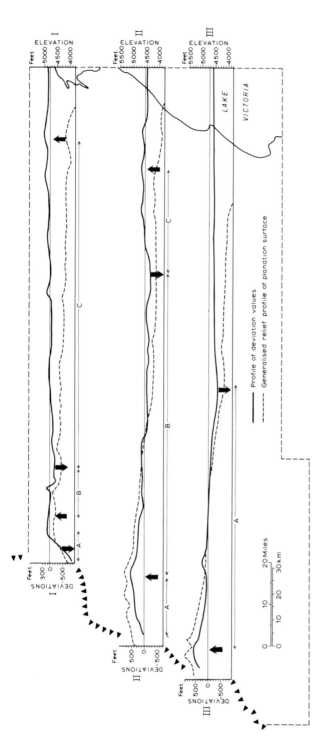

Fig. 9.11 Cross-sectional profiles of deviation values and generalised relief along the base-line positions, on the Lowland Landscape. The arrows indicate the inferred zones of warping. The gradients along each of the marked sections are listed in table 9.3.

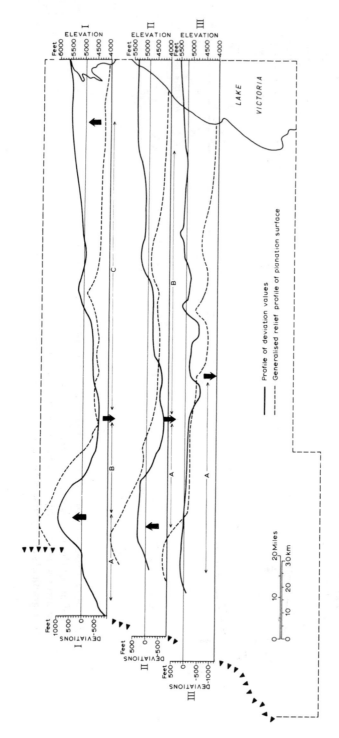

Fig. 9.12 Cross-sectional profiles of deviation values and gener-
alised relief along the base-line positions, on the Upland Landscape.
The arrows indicate the inferred zones of warping.
The gradients along each of the marked sections are listed in table 9.3.

—— Profile of deviation values
------ Generalised relief profile of planation surface

a means of isolating the tilt component, while the deviations could be used to identify the main zones of warping. The gradient of the tilt component for the Lowland Landscape is 9 ft/mile (1·71 m/km) (from fig. 9.3) while that for the Upland Landscape is 21 ft/mile (3·98 m/km) (from fig. 9.4). To these absolute values there needs to be added the original westward gradient of the planation surfaces if the total amount of tilt is to be calculated (Doornkamp and Temple 1966).

The planation surfaces are locally more distorted, however, than these values suggest because of the warping component of the tectonic influences. The profiles in figs. 9.11 and 9.12 can be subdivided into units coincident with the flanks of the individual deformations recognised above. The gradients of the units are listed in table 9.3. Two values are

Table 9.3 *Local generalised gradients of the deformed planation surfaces* (obtained from figs 9.11 and 9.12)

Profile	Unit	Gradient from generalised contours		Gradient of deviations	
		ft/ml	m/km	ft/ml	m/km
Upland Landscape (*see fig. 9.12*)					
I	A	+85·7	+16·23	+81·25	+15·39
	B	−84·62	−16·03	−64·00	−12·12
	C	—	—	+17·14	+3·25
II	A	−54·55	−10·33	−40·00	−7·58
	B	—	—	+16·67	+3·16
III	A	−33·33	−6·31	−15·56	−2·95
Lowland Landscape (*see fig. 9.11*)					
I	A	+83·3	+15·78	+160·00	+30·31
	B	−50·00	−9·47	−33·33	−6·31
	C	—	—	+6·67	+1·26
II	A	+25·00	+4·74	+33·33	+6·31
	B	−17·14	−3·25	−10·00	−1·89
	C	+5·56	+1·05	+28·00	+5·30
III	A	−20·00	−3·79	−15·00	−2·84

+ indicates rise from W to E − indicates rise from E to W

necessary. That which is derived from an analysis of the generalised contours (figs. 9.6 and 9.7) is the overall gradient of the planation surface remnants *as they occur* in the landscape, and include the tilt component. The gradients derived from the deviation values (i.e. those of figs. 9.8 and 9.9) are a measure of the amount of surface distortion, and have the tilt component removed. All of these values are greater than those previously advanced for this area (Bishop and Posnansky 1960; Spurr 1955; Bishop 1966; Pallister 1956; de Swardt 1964; Doornkamp and Temple 1966; Doornkamp 1970).

Conclusion

Trend-surface analysis may usefully be applied to the analysis of planation surfaces so long as the data are in a form which is compatible with the use of this technique. Not only does this imply a numerical examination of the data, it also means that field evidence and not cartographic evidence alone, should be used in obtaining the data. Applications should relate to the testing of a model which can be substantiated on geomorphological grounds. In the case study presented it has been possible to test and confirm the validity of a model of planation surface deformation resulting from rift valley tectonics in southern Uganda. At the same time absolute values are suggested which indicate that surface deformation may have been considerably more intense than had previously been recognised. The application of trend-surface analysis in this case provides a valuable method of measuring the scale of warping achieved by post-oligocene tectonics on the flank of a rift valley.

Acknowledgements

The fieldwork for this study was generously supported by a grant from Makerere University College, Uganda. The computer program was written by M. J. McCullagh.

Bibliography

This bibliography includes not only those works specifically referred to in the text but also some of the others which are particularly useful contributions to the subject.

AGOCS, W. B. (1951) Least squares residual anomaly determination; *Geophysics* 16, 686–96.

AGTERBERG, F. P. (1964) Methods of trend surface analysis; *Colorado School of Mines Quarterly* 59, 111–30.

AGTERBERG, F. P. (1967) Computer techniques in geology; *Earth Science Reviews* 3, 47–77.

ALLEN, P. and KRUMBEIN, W. C. (1962) Secondary trend components on the Top Ashdown Pebble Bed: a case history; *Journal of Geology* 70, 507–38.

ANDREWS, J. T. (1970) A geomorphological study of post-glacial uplift with particular reference to Arctic Canada; *Institute of British Geographers, Special Publication* 2, 156p.

BATCHA, J. P. and REESE, J. R. (1964) Surface determination and automatic contouring for mineral exploration, extraction, and processing; *Colorado School of Mines Quarterly* 59(4), 1–14.

BEAUMONT, P. (1970) Geomorphology; In *Durham County and City with Teesside* (British Association, Durham 1970), 26–45.

BISHOP, W. W. (1966) Stratigraphical geomorphology; In Dury, G. H. (Ed.), *Essays in Geomorphology* (Heinemann, London), 139–76.

BISHOP, W. W. and POSNANSKY, M. (1960) Pleistocene environments and early man in Uganda; *Uganda Journal* 24, 44–61.

BISHOP, W. W. and TRENDALL, A. F. (1967) Erosion surfaces, tectonics and volcanic activity in Uganda; *Quarterly Journal of the Geological Society of London* 122, 385–420.

BOON, J. D. (1968) Trend surface analysis of sand tracer distributions on a carbonate beach, Bimini, B.W.I.; *Journal of Geology* 76(1), 71–87.

BROWN, E. H. (1960) *The Relief and Drainage of Wales* (University of Wales Press, Cardiff), 186p.

CHAYES, F. (1970) On deciding whether trend surfaces of progressively higher order are meaningful; *Bulletin of the Geological Society of America* 81(4), 1273–8.

CHORLEY, R. J. (1964) An analysis of the areal distribution of soil size facies on the Lower Greensand rocks of East Central England by the use of trend surface analysis; *Geological Magazine* 101(4), 314–21.

CHORLEY, R. J. and HAGGETT, P. (1965) Trend-surface mapping in geographical research; *Transactions of the Institute of British Geographers* 37, 47–67.

CHORLEY, R. J., STODDART, D. R., HAGGETT, P. and SLAYMAKER, H. O. (1966) Regional and local components in the areal distribution of surface sand facies in the Breckland, eastern England; *Journal of Sedimentary Petrology* 36, 209–20.

COMBE, A. D. (1932) The geology of south-west Ankole and adjacent territories with special reference to the tin deposits; *Memoir Geological Survey of Uganda* No. 2.

DAVIS, J. C. (1967) Application of response-surface analysis to sedimentary petrology; In Merriam, D. F. and Cocke, H. C. (Eds.), Colloquium on trend analysis; *Kansas Computer Contribution* No. 12, 57–62.

DAVIS, J. C. (1969) Distribution of hydrocarbons in three dimensions; In *Kansas Computer Contribution* No. 40, 34–40.

DILLON, E. L. (1967) Expanding role of computer in geology; *Bulletin of the American Association of Petroleum Geology* 51, 1185–202.

DOORNKAMP, J. C. (1970) The geomorphology of the Mbarara area (Sheet SA-36-1); *Geomorphological Report* No. 1 (Department of Geography, University of Nottingham and Geological Survey and Mines Department, Uganda), 78p.

DOORNKAMP, J. C. and TEMPLE, P. H. (1966) Surface, drainage and tectonic instability in part of southern Uganda; *Geographical Journal* 132. 238–52.

ESSELAAR, P. A. (1969) An introduction to polynomial, Fourier, and moving average techniques of trend surface analysis of geological data (Abs.); In

Colloquium on trend surface analysis in economic geology (Geological Society of South Africa, University of Witwatersrand, Economic Geology Research Unit, Johannesburg), p. 3.

FORGOTSON, J. M. and IGLEHART, C. F. (1967) Current uses of computers by exploration geologists; *Bulletin of the American Association of Petroleum Geologists* 51, 1202–24.

GRANT, F. A. (1954) A theory for the regional correction of potential field data; *Geophysics* 19, 23–45.

GRANT, F. A. (1957) A problem in the analysis of geophysical data; *Geophysics* 22, 309–44.

GRISWOLD, W. T. and MUNN, M. J. (1907) Geology of oil and gas fields in Steubenville, Burgettstown, and Claysville Quadrangles Ohio, West Virginia and Pennsylvania; *United States Geological Survey Bulletin* 318, 1–194.

HARBAUGH, J. W. and MERRIAM, D. F. (1968) *Computer Applications in Stratigraphic Analysis* (Wiley, New York) (see 61–112).

HOWARTH, R. J. (1967) Trend-surface fitting to random data – an experimental test; *American Journal of Science* 265, 619–25.

JAMES, W. R. (1966A) The Fourier series model in map analysis; *Office of Naval Research, Geography Branch, Task No. 388–078, Contract Nonr 1228(36), Technical Report No. 1, Department of Geology, Northwestern University.*

JAMES, W. R. (1966B) Fortran IV program using double Fourier series for surface fitting of irregularly spaced data; *Kansas Computer Contribution No. 5.*

JAMES, W. R. (1967) Nonlinear models for trend analysis in geology; In Merriam, D. F. and Cocke, N. C. (Eds.), Colloquium on trend analysis; *Kansas Computer Contribution* No. 12, 26–30.

KING, C. A. M. (1969A) Glacial geomorphology and chronology of Henry Kater Peninsula, east Baffin Island, N.W.T.; *Arctic and Alpine Research* 1, 195–212.

KING, C. A. M. (1969B) Trend-surface analysis of central Pennine erosion surfaces; *Transactions of the Institute of British Geographers* 47, 47–59.

KRIDGE, D. C. (1964) Recent developments in South Africa of trend surface and multiple regression techniques to gold ore valuation; *Colorado School of Mines Quarterly* 59, 795–809.

KRIDGE, D. C. (1966) Ore value trend surfaces for the South African gold mines based on a weighted moving average; *Proceedings of the Symposium on Computers Operation Research in Mineral Industry*, Pennsylvania State University, 1, G1–29.

KRUMBEIN, W. C. (1956) Regional and local components in facies maps; *Bulletin of the American Association of Petroleum Geologists* 40, 2163–94.

KRUMBEIN, W. C. (1960) Stratigraphic maps from data observed at outcrop; *Proceedings of the Yorkshire Geo   ical Society* 32, 353–66.

KRUMBEIN, W. C. (1963) Confidence intervals in low-order polynomial trend surfaces; *Journal of Geophysical Research* 68, 5869–78.

KRUMBEIN, W. C. (1964) A geological process-response model for analysis of beach phenomena; *Annual Bulletin of the Beach Erosion Board* 17, 1–15.

KRUMBEIN, W. C. (1966A) A comparison of polynomial and Fourier models in map analysis; *Office of Naval Research, Geography Branch, Task No. 388–078, Contract Nonr 1228(36), Technical Report No. 2, Northwestern University.*

KRUMBEIN, W. C. (1966B) Classification of map surfaces based on the structure of polynomial and Fourier coefficient matrices; *Kansas Computer Contributions* No. 7, 12–18.

KRUMBEIN, W. C. (1967) The general linear model in map preparation and analysis; In Merriam, D. F. and Cocke, N. C. (Eds.), Colloquium on trend analysis; *Kansas Computer Contribution* No. 12, 38–44.

KRUMBEIN, W. C. and GRAYBILL, F. A. (1965) *An Introduction to Statistical Models in Geology* (McGraw-Hill, New York), 319–57.

KRUMBEIN, W. C. and IMBRIE, J. (1963) Stratigraphic factor maps; *Bulletin of the American Association of Petroleum Geologists* 47, 698–701.

KRUMBEIN, W. C. and SLOSS, L. L. (1963) *Stratigraphy and Sedimentation,* 2nd Edn (Freeman & Company, San Francisco).

LEE, P. J. and MIDDLETON, G. V. (1967) Application of canonical correlation to trend analysis; In Merriam, D. F. and Cocke, N. C. (Eds.), Colloquium on trend analysis, *Kansas Computer Contribution* No. 12, 19–21.

LINK, R. F. and KOCH, G. S. (1962) Quantitative areal modal analysis of granitic complexes: discussion; *Bulletin of the Geological Society of America* 63, 411–14.

MANDELBAUM, H. (1963) Statistical and geological implications of trend mapping with nonorthogonal polynomials; *Journal of Geophysical Research* 68, 505–19.

MANDELBAUM, H. (1966) Comments on a paper by Daniel F. Merriam and Peter H. A. Sneath; 'Quantitative comparison of contour maps'; *Journal of Geophysical Research* 71(18), 4431–2.

MCCANN, S. B. (1966) The main Post-glacial raised shoreline of Western Scotland from the Firth of Lorne to Loch Broom; *Transactions of the Institute of British Geographers* 39, 87–99.

MCCANN, S. B. and CHORLEY, R. J. (1967) Trend-surface mapping of raised shorelines; *Nature* 215, 611–12.

MCINTYRE, D. B. (1967) Trend-surface analysis of noisy data; In Merriam, D. F. and Cocke, N. C. (Eds.), Colloquium on trend analysis, *Kansas Computer Contribution* No. 12, 45–56.

MERRIAM, D. F. and HARBAUGH, J. W. (1964) Trend-surface analysis of regional and residual components of geologic structure in Kansas; *State Geological Survey, Kansas Special Distribution Publication* 11.

MERRIAM, D. F. and SNEATH, P. H. (1966) Quantitative comparison of contour maps; *Journal of Geophysical Research* 71(4), 1105–15.

MIESCH, A. T. and CONNOR, J. J. (1967) Stepwise regression in trend analysis; In Merriam, D. F. and Cocke, N. C. (Eds.), *Kansas Computer Contribution* No. 12, 16–18.

MIESCH, A. T. and CONNOR, J. J. (1968) Stepwise regression and non-polynomial models in trend analysis; *Kansas Computer Contribution* No. 27, 40p.

MILLER, R. L. (1956) Trend surfaces, their application to analysis and description of environments of sedimentation; *Journal of Geology* 64, 425–46.

MILLER, R. L. and ZIEGLER, J. M. (1958) A model relating dynamics and sediment pattern in equilibrium in the region of shoaling waves, breaker zone and foreshore; *Journal of Geology* 66, 417–41.

MIRCHINK, M. F. and BUKHARTSEV, V. P. (1959) The possibility of a statistical study of structural correlations; *Dokl., Earth Science Section, English Translation* 126(5), 1062–5.

NETTLETON, L. L. (1954) Regionals, residuals and structures; *Geophysics* 19, 1–22.

NORCLIFFE, G. B. (1969) On the use and limitations of trend surface models; *Canadian Geographer* 13, 338–48.

OLDHAM, C. H. G. and SUTHERLAND, D. B. (1955) Orthogonal polynomials: their use in estimating the regional effect; *Geophysics* 20, 295–306.

PALLISTER, J. W. (1956) Slope form and erosion surfaces in Uganda. *Geological Magazine* 93, 465–72.

PETERSON, J. A. and ROBINSON, G. (1968) Partial trend-surface analysis of the distribution of cirques in Tasmania; *Conference papers of the Institute of Australian Geographers* (Quoted in Robinson, G., 1970, *Area* 2(3), 31–6).

PETERSON, J. A. and ROBINSON, G. (1969) Trend-surface mapping of cirque floor levels; *Nature* 222(5188), 75–6.

ROBERTS, M. C. and MARK, D. M. (1970) The use of trend surfaces in till fabric analysis; *Canadian Journal of Earth Sciences* 7(4), 1179–84.

ROBINSON, G. (1970) Some comments on trend-surface analysis; *Area* 2(3), 31–6.

ROBINSON, G. and FAIRBAIRN, K. J. (1969) An application of trend surface mapping to the distribution of residuals from a regression; *Annals of the Association of American Geographers* 59, 158–70.

RODDA, J. C. (1970) A trend-surface analysis trial for the planation surfaces of north-Cardiganshire; *Transactions of the Institute of British Geographers* 50, 107–14.

ROMANOVA, M. A. (1970) Trend analysis of geologic data (basic literature); In *Topics in Mathematical Geology* 273–8.

SIMPSON, S. M. (1954) Least square polynomial fitting to gravitational data and density plotting by digital computers; *Geophysics* 19, 255–69.

SMITH, D. E., SISSONS, J. B. and CULLINGFORD, R. A. (1969) Isobases for the Main Perth Raised Shoreline in south-east Scotland as determined by trend-surface analysis; *Transactions of the Institute of British Geographers* 46, 45–52.

SPURR, A. M. N. (1955) The Pleistocene deposits of part of the Kagera Valley, Bukoba District; *Report of the Geological Survey of Tanganyika* (unpublished).

SVENSSON, H. (1956) Method for exact characterizing of denudation surfaces, especially peneplains, as to the position in space; *Lund Studies in Geography, Series A*, No. 8, Stockholm.

SWARDT, A. M. J. DE (1964) Lateritisation and landscape development in parts of equatorial Africa; *Zeitschrift für Geomorphologie* 8(3), 313–33.

TARRANT, J. R. (1970) Comments on the use of trend-surface analysis in the study of erosion surfaces: *Transactions of the Institute of British Geographers* 51, 221–2.

THORNES, J. B. and JONES, D. K. C. (1969) Regional and local components in the physiography of the Sussex Weald; *Area* 1(2), 13–19.

TINKLER, K. J. (1969) Trend surface with low 'explanations'; the assessment of their significance; *American Journal of Science* 267, 114–23.

UNWIN, D. J. (1970) Percentage RSS in trend surface analysis; *Area* 2(1), 25–8.

UNWIN, D. J. and LEWIN, J. (1971) Some problems in the trend analysis of erosion surfaces; *Area* 3(1), 13–14.

WAYLAND, E. J. (1929) Rift valleys and Lake Victoria; *International Geological Congress*, Session IV, Vol. II, Sec. VI, Pretoria, 323–53.

WAYLAND, E. J. (1934A) Rifts, rivers, rains and early man in Uganda; *Journal of the Royal Anthropological Institute* 64, 333–52.

WAYLAND, E. J. (1934B) Pleistocene geology and prehistory – Pleistocene history and the Lake Victoria region; *Report and Bulletin, Geological Survey of Uganda, 1933*, 71–7.

WHITTEN, E. H. T. (1959) Composition trends in a granite: modal variation and ghost stratigraphy in part of the Donegal granite, Eire; *Journal of Geophysical Research* 64, 835–49.

WHITTEN, E. H. T. (1961A) Quantitative areal model analysis of granitic complexes; *Bulletin of the Geological Society of America* 72, 1331–59.

WHITTEN, E. H. T. (1961B) Systematic quantitative areal variation in five granite masses from India, Canada, and Great Britain; *Journal of Geology* 69, 619–46.

WHITTEN, E. H. T. and BOYER, R. E. (1964) Process-response models based on heavy-mineral content of the San Isabel Granite, Colorado; *Bulletin of the Geological Society of America* 75, 841–62.

K

10 The application of harmonic and spectral analysis to the study of terrain

JOHN N. RAYNER

Department of Geography, The Ohio State University

In 1965 Chorley and Haggett noted that although their traditional concern was with areal distributions, geographers generally had not been involved in the development of techniques in that field. The authors went on to discuss the use of trend-surface mapping as an example. In similar vein this chapter addresses another group of techniques, which bear the name of the nineteenth-century mathematician and developer of the basic underlying theory, Fourier. In particular it deals with two-dimensional Fourier techniques which, despite their relatively long history in allied fields, are little used in geomorphology.

Major development in Fourier techniques for two-dimensional patterns followed closely the basic work on one-dimensional analysis by Bartlett (1948) and Tukey (1949). The initial impetus came from oceanography (e.g. Cox and Munk 1954; Pierson 1960) where there was an obvious correspondence between water waves and theoretical sinusoidal waves. Furthermore, the output had an immediate physical interpretation. The widespread introduction of digital computers and the enormous increase in two-dimensional data from remote sensing devices spurred further development. The same general ideas were applied in geophysics (Horton *et al.* 1962), meteorology (Leese and Epstein 1963) and geology (Preston and Harbaugh 1965). Paralleling the digital analysis optical methods were developed using the laser and a general theory which had been set down by Abbe in 1873 (O'Neill 1956; Pincus and Dobrin 1966). In geography Barton and Tobler (1971) have analysed optically settlement patterns and Rayner (1971) has dealt digitally with topographic features.

The nature of two-dimensional Fourier techniques

Harmonic and spectral analysis techniques are nothing more than a set of procedures based upon a particular transformation of data. Common to any transformation of this type the purpose is to rearrange the original information in such a way that it can be more easily manipulated or more easily interpreted.

In its simplest and basic form the Fourier transformation of a two-dimensional field involves the fitting by least squares, of a set of parallel sinusoidal waves of varying wavelength and orientation. The coefficients, or transformed output, define the amplitude, phase (point at which the particular wave intersects the origin) and orientation of each wave. The original data array is replaced therefore by a new array of the same size and information content. However, now the co-ordinates are size or length of wave rather than distance from an origin. If the amplitude of the wave is considered as a measure of the variability at that wavelength then the transformed array presents a ranking of the variability as a function of magnitude and direction of scale.

An important characteristic of the Fourier transform is that the reverse process is possible using the same procedure (and therefore same computer subroutine). Consequently the ranked variabilities are simply transformed back to data. Also, because the initial transform produces a function which describes the two-dimensional surface it may be used for interpolation purposes. An additional advantage springs from the fact that certain calculations, which are relatively complicated in the data domain, are made simple in the spectral domain due to the convolution theorem which will be discussed later. An obvious example is that of scale removal from the data array. The frequently used technique of calculating a moving average is tedious and in fact a poor method of removing high frequencies from data. In the spectral domain the high-frequency coefficients may be set to zero and the remaining set retransformed to produce a trend map, which contains only the broad regional features (the low frequencies).

Other more complicated filters may be applied in the same way. For example, by removing certain orientations from an aerial photograph of part of the Greenland ice cap Bauer *et al.* (1967) were apparently able to find crevasses which were masked in the original photograph.

Another closely related example is pattern recognition. *One* method might be to take a known landform, called the target, and to move it in steps across the landscape and to calculate at each step the correlation between the two patterns (Rosenfeld 1962). High correlation would indicate the likelihood of a similar feature in the landscape. Again such a procedure is accomplished simply in the spectral domain through the straightforward multiplication of the landscape spectrum with the target spectrum. Inverse transformation produces a map of covariances. Of course, the brain could perform the task almost as quickly over a small area. The advantage of the mathematical-computer technique comes in the initial search for features from many aerial photographs. It can identify possible locations which may then be subject to closer analysis by more sophisticated techniques. Furthermore, a mechanical-electrical device does not tire of monotonous procedures like the brain and it may be operated successfully by a technician.

A particular type of filter is the 1, −1 finite difference operator which may be used to calculate slopes. This hardly needs the Fourier transform for its application but a more refined operator is the partial derivative which can be applied in the spectral domain and used directly in a number of slope development models. For example, both the first and second derivatives of slope are required to test some aspects of Culling's (1965) model.

Slope appears frequently because it is both a factor in and a result of erosion. A less familiar influence occurs through its effect upon weathering by controlling the radiation balance. Soons and Rayner (1968) found that sediment yield on slopes was strongly controlled by surface loosening through frequent ice needle growth and melting which had its highest frequency on sun-facing slopes: the shadier slopes were permanently frozen.

In these examples the use of the discrete spectrum may be likened to the use of logarithms. Data are transformed into a new state or plane in which certain numerical manipulations are relatively simple and then they are retransformed back into the original state. For example, powering a number, by say 10, involves nine multiplications by direct calculations, but requires only one multiplication and two transformations when logarithms are used. Like logarithms the spectrum contains exactly the same information as the original data and one state is equally easily obtained from the other. However, unlike logarithms the Fourier transformation for one term in one state involves all the terms in the other state.

Despite its description so far as a purely manipulative device, the Fourier transform does provide information which should not be overlooked. For instance, in an aggregation process seldom does the geographer consider thoroughly the effects upon his data in terms of scale (Tobler 1969). The choice of an aggregation procedure, such as summing in rectangles, *may* lead to erroneous final results because the low frequencies, which were thought to be removed, are in fact retained with reversed sign in the data domain.

Concern with the scale content of a two-dimensional field then focuses attention upon the transformed output itself and the question may be asked whether the spectrum of the coefficients as so far described provides the best description of that information. It does not for two major reasons, both of which are related to the periodic nature of trigonometric functions. First, for practical purposes the number of coefficients (different scale components) is finite with a distinct scale separation between them, yet it would seem extremely unlikely that those and only those scales existed in the original data. Therefore, it is more reasonable to assign the variance estimate for a particular frequency to a band or block of scales rather than to a discrete scale. Also from a statistical standpoint, because each estimate has at the most two degrees of freedom and presents an erratic spectrum, adjacent blocks are usually summed. Secondly, since the original data represent a sample region from a continuum, the edges of the array present a problem, for they must be described by the same function which describes the fluctuations within the sample. Hence edge smoothing, known as tapering or window application, is necessary. In addition, because in the calculations the sample repeats itself, linear and nonlinear trends, even with edge smoothing, will appear as components increasing in magnitude towards the larger scale. Consequently, some initial trend removal, say by a polynomial, may be necessary.

The result of these modifications is a continuous estimate of the variability as a function of scale or frequency (the number of scale lengths in a given interval). The spectrum may be plotted in a number of different ways. If variance is the spectral ordinate it may be considered as a form of variance analysis or component analysis where the components are scale values. Alternatively, if the variance elements are divided by the total variance the ordinate becomes a probability density function describing the expected distribution of scales within the data. The latter is often a necessary statistical characteristic to be put alongside the mean and total variance in the description of nonperiodic spatial data. In other words, it expresses the spatial aspects of variance. Thus, with a relatively fine grid if one observation is made in a valley floor an adjacent one is more likely also to be in the valley rather than close to the mean elevation or ridge top. With randomly selected data the mean elevation would have the highest probability of being observed next.

The user may be interested in describing and measuring the amount of variance in a particular direction and/or at a particular scale. If no clear orientation is present he may average in arcs of constant frequency to remove direction and to produce the averaged vector spectrum. This expresses the probability of observing a particular scale in the landscape from a traverse regardless of the direction of that traverse. Clearly some measure of surface variability and therefore roughness is necessary although there is no accepted standard method. For terrain

Gould (1967) suggests the use of eigenvalues, whereas Rozema (1969) uses the spectrum.

As yet no theory exists for a given shape of the relief spectrum, but it is suggested here that different processes may produce characteristic forms and that in analogy with other fields such a possibility should be investigated. For example, in meteorology the Kolmogorov theory relates the spectrum of turbulence to the physical processes (e.g. Munn 1966), and in geophysics the theory of Bhattacharyya relates the shape of the spectrum of aeromagnetic data to the thickness and shape of the magnetic anomaly source (e.g. Spector and Grant 1970).

One frequently employed and successful method of seeking association between variables is regression. The usual technique does not take into account spatial arrangement or scale. Consequently, whereas there may be high correlation at one scale its true significance may be masked in the overall relationship through low correlation at other scales. This effect may be compounded in multiple regression if the independent variables influence the dependent variable at different scales. If such a scale dependence exists it can be identified by spectral regression or cross spectral analysis, which supplies all the statistics of regular regression as functions of frequency (Rayner 1967).

These then are some of the possible uses of harmonic and spectral analysis in geomorphology and in spatial studies generally. The next section deals with the basic mathematical formulae used in the application of the techniques but, since these are not necessary for a superficial understanding of the results, the reader may wish to pass directly to the last section which presents some computed examples.

The mathematical relationships

The Fourier transform

The equations for the spectral (or Fourier) transformation and its inverse for a two-dimensional surface may be written (e.g. Rayner 1971)

$$X[f1, f2] = \int_{-\infty}^{\infty} \int_{-\infty}^{\infty} x[t1, t2] \, e^{-i2\pi(f1\,t1+f2\,t2)} \, dt1 \, dt2 \qquad (1)$$

$$x[t1, t2] = \int_{-\infty}^{\infty} \int_{-\infty}^{\infty} X[f1, f2] \, e^{i2\pi(f1\,t1+f2\,t2)} \, df1 \, df2 \qquad (2)$$

where the data surface x has co-ordinates $t1, t2$, and the spectral array X, which is complex, has co-ordinates $f1, f2$. Sometimes x may be complex. Often in practice the data are a finite equally spaced $N1 \times N2$ array and (1) and (2) become

$$a[k1, k2] - ib[k1, k2] =$$

$$\frac{1}{n1 \cdot n2} \sum_{j2=0}^{N2-1} \sum_{j1=0}^{N1-1} x[j1, j2] \, e^{-i2\pi\left(\frac{j1\,k1}{N1} + \frac{j2\,k2}{N2}\right)} \qquad (3)$$

$$x[j1, j2] =$$

$$\sum_{k2=0}^{N2-1} \sum_{k1=0}^{N1-1} (a[k1, k2] - ib[k1, k2]) \, e^{i2\pi\left(\frac{j1\,k1}{N1} + \frac{j2\,k2}{N2}\right)} \quad (4)$$

where the j's and k's are integers of position in the data and spectral planes respectively and i is the complex number equal to $(-1)^{1/2}$. The a's and b's in (3) may be separated by equating real (without i) and imaginary (with i) parts. The output spectrum is limited by the k's which in turn vary between 0 and $N - 1$ or $-N/2$ and $+N/2$. The reader should note that the a's and b's are two-sided functions and that frequently they are defined for only one side when the right-hand side of (3) is doubled. If this is the case only two quadrants of the output spectrum need be retained. Any adjacent pair may be chosen. Here in the examples, the lower two will be used so that the zero frequency is half-way along the top row. The frequency of any entry $[k1, k2]$ is then given by

$$f = \left\{ \left(\frac{k1}{N1\,\Delta t1}\right)^2 + \left(\frac{k2}{N2\,\Delta t2}\right)^2 \right\}^{1/2}$$

or, if $N1 = N2$ and $\Delta t1 = \Delta t2$

$$k = (k1^2 + k2^2)^{1/2}.$$

A note of caution should be injected here about the spacing of the data, $\Delta t1$, $\Delta t2$. If the grid is so large that significant variation occurs between grid points some of this variation will be contained in the data yet cannot be assigned to the correct frequency. Nevertheless, because the Fourier series is made to fit the data exactly this high-frequency variation is incorporated into and may confuse the lower calculable frequencies. This is known as aliasing.

Convolution

Many of the calculations performed on spatial data such as the application of a weighting function and the comparison of patterns for identification involve a convolution. That is, each element in a new array is produced from the sum of adjacent elements in the old array multiplied by some weighting array, i.e.

$$\begin{matrix} \text{New} \\ \text{filtered} \\ \text{array} \end{matrix} = \int_{-\infty}^{\infty} \int_{-\infty}^{\infty} x[t1' + t1, t2' + t2] \, w[t1, t2] \, \mathrm{d}\,t1 \, \mathrm{d}\,t2 \quad (5)$$

where $x[t1, t2]$ and $w[t1, t2]$ are the data and weights respectively.

The importance of the Fourier transform in this situation results from the fact that the spectrum of (5) is the equivalent of the product of the individual spectra, i.e.

The Fourier
transform of
$$\int_{-\infty}^{\infty} \int_{-\infty}^{\infty} x[t1' + t1, t2' + t2] \, w[t1, t2] \, d \, t1 \, d \, t2$$

$$= X[f1, f2] \, W^*[f1, f2] \quad (6)$$

where $X[f1, f2]$ and $W[f1, f2]$ are the transforms of $x[t1, t2]$ and $w[t1, t2]$ and the asterisk refers to the conjugate of $W[f1, f2]$ which is obtained by changing the sign of i.

Therefore

$$
\begin{matrix}
\text{Filtered} \\
\text{array} \\
\text{(5)}
\end{matrix}
=
\begin{matrix}
\text{Inverse} \\
\text{Fourier} \\
\text{transform}
\end{matrix}
\text{of}
\left(
\begin{matrix}
\text{Fourier} \\
\text{transform} \\
\text{of } x[t1, t2]
\end{matrix}
\times
\begin{matrix}
\text{Fourier} \\
\text{transform} \\
\text{of } w[t1, t2]
\end{matrix}
\right)
\quad (7)
$$

Here the analogy with logarithms is clear. In a particular situation the weight might form a complicated function and the manipulations on the left-hand side of (7) could be lengthy even for a modern digital computer. With recently developed algorithms the transformation and its inverse are fairly simple and straightforward. An advantage of this technique is that the data need be transformed only once and stored in the transformed state. Then they may be recalled for many different weighting functions as well as for pattern recognition and differentiation. Furthermore, as each weighting function spectrum is calculated explicitly its effect upon the various scales within the data is immediately apparent. In fact, in some situations the weighting function itself need not be specified and only its spectral equivalent defined. This, for instance, is what may be done in some forms of trend analysis. Some cut-off frequency is defined beyond which the spectral weighting function $W[f1, f2]$ is zero. Thus on retransformation only the lowest frequencies (largest scales) remain.

Differentiation

Although the array x may be in discrete form differentiation is relatively simple since equation (2) provides a functional equivalent in the spectral plane. Therefore

$$\frac{\partial}{\partial t1}(x[t1, t2]) = \int_{-\infty}^{\infty} \int_{-\infty}^{\infty} \frac{\partial}{\partial t1}(X[f1, f2] \, e^{i2\pi(f1\,t1+f2\,t2)}) \, d \, f1 \, d \, f2$$

$$= \int_{-\infty}^{\infty} \int_{-\infty}^{\infty} i2\pi f1 \, X[f1, f2] \, e^{i2\pi(f1\,t1+f2\,t2)} \, d \, f1 \, d \, f2 \quad (8)$$

which in the form of equation (4) becomes

$$\frac{\partial}{\partial t1}(x[t1, t2]) \text{ at } [j1, j2] =$$

$$\sum_{k2=0}^{N2-1} \sum_{k1=0}^{N1-1} \frac{i2\pi \, k1}{N1 \, \Delta t1} \left(a[k1, k2] - ib[k1, k2] \right) e^{i2\pi \left(\frac{j1\,k1}{N1} + \frac{j2\,k2}{N2} \right)} \quad (9)$$

where $\Delta t1$ is the spacing of the original data in the $t1$ direction.

Similarly $\dfrac{\partial}{\partial t2}(x[t1, t2])$ may be calculated in like manner. It will be noted that the change from (4) to (9) involves the multiplication by the complex factor of frequency $i2\pi\, k1/N1\,\Delta t1$. Besides other things this has the effect of emphasising the larger frequencies at the expense of the smaller. A little thought will reveal that this is truly the case: that the slope of a surface depends very much upon the smaller-scale fluctuations (higher frequencies) and only to a limited extent on the larger scale. Consequently the accuracy of the results will depend on how well the high frequencies have been recorded.

The two partial derivatives may be combined to give the magnitude and direction of maximum slope, i.e.

$$\text{maximum slope} = \left(\left\{\frac{\partial}{\partial t1}(x[t1, t2])\right\}^2 + \left\{\frac{\partial}{\partial t2}(x[t1, t2])\right\}^2\right)^{1/2}, \tag{10}$$

and $\quad$ $$\text{direction} = \arctan\left\{\frac{\partial}{\partial t2}(x[t1, t2]) \Big/ \frac{\partial}{\partial t1}x[t1, t2])\right\}, \tag{11}$$

which will relate to the way in which the original $t1, t2$ co-ordinate system was arranged.

The statistical spectrum

As previously indicated certain modifications to $x[t1, t2]$ and $X[f1, f2]$ are necessary to produce the statistical spectrum. First, $x[t1, t2]$ should have a zero mean and no general slope (e.g. no lower degree polynomial surface). Then, the edges of the $D1 \times D2$ array should be smoothed, say with the cosine bell $h[j1, j2]$ of Tukey (1967),

$$x'[j1, j2] = x[j1, j2]\, h[j1, j2] \tag{12}$$

where on one edge, for example, $h[j1, j2] = \tfrac{1}{2}(1 - \cos \pi\, j1/G1)$, $0 \leqslant j1$ $\leqslant G1$ and $G1$ is by rule of thumb between $D1/20$ and $D1/4$. Other edges will be smoothed in a similar manner.

Next, rows and columns of zeros are added to $x'[j1, j2]$ to create an $N1 \times N2$ array which increases resolution and makes easy factoring, a necessary element in the present computational algorithms. This is followed by the application of equation (3) to produce $\hat{a}$ and $\hat{b}$ coefficients. These are then combined and summed over rectangles, $(2\, z1 + 1) \times (2\, z2 + 1)$, to give an estimate of the variance in a band. Variance estimate,

$$\widehat{XX}[r1, r2] = \sum_{k2=r2(2\, z2+1)-z2}^{r2(2\, z2+1)+z2} \sum_{k1=r1(2\, z1+1)-z1}^{r1(2\, z1+1)+z1} \frac{(\hat{a}^2[k1, k2] + \hat{b}^2[k1, k2])}{2}, \tag{13}$$

for $0 < r1 < M1$, $0 < r2 < M2$. Edge rows and columns will be only $z1 + 1$ and $z2 + 1$ wide.

Alternatively the combined $\hat{a}$'s and $\hat{b}$'s may be averaged in arcs to give the averaged vector spectrum.

Spectral regression

The spectral relationship between two arrays x and y may be given in terms of their individual elementary $\hat{a}$'s and $\hat{b}$'s. Thus, for example, the scale breakdown of covariance, known as the cospectrum, is given by

$$C\hat{X}Y[r1, r2] = \sum\sum_{\substack{\text{same summation}\\\text{limits as in (13)}}} \frac{(\hat{a}_x[k1, k2]\,\hat{a}_y[k1, k2] + \hat{b}_x[k1, k2]\,\hat{b}_y[k1, k2])}{2} \tag{14}$$

An associated spectrum is the quadrature spectrum,

$$Q\hat{X}Y[r1, r2] = \sum\sum_{\substack{\text{same summation}\\\text{limits as in (13)}}} \frac{(\hat{a}_x[k1, k2]\,\hat{b}_y[k1, k2] - \hat{a}_y[k1, k2]\,\hat{b}_x[k1, k2])}{2} \tag{15}$$

Together (13), (14) and (15) supply the basic input for the other statistics of spectral regression.

The coherence, corresponding to the coefficient of determination, is

$$\text{coh}\,[r1, r2] = \frac{C\hat{X}Y^2[r1, r2] + Q\hat{X}Y^2[r1, r2]}{X\hat{X}[r1, r2]\,Y\hat{Y}[r1, r2]} \tag{16}$$

The phase, or relative displacement of the average waves in y with respect to x is given by

$$\Phi[r1, r2] = \arctan\,(Q\hat{X}Y[r1, r2]/C\hat{X}Y[r1, r2]) \tag{17}$$

Similar to the slope of the linear regression line is the gain of y upon x,

$$G\hat{X}Y[r1, r2] = \frac{|(C\hat{X}Y^2[r1, r2] + Q\hat{X}Y^2[r1, r2])^{1/2}|}{X\hat{X}[r1, r2]}. \tag{18}$$

An excellent discussion of the confidence intervals for the estimates is given in Jenkins and Watts (1968) for one-dimensional analysis and these may be applied in the two-dimensional situation given the number of degrees of freedom (Rayner 1971). In all, the data present $D1 \times D2$ degrees of freedom. Harmonic analysis produces $D1 \times D2$ $\hat{a}$'s plus $\hat{b}$'s. Therefore, each has one degree of freedom. In spectral analysis the addition of zeros and the smoothing of the edges reduces this to

$$\nu12 \approx \left(\frac{D1 - G1}{N1}\right) \times \left(\frac{D2 - G2}{N2}\right) \tag{19}$$

Therefore, for each elementary estimate $[k1, k2]$ which combines an $\hat{a}$ and $\hat{b}$, $\nu12$ must be doubled. In the averaged vector spectrum this must be multiplied again by the number of terms summed. Similarly, for the two-dimensional spectral estimates [$r1$, $r2$] (19) must be multiplied by 2. (2 $z1$ + 1)(2 $z2$ + 1) excepting at the edges. Alternatively ν_r may be written

$$\nu_r \approx \left(\frac{D1 - G1}{2\,M1}\right) \times \left(\frac{D2 - G2}{2\,M2}\right) \tag{20}$$

where $M = (N/2)/(2z + 1)$ is the number of separate estimates along a given axis of the spectrum.

Some examples

The original relief data were kindly supplied by Dr Tobler (1967) and later modified by the author. The area chosen for analysis came from the 1/62,500 U.S.G.S. Quadrangle sheet for Alma, Wisconsin. A rectangle with map co-ordinates (X, Y) (1·6, 10·4), (8·5, 10·4), (8·5, 17·3) and (1·6, 17·3) in inches measured from the borders was digitised every $\frac{1}{10}$ inch on the map (520·8 ft on the ground = $\Delta\,t1 = \Delta\,t2$) which yielded

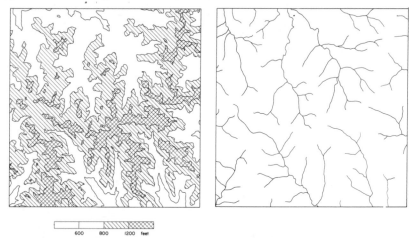

Fig. 10.1 Relief from a portion of the U.S. Geological Survey Quadrangle map, Alma, Wisconsin.

Fig. 10.2 Rivers corresponding to the area shown in fig. 10.1.

a matrix of 70 rows and 70 columns. Because the digitisation process is a form of smoothing not all the features of the relief are recorded. Figure 10.1 is a computer-drawn map of the input data which may be compared with the original topographic sheet. It should be remembered also that the topographic map itself is a filtered version of reality and detail is lost through the use of 20-ft interval contours.

The corresponding stream map was digitised according to stream order (fig. 10.2). Since no orders greater than 3 were present each grid point was assigned a digit 0 through 3. Consequently, as most of the map is composed of zeros, the output magnitudes are small. Furthermore, they have no absolute value and are useful only for comparative purposes.

Smoothing

Figure 10.3 presents a smoothed version of fig. 10.1 and is a reproduction of a similar calculation given in Rayner (1971). In this case the data of fig. 10.1 were transformed and then only frequencies of $k \leqslant 9$ were retained for retransformation. Selection of the limiting value of k was made on the basis of running several examples and a preliminary analysis of the spectrum (see later section) which showed that over 90% of the variance in the data could be accounted for by the lower 10 frequencies. Closer examination of the spectrum suggests that perhaps $k \leqslant 7$ would have been a better choice. However, the present example

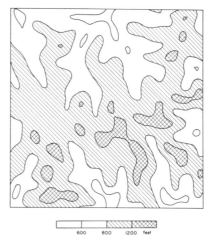

Fig. 10.3 Smoothed version of fig. 10.1. Frequencies greater than $k = 9$ removed.

suffices as an illustration of frequency rejection as a method of trend-surface mapping.

It will be seen that fig. 10.3 does present an acceptable smoothed image of fig. 10.1 excepting at the edges. Here, because of the periodicity of the Fourier series, the edge effect does create serious error particularly one-third of the way along the southern and northern edges where a ridge is met by a valley. Fortunately elsewhere lowland meets lowland and highland meets highland. Nevertheless several closed lowland depressions occur.

As Chorley and Haggett (1965) have already outlined the major uses of trend-surfaces as descriptive and interpretative devices, a review relating to this aspect of filtering will not be attempted here. However, a few comments are pertinent with regard to the double Fourier series. In terms of scale it cannot compete with the lower-degree polynomial in isolating the broad regional change. Next to the horizontal plane the simplest Fourier surface must contain one maximum and one minimum. A first-degree polynomial is a sloping plane and a second-degree has at the most either a maximum or a minimum. Thus, for instance, many trigonometric terms will be required to describe the least-squares-fitted linear polynomial of the Alma map,

$$x_{\text{plane}}[j1, j2] = 976 \cdot 82 + 0 \cdot 14\, ji + 1 \cdot 42\, j2,$$

where $j1$ and $j2$ increase towards the east and south respectively.

Normally a polynomial surface is not calculated beyond the sixth degree whereas the Fourier surface may contain complete oscillations numbering up to $N1/2$, $N2/2$ and $(N1^2 + N2^2)^{1/2}/2$ in the west to east, north to south and two diagonal directions respectively. Consequently the latter is much more sensitive, even at the lower frequencies, to local variation within a map.

If residuals from trend are required they may be obtained directly from the transform by the removal of the low frequencies. On the other hand it may be more revealing to isolate particular bands of frequencies rather than to take one and/or other end of the spectrum. This is the kind of procedure followed in one-dimensional time series analysis where, after an initial calculation of the spectrum, any peaks in the variance, such as the diurnal, will be analysed separately.

The example presented here is the simplest form of filtering. Besides the importance of the technique in applying filtering functions in general, it should be remembered that it also provides a means of analysing the scale effects of other smoothing procedures such as aggregation over rectangles.

Differentiation

As indicated in the discussion of the formulae for differentiation the results are very dependent upon the higher frequencies. This usually means that, for undulating relief, nearly every grid point will have a slope which is significantly different from the adjacent ones. Consequently, although derivative vectors are calculable, they are very difficult to combine into a coherent map. One way around this problem is to average adjacent slope elements, but this involves a convolution which is more easily handled in the spectral domain. The two processes, differentiation and smoothing, therefore may be handled together.

In this example, to reduce the number of maps a smoothing with $k \leqslant 9$ was used. Therefore either fig. 10.1 or fig. 10.3 may be taken as the surface being differentiated. If fig. 10.1 is used then the equivalent of final smoothing of the derivative vectors must be visualised. The results

are given in figs. 10.4 and 10.5. The magnitude of the slope is given by
the tangent ratio with isopleths drawn at 0·1 and 0·2 (angles of 5·7°
and 11·3°). For comparison it might be mentioned that the extreme
finite difference estimation of the slope was 0·4 with an average value of
less than 0·1 over most of the region. An interesting feature of fig. 10.4
is that the areas of highest slope are aligned mainly in a north-north-
west to south-south-east direction although the original map only gives
a vague suggestion of this finding.

Even with the smoothing the direction of slope was still very variable

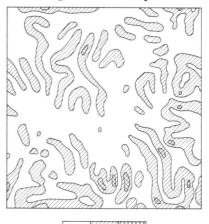

Fig. 10.4 Slope magnitude of fig. **10.3.**

Fig. 10.5 Direction of slope to
the nearest 90° in fig. 10.3.

and for cartographic purposes classification had to be limited to four
points of the compass. Variability was highest, and consequently
accuracy was the lowest, where the magnitude of the slope was a
minimum. Therefore fig. 10.5 should be analysed in association with
fig. 10.4.

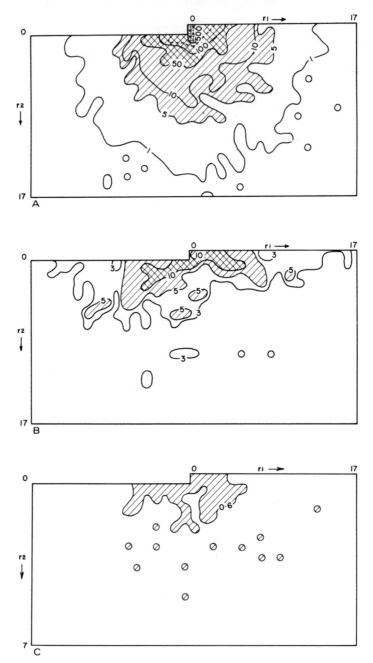

Fig. 10.6 A. Variance spectrum of fig. 10.1.
   B. Variance spectrum of fig. 10.2.
   C. Coherence spectrum of figs 10.1 and 10.2.

Whereas maps of this type may be very useful to the geomorphologist the originally computed partial derivatives would seem to have more potential. As noted in the introduction these values may be fed directly into theoretical models. For example, it is a simple matter to calculate the solar radiation receipt for all slopes in a region given one fully equipped recording station.

Spectral and cross spectral analysis

As elements in the analysis of the Alma map, filtering and gradient calculations have been presented, but the first step in scientific enquiry,

Table 10.1   *Calculated statistics from Alma relief and rivers*

	Relief		Rivers
Mean	1031 ft		0·2467
Total variance	20,480 ft^2		0·4013
Covariance		−34·30	
Correlation		−0·38	

that of objective description, has yet to be performed. This requires at least the calculation of the mean and variance, which are given in table 10.1, and the two-dimensional variance spectrum, which is plotted

Table 10.2   *Control parameters used in the two-dimensional spectral and cross spectral analyses*

$D1 = D2 = 70$	$G1 = G2 = 7$
$\triangle t1 = \triangle t2 = 520 \cdot 8$	$M1 = M2 = 17$
$N1 = N2 = 102$	$Z1 = Z2 = 1$

Approximate Number of degrees of freedom per
$\begin{cases}\text{elementary } k \text{ estimate} = 1\cdot52 \\ r \text{ block } (3 \times 3 \text{ elementary estimates}) = 13\end{cases}$

in fig. 10.6a. It should be remembered that the spectra in fig. 10.6 are one-sided and that the upper antisymmetrical parts are the inverted mirror images of the lower quadrants. Table 10.2 lists the parameters used in the calculations and table 10.3 includes a list of relationships between the radial frequency integer in fig. 10.6 and the corresponding *average* scale distance (wavelength).

Figure 10.6a reveals that in the Alma region the larger-scale elements of relief are by far the most important since the spectrum declines rapidly as scale decreases (increasing r). Furthermore, the larger-scale

Table 10.3 *Frequency–period relationships and number of terms summed in the averaged vector spectrum*

Terms summed in averaged vector spectrum	Integer frequency		Wavelength		Terms summed in averaged vector spectrum	Integer frequency		Wavelength	
	k	r	feet	miles		k	r	feet	miles
1	0	0	∞	∞	82	26		2043	
4	1		53122	10	80	27	9	1967	
6	2		26561	5	92	28		1897	
8	3	1	17707		86	29		1832	
16	4		13280		100	30	10	1771	$\frac{1}{3}$
14	5		10623	2	96	31		1714	
20	6	2	8854		94	32		1660	
20	7		7589		104	33	11	1610	
24	8		6640		112	34		1562	
34	9	3	5902		112	35		1518	
28	10		5312	1	114	36	12	1476	
36	11		4829		112	37		1436	
34	12	4	4427		124	38		1398	
44	13		4086		118	39	13	1362	
44	14		3794		132	40		1328	$\frac{1}{4}$
42	15	5	3541		124	41		1296	
56	16		3320		132	42	14	1265	
56	17		3125		138	43		1235	
56	18	6	2951		132	44		1207	
58	19		2796		144	45	15	1180	
56	20		2656	$\frac{1}{2}$	138	46		1155	
72	21	7	2530		152	47		1130	
70	22		2415		152	48	16	1107	
72	23		2310		156	49		1084	
72	24	8	2213		158	50		1062	
84	25		2125			51	17	1042	$\frac{1}{5}$

features have a clear orientation in the north-north-west to south-south-east direction (at right angles to the major axis of the dashed elliptical curve). This has already been noted in the magnitude of the slope field (fig. 10.4). At scales less than 3500 ft ($r > 5$) no distinct orientation exists and the deviation of the isopleths of constant variance from semicircles of constant scale can be accounted for by random sampling errors. The average rate at which the spectrum declines with decreasing scale is clearly evident in the averaged vector spectrum of fig. 10.7.

Together, the mean, total variance, two-dimensional spectrum and averaged vector spectrum may be used to characterise the shape and roughness of the terrain. However, much further work will be necessary in other types of region before it is possible to ascertain whether these statistics are sufficient to set up useful classifications. More important is the possibility that the spectrum may lead to a recognition of different

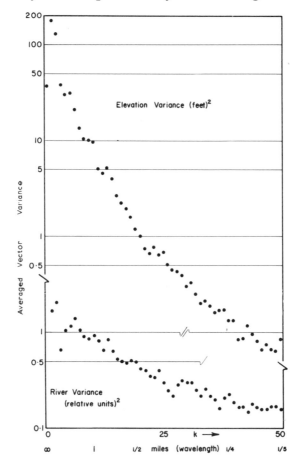

Fig. 10.7 Averaged vector variance spectra for figs. 10.1 and 10.2.

generating processes. For example, the larger-scale features of the Alma map have an orientation whereas the smaller scales do not. Does this finding lead to new theories about the composite relief? In the averaged vector spectrum there is a suggestion of a break in the plot at about $k = 23$. In fact, two quite separate straight lines may be drawn through the logarithmic plot, one from $k = 4$ to $k = 22$ and the other from $k = 23$ to $k = 49$. Does this indicate a truly different process or set of features, or is this just playing with data? Most of the time the former ideas will prove to be unfruitful but this should not deter the researcher: the one successful hypothesis will be worth the thousand discarded ones.

To pursue explanation further the distribution under investigation may be related statistically to other influencing distributions. An obvious choice for relief seems to be the river pattern. The mean and total variance of the Alma rivers would seem to have little immediate meaning in themselves but the correlation coefficient might be more useful. The negative value is to be expected because the presence of a river is indicated by a positive number. The corresponding zero mean $x[j1, j2]$ of the relief is a valley floor which is probably negative. The magnitude of the correlation coefficient 0·38 (coefficient of determination $= 0·14$), is not very revealing. Its relatively small size indicates that the relief as a whole is not closely related to the river pattern as shown in fig. 10.2. On the other hand spectral regression presents a different picture.

First, the two-dimensional variance spectrum (fig. 10.6b) contains a strong orientation which matches the large-scale features of the relief. This is confirmed by the coherence, the spectral equivalent of the coefficient of determination (fig. 10.6c), which shows a high relationship at large scale. Also, a similar, although less distinct, break in the averaged vector spectrum (fig. 10.7) appears in the river variance. Of course, it should be remembered that only the permanent streams were drawn on the original topographic map and that digitisation will have removed some small-scale features. Nevertheless the present analysis does pose some interesting questions concerning the relationship between streams and their slopes, which may have been inherited from processes which are no longer present.

It is hoped that these examples will encourage geomorphologists to apply some of these techniques themselves. Computers and programmes are generally easily accessible, so the field worker will not have to concern himself too much with the detailed mathematics or computational algorithms (see, for example, Rayner and McCalden 1972). On the other hand he must make himself aware of the assumptions and limitations of the technique. In return he will have much to contribute to the interpretation step from his intimate knowledge of the processes and forms in the field.

References

BARTLETT, H. G. (1948) Smoothing periodograms from time series with continuous spectra; *Nature* 161, 686–7.

BARTON, B. and TOBLER, W. (1971) A spectral analysis of innovative diffusion; *Geographical Analysis* 7, 182–6.

BAUER, A., FONTANEL, A. and GRAU, G. (1967) The application of optical filtering in coherent light to the study of aerial photographs of Greenland glaciers; *Journal of Glaciology* 6, 781–93.

CHORLEY, R. J. and HAGGETT, P. (1965) Trend-surface mapping in geographical research; *Transactions of the Institute of British Geographers* 37, 47–67.

COX, C. and MUNK, W. (1964) Statistics of the sea surface derived from sun glitter; *Journal of Marine Research* 13, 198–227.

CULLING, W. E. H. (1965) Theory of erosion on soil-covered slopes; *Journal of Geology* 73, 230–54.

GOULD, P. R. (1967) On the geographical interpretation of eigenvalues; *Transactions of the Institute of British Geographers* 42, 53–86.

HARBAUGH, J. W. and PRESTON, F. W. (1968) Fourier series analysis in geology; In *Spatial Analysis*, Berry, B. J. L. and Marble, D. F. (Eds.) (Prentice-Hall, Englewood Cliffs), 218–38.

HORTON, C. W., HEMPKINS, W. B. and HOFFMANN, A. A. J. (1964) A statistical analysis of some aeromagnetic maps from the northwestern Canadian shield; *Geophysics* 29, 582–601.

JENKINS, G. M. and WATTS, D. C. (1968) *Spectral Analysis and its Application* (Holden Day, San Francisco).

LEESE, J. A. and EPSTEIN, E. S. (1963) Application of two-dimensional spectral analysis to the quantification of satellite cloud photographs; *Journal of Applied Meteorology* 2, 629–44.

MUNN, R. E. (1966) *Descriptive Micrometeorology* (Academic Press, New York).

O'NEILL, E. L. (1956) Spatial filtering in optics; *Institute of Radio Engineers, Transactions on Information Theory* 2, 56–65.

PIERSON, W. J. (Ed.) (1960) The directional spectrum of a wind generated sea as determined from data obtained by the stereo wave observation project; *Meteorology Papers*, 2, 6 (N.Y. University, Coll. of Engineering).

PINCUS, H. and DOBRIN, M. B. (1966) Geological applications of optical data processing; *Journal of Geophysical Research* 7, 4861–9.

PRESTON, F. W. and HARBAUGH, J. W. (1965) BALGOL program and geologic applications for single and double Fourier series using IBM 7090/7094 computers; *Kansas Geological Survey, Special Distribution Publication* No. 24, 72p.

RAYNER, J. N. (1967) Correlation between surfaces by spectral methods; *Kansas Computer Contribution* No. 12, 31–7.

RAYNER, J. N. (1971) *An Introduction to Spectral Analysis* (Pion, London).

RAYNER, J. N. and MCCALDEN, G. (1972) Programs for two dimensional spectral and cross spectral analysis; *Papers, Department of Geography, The Ohio State University.*

ROSENFELD, A. (1962) Automatic recognition of basic terrain types from aerial photographs; *Photogrammetric Engineering* 27, 115–32.

ROZEMA, W. J. (1969) The use of spectral analysis in describing lunar surface roughness; *U.S. Geological Survey Professional Paper* 650-D, 180–8.

SOONS, J. M. and RAYNER, J. N. (1968) Micro-climate and erosion processes in the Southern Alps, New Zealand; *Geografiska Annaler* 50, 1–15.

SPECTOR, A. and GRANT, F. S. (1970) Statistical models for interpreting aeromagnetic data; *Geophysics* 35, 293–302.

TOBLER, W. R. (1968) A digital terrain library; *Technical Report, University of Michigan* (under contract with U.S. Army Research Office, Durham).

TOBLER, W. R. (1969) Geographical filters and their inverses; *Geographical Analysis* 1, 234–53.

TUKEY, J. W. (1949) The sampling theory of power spectrum estimates; *Symposium on applications of auto correlation analysis to physical problems, Office of Naval Research, Woods Hole NAVEXOS-P-735*, 47–67.

TUKEY, J. W. (1967) Spectrum calculations in the new world of the fast Fourier transform; In Harris, B. (Ed.), *Advanced Seminar on Spectral Analysis of Time Series* (Wiley, New York), 25–46.

PART V
Space partitioning

11 Areal classification in geomorphology

PAUL M. MATHER

Department of Geography, University of Nottingham

The most common form of spatial analogue model is that in which adjacent contiguous areas are grouped together on the assumption that each unit can be better understood in terms of generalizations about some larger region of which it forms a part. (R. J. CHORLEY 1967, 62)

Introduction

Classification is one of the fundamental procedures in any scientific discipline. Gilbert (1895), for example, suggested that the first stage in a scientific investigation involves the grouping of facts according to common characteristics; the resulting classification serves as a framework for observation and acts as a preliminary to a 'more logical relational classification'. The history of disciplines such as biology, geology and zoology shows Gilbert's prognostications to have been correct. These disciplines have developed from the stage of cataloguing observed facts to the detection of order in the observed data and the searching out of the processes or forces responsible for the generation of this order. Central to this progress have been the improvements in the methods of classification. Automated techniques, developed for digital computers, have allowed biological taxonomists to reconsider the Linnean classification and to use these new taxonomic schemes in attempts to unravel the ancestry of the species of interest (Sokal and Sneath 1963; Sokal 1966; Dupraw 1965; Crowson 1965).

Geographers have traditionally been concerned with the description and differentiation of the earth's surface (Hartshorne 1939, 1959; Grigg 1967; Wrigley 1965). The role of classification was not immediately understood, for the 'regional method' was seen by many as an art form (viz. Gilbert 1957, 346–7) rather than as the manifestation of a technique common to all sciences. The controversy over the uniqueness or otherwise of geographical phenomena diverted attention from the problems of regional definition, and it was not until comparatively recently that progress began to be made. Schaefer (1953), in an attack on what he termed 'exceptionalism' in geography, placed the subject firmly among the sciences, giving it the same aim – explanation – and the same logical procedures as established disciplines such as physics and chemistry. Thus, in 1956, it was possible for Reynolds to state that 'the delineation of regions is essentially a classification process'. With the availability of automatic procedures, geographers were not slow to realise their potential (Berry 1961) and their use is now widespread (Spence and Taylor 1971 give a bibliography).

Areal studies have not been prominent in post-Davisian geomorphology in Britain. Wooldridge and Linton (1955), Sparks (1949) and Brown (1960) have produced studies of south-eastern England and Wales which present the landscape as the product of several periods of planation (the 'denudation chronology' approach) while Sissons (1967) follows Fenneman (1935) in his reliance on geological rather than morphological boundaries; these latter are frequently of dubious significance. In recent years the morphological mapping approach, described by Savigear (1965), has tended to express the spatial aspect of geomorphology while the main stream of writing and research has concentrated on processes rather than regional analysis. (Schaefer's remark that 'in physical geography it has been felt that regional work, since it did not lead directly to the formulation of laws, was not worth doing and had better be abandoned' (1953) is pertinent in this context.) This concentration on process is possibly a reaction against denudation chronology, in which relatively few, and often minor, landscape features were considered to be of paramount importance (Chorley 1965). The concentration of morphological mapping on a few selected landscape features carries on this tradition, which is in strong contrast to the feeling in other disciplines that it is essential to consider the object of study as a unit and 'not as a haphazard jumble of piecemeal measurements' (Bronowski and Long 1951).

The special nature of geographical data

The objects or entities (O.T.U.'s) which are classified in numerical taxonomy are generally discrete, independent and unambiguously identifiable. When these techniques are applied to areal data several problems arise. These problems concern (i) the arbitrariness involved in defining a geographical individual; (ii) the effects of variation in size and shape of the individual areal units; (iii) the nature and measurement of location.

The problems involved in the definition of the geographical individual are treated by Grigg (1967). He points out the difficulty encountered in picking out separate, individual areal units from a continuum (the earth's crust), and concludes that 'there are no easily recognizable entities which form a mosaic whose interpretation and grouping is the business of regionalization,' (1967, 484). There is no easy solution to this problem, though its impact can be reduced. The most common method among geomorphologists is the selection of grid squares as the basic areal unit, characteristics of topography, hydrology and climate being averaged out for each grid square.

The selection of grid squares as the basic unit overcomes, at least partially, the problem of recognising the O.T.U.'s in the regional scheme. Disadvantages result from the fact that an infinite number of orthogonal grids can be placed over a given area, and differing results can be expected for each. Again, grid squares are not 'pure' individuals in that they do not necessarily consist of one landscape type only; two or more contrasting types may be present, especially in boundary areas. However, since grid squares are all of the same size and shape, their use does eliminate variability in these properties. This can be especially valuable in fluvial geomorphology where, according to Anderson (1957), area is 'the devil's own variable', correlating with virtually every other characteristic of drainage basins.

The third problem, measurement of location, is peculiar to geography. Geographers do not seem to have identified the problem sufficiently clearly nor have any convincing solutions been obtained. Location, unlike other variables, is two-dimensional, requiring two parameters in its specification whether polar or cartesian co-ordinates are used. If location is treated like other variables (as Bunge (1966) suggests) then discontinuous regions are likely, and the interpretation of locational variation within each such region might prove difficult. Berry (1966) includes latitude and longitude as separate variables in his factor analysis of the Indian economy, though he does not proceed to a classification. It is interesting to consider whether the selection of reference axes other than the conventional N–S/E–W would have any effect upon the outcome. The most common solution is to make relative location (as measured by spatial contiguity) the dominant variable in the analysis, allowing individual areas to become members of a region only if they are contiguous to an existing member of that region. An interesting discussion of this topic is given by Taylor (1969) and reference should also be made to Grigg (1967, 484–5).

Methods of numerical taxonomy

In general, grouping algorithms are of two kinds – hierarchical and nucleated. The former results in the familiar linkage tree or dendrogram. The advantages or disadvantages of the two approaches depend upon the nature of the investigation. A full description of the use of hierarchical and non-hierarchical schemes is given by Johnson (1968), Spence and Taylor (1971), Pocock and Wishart (1969) and Lankford (1969). Papers by Sneath (1961, 1965 and 1967), Williams and Dale (1965), Gengerelli (1965), Ball and Hall (1967) and Kendall (1965) are also relevant.

The merits of each technique will not be argued here. Instead, one algorithm will be examined and applied to a simple example. The algorithm leads to a hierarchical solution and is a modification of the group average method of Sokal and Michener (1958). This method has already been used successfully in fields as diverse as marketing research (Frank and Green 1968), paleoecology (Mello and Buzas 1968), sedimentology (Imbrie and Purdy 1962) and geography (Berry 1961). Its advantages lie in its comparative simplicity, ease of interpretation and general reliability. With data that do not exhibit fairly strong patterns, spurious results may be generated due to a process termed 'chaining' by Lankford (1969, 198). Furthermore, demands on computer time and storage space restrict its use to problems involving 200 or less objects. (An example with 195 individual areas took 19·1 minutes on the KDF9 computer.) The algorithm is, then, limited to applications in which the data are expected to be well ordered, and where the number of objects to be classified is less than 200. This description fits adequately many geomorphological situations.

The details of the calculations involved in the technique are well known (Berry 1961; Sokal and Sneath 1963; Mather 1969A). Briefly, the steps involved are:

(i) calculation of the similarity matrix (the euclidean distance coefficient is used in this study);
(ii) modification of the matrix by placing a negative sign before elements representing similarity between non-contiguous areas;
(iii) search for the smallest non-negative entry (say d_{ij}) and subsequent replacement of the entries for area i by the weighted average of the similarity values of the remaining areas and the centroid of the cluster now containing i and j. The following formula is used:

$$d^2_{ca} = \frac{Wt(i)}{Wt(c)} . d^2_{ia} + \frac{Wt(j)}{Wt(c)} . d^2_{ja} - \frac{Wt(i) \times Wt(j)}{Wt(c)^2} . d^2_{ij}$$

where　c　　= centroid of group holding i and j
a　　= any other area
d^2_{ij}　= square of distance coefficient for i and j
$Wt(j)$　= weight given to area j
$Wt(c)$　= weight given to centroid c

Each area is originally allocated a weight of unity; as clusters develop, their centroids are given a weight equal to the sum of the weights of their members.

(iv) updating of the similarity matrix firstly by the elimination of the entries for j and secondly, by the modification of the signs attached to the distance coefficients in view of changes in contiguity brought about by the amalgamation of i and j at the preceding step.

 (v) Steps (iii) and (iv) are repeated $(n-1)$ times, where n is the number of individual areas. At this stage all areas have been assigned to groups.

(vi) The 'linkage order' is derived from the sequence of (i, j) pairings.

The adoption of a method described by Parks (1969) allows the introduction of non-quantitative multistate data and its consideration together with continuously measured variables. Multistate data, defined as data which is expressed in a form such as 'yes–no', 'present–absent' or 'abundant - rare - absent', can be given numerical values in the range 0–1. (The preceding examples would take the respective values 0–1, 0–1, and $0 - 0 \cdot 5 - 1$.) The continuously measured data is transformed on to the range 0–1 by the formula

$$x_i = \frac{x_i - x_{\min}}{\text{range}} \qquad (i = 1, 2, \ldots, n)$$

where n is, as usual, the number of individual areas. The use of multistate data allows the inclusion in the regional scheme of information that is normally overlooked or ignored in quantitative analyses.

A major assumption of the scheme is that of orthogonality, or lack of correlation between variables. This is normally achieved via principal components analysis (for continuous data). Omission of this step leads to error in the distance coefficients proportional to the inter-variable correlation. This step has been ignored in some published work, especially in biology and zoology, and it is not clear whether results are adversely affected. Furthermore, it is frequently difficult to attach a meaning to principal components beyond the first few, and the tendency to include only 'meaningful' components is tantamount to throwing away information in order that what remains satisfies the orthogonality restriction. Neither is it necessarily true that components having a correspondingly eigenvalue greater than one are error-free – error is not concentrated in the last components.

A second assumption is that all variables are measured on the same scale, since the similarity measure used (the euclidean distance coefficient) is sensitive to scale. The most common standardisation technique is that which produces transformed variables with a mean of 0 and a variance of 1 (the 'Z scores' of Krumbein and Graybill 1965). The

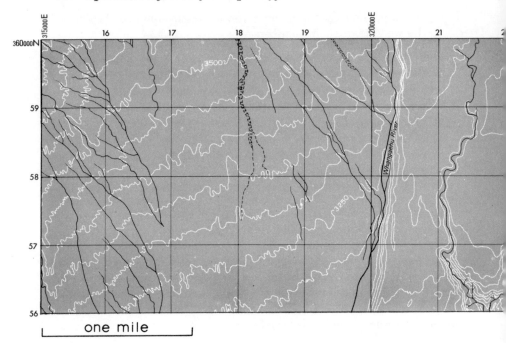

Fig. 11.1 Extract from New Zealand Survey Sheet number N 122/5 (Te Rotete) showing the example area.

alternative method is to transform on to a 0–1 scale. Neither method is completely satisfactory (See Horst 1965 for a discussion.)

An example

To illustrate the use of the algorithm an example is given. The data used, and the results obtained, should not be treated as anything more than an attempt to illustrate, in a naive way, the procedure involved. An area of 4 × 15 km was selected from the New Zealand Survey 1/25,000 sheet N 122/5 (Te Rotete) which includes, in its eastern part, a section of the Kaimanawa Mountains (fig. 11.1). The western area of the map is less rugged, the drainage pattern more tenuous and the slopes smoother than the Kaimanawa Mountains. Ten continuous variables were measured or estimated for each of the 60 grid squares; they are:

1. Maximum height;
2. Minimum height;
3. Relative relief;
4. Maximum number of streams crossed by one straight line;

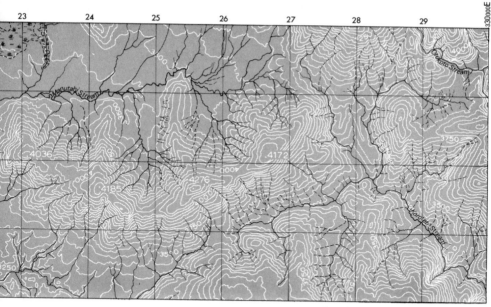

Contour interval 50 feet

5. Maximum number of contour lines crossed by one straight line;
6. Number of closed contours;
7. Number of stream sources;
8. Percentage of swamp;
9. Number of stream sinks;
10. Percentage of scrub and bush (darker shading on map, Fig. 11.1).

The data were recorded variable by variable (table 11.1) from grid squares numbered according to the following convention:

$$01 \rightarrow 15$$
$$30 \leftarrow 16$$
$$31 \rightarrow 45$$
$$60 \leftarrow 46$$

Contiguity was defined by the presence of a common grid square boundary. Grid squares having only one corner in common were not counted as contiguous. An extract from the unmodified distance matrix is shown in table 11.3 while the pairing sequence and linkage tree are shown in table 11.4 and fig. 11.2. Inspection of the pairing sequence

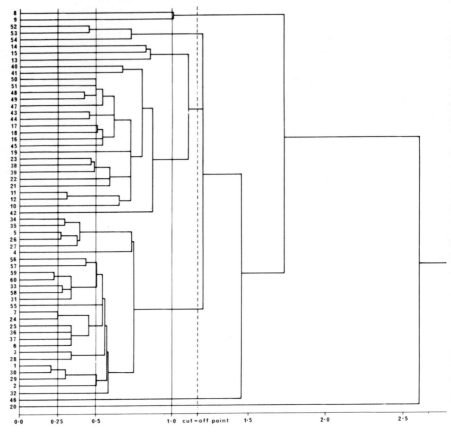

Fig. 11.2 Linkage tree.

shows that backward jumps in the value of the minimum non-negative distance coefficient are occasionally present; this is a result of mutually similar areas being prevented from linking by the contiguity constraint, due to the presence of an intervening area with lower similarity. To prevent confusion in the drawing up of the linkage tree, these 'backward' jumps are replaced by the value of the immediately preceding distance coefficient in the pairing sequence. Group membership is not affected.

Inspection of fig. 11.2 shows that two major regions are present. The first is composed of grid squares 1–7, 24–37 and 55–60, inclusive. This is the western region and, from consideration of the data matrix (table 11.1) or the table of standard (Z) scores (table 11.2), it is seen to be characterised by low relative relief, lower than average maximum altitude, a low number of stream sources, low 'terrain roughness' (as expressed by variable 5) and below-average intensity of dissection (variable 4). The hierarchical nature of the classification allows closer inspection of the internal structure of the region. Several sub-regions can be identified, for example the cluster consisting of grid squares 4, 5, 26, 27, 34 and 35, which have an above-average value on variable 10. The standard scores for squares 5 and 55 are given in table 11.5 for comparative purposes.

The second major regional grouping covers grid squares 11–19, 21–23, 38–45 and 47–51. This is the eastern region, the Kaimanawa Mountains area, to which reference was made earlier. It has above-average values for maximum altitude, relative relief, terrain roughness, number of stream sources and intensity of dissection. Again, sub-regional groupings occur and these can be picked out from the linkage diagram. The standard scores for squares 18 and 43 are shown in table 11.5.

Apart from these two major groups, two smaller but still separate areas can be identified. The first (grid squares 8 and 9) is characterised by marshy conditions, while the second (grid squares 52, 53 and 54) is seen to have a greatly above-average number of closed contours, as well as a lower maximum height than the eastern region in which it is embedded. Square 20 remains unallocated at the cut-off level chosen, as does square 46. Square 20 has 10 stream 'sinks' and is very dissimilar in this respect to any other grid square, while square 46 has a very large relative relief, low minimum height and above-average number of closed contours compared with adjacent squares.

Conclusion

The purpose of regionalisation is to classify the surface of the earth with respect to the properties of interest, in order that the processes responsible for variation and covariation in these properties can be isolated and examined. In the past the main difficulty experienced in the building-up of a regional scheme has been the lack of objective methods of defining regional boundaries. The technique described above goes some way towards resolution of this difficulty. Selection of the properties to be included in the analysis is still the responsibility of the individual; the outcome will ultimately depend upon the variables used. While computer-based techniques can provide more objective methods of data analysis, they cannot make up for improper selection or measurement of data.

The results of a classification analysis, such as the one described, are open to interpretation. The position of the cut-off point in fig. 11.2 is, initially, a matter of opinion – there is no general automatic method of deriving the 'optimum' cut-off point. Since it is well known that hierarchical clustering schemes will produce linkage trees even where the data is random (McCammon 1968) it is evident that some method

L

Table 11.1   *Raw data, for 10 variables measured over the 60 grid squares shown in fig. 11.1 (Note that each block constitutes the 60 measurements on a single variable)*

```
3650.0000 3650.0000 3450.0000 3510.0000 3455.0000 3410.0000
4400.0000 4400.0000 4350.0000 3950.0000 3950.0000 4050.0000
3460.0000 3300.0000 3440.0000 3450.0000 3500.0000 3600.0000
3420.0000 3900.0000 4070.0000 4185.0000 4276.0000 4150.0000
4000.0000 3950.0000 3850.0000 3550.0000 3400.0000 3400.0000
             INPUT FORMAT = (10F4.0)


3500.0000 3450.0000 3440.0000 3400.0000 3300.0000 3290.0000
3650.0000 3500.0000 3650.0000 3350.0000 3300.0000 3330.0000
3250.0000 3250.0000 3340.0000 3350.0000 3390.0000 3450.0000
3250.0000 3300.0000 3400.0000 3555.0000 3570.0000 3390.0000
3450.0000 3370.0000 3310.0000 3220.0000 3100.0000 3050.0000
             INPUT FORMAT = (15F3.0)


 150.0000  200.0000   10.0000  110.0000  155.0000  120.0000
 750.0000  900.0000  700.0000  600.0000  650.0000  720.0000
 210.0000   50.0000  100.0000  100.0000  110.0000  150.0000
 170.0000  600.0000  670.0000  630.0000  706.0000  760.0000
 550.0000  580.0000  540.0000  330.0000  300.0000  250.0000
             INPUT FORMAT = (20F2.0)


   6.0000    4.0000    0.0000    3.0000    4.0000    4.0000
   4.0000   10.0000   12.0000   10.0000    8.0000    5.0000
   2.0000    4.0000    4.0000    0.0000    5.0000    6.0000
   1.0000    5.0000    7.0000    5.0000    7.0000   12.0000
   5.0000    5.0000    5.0000    3.0000    2.0000    0.0000
             INPUT FORMAT = (20F2.0)


   3.0000    3.0000    2.0000    3.0000    3.0000    7.0000
  20.0000   21.0000   26.0000   20.0000   21.0000   17.0000
   6.0000    3.0000    2.0000    3.0000    3.0000    4.0000
   7.0000   16.0000   20.0000   25.0000   25.0000   21.0000
  17.0000   18.0000   12.0000   11.0000   17.0000   14.0000
             INPUT FORMAT = (30F1.0)


   0.0000    0.0000    0.0000    0.0000    0.0000    0.0000
   0.0000    1.0000    1.0000    2.0000    1.0000    1.0000
   0.0000    0.0000    0.0000    0.0000    0.0000    0.0000
   0.0000    1.0000    1.0000    2.0000    1.0000    0.0000
   2.0000    0.0000    1.0000    4.0000    4.0000    2.0000
             INPUT FORMAT = (20F2.0)


   4.0000    0.0000    0.0000    2.0000    1.0000    1.0000
  12.0000   13.0000   17.0000   16.0000   19.0000   11.0000
   0.0000    1.0000    2.0000    0.0000    0.0000    3.0000
   0.0000    3.0000    8.0000   12.0000   25.0000   21.0000
  11.0000    9.0000    7.0000    4.0000    8.0000    9.0000
```

of testing the validity of the outcome should be employed unless there are prior grounds for believing that some structure is present within the data. Techniques of multivariate analysis provide a means of testing hypotheses relating to the significance of differences between the multivariate means of the regions derived from study of the linkage tree (multivariate analysis of variance, Hotelling's T^2) and of allocating additional or previously unclassified areas to the regional

Table 11.1 (*continued*)

```
3460.0000  3460.0000  3600.0000  3650.0000  3700.0000  3710.0000
4172.0000  3950.0000  3950.0000  4036.0000  4000.0000  3460.0000
3510.0000  3440.0000  3380.0000  3350.0000  3320.0000  3425.0000
3750.0000  3600.0000  4000.0000  3900.0000  3950.0000  3900.0000
3400.0000  3240.0000  3260.0000  3300.0000  3350.0000  3450.0000


3400.0000  3440.0000  3440.0000  3440.0000  3470.0000  3530.0000
3540.0000  3440.0000  3440.0000  3440.0000  3430.0000  3400.0000
3350.0000  3300.0000  3280.0000  3250.0000  3200.0000  3190.0000
3300.0000  3230.0000  3230.0000  3000.0000  3250.0000  3400.0000
3140.0000  3160.0000  3170.0000  3200.0000  3210.0000  3260.0000


260.0000   20.0000    160.0000   210.0000   230.0000   170.0000
632.0000   510.0000   510.0000   596.0000   570.0000   60.0000
160.0000   140.0000   100.0000   100.0000   120.0000   235.0000
450.0000   370.0000   770.0000   900.0000   700.0000   500.0000
260.0000   80.0000    90.0000    100.0000   140.0000   190.0000


2.0000     2.0000     3.0000     5.0000     8.0000     10.0000
8.0000     11.0000    5.0000     8.0000     4.0000     2.0000
3.0000     6.0000     2.0000     3.0000     2.0000     4.0000
9.0000     7.0000     9.0000     6.0000     7.0000     6.0000
0.0000     6.0000     4.0000     4.0000     0.0000     0.0000


2.0000     4.0000     3.0000     6.0000     7.0000     4.0000
21.0000    11.0000    11.0000    15.0000    13.0000    3.0000
3.0000     3.0000     2.0000     3.0000     3.0000     10.0000
13.0000    13.0000    21.0000    27.0000    18.0000    14.0000
8.0000     6.0000     2.0000     2.0000     3.0000     4.0000


0.0000     0.0000     0.0000     1.0000     0.0000     0.0000
2.0000     1.0000     0.0000     2.0000     0.0000     0.0000
0.0000     0.0000     0.0000     0.0000     0.0000     0.0000
3.0000     2.0000     1.0000     3.0000     2.0000     1.0000
1.0000     0.0000     0.0000     0.0000     0.0000     0.0000


1.0000     3.0000     2.0000     0.0000     10.0000    9.0000
21.0000    30.0000    7.0000     9.0000     4.0000     1.0000
0.0000     0.0000     0.0000     1.0000     0.0000     0.0000
11.0000    14.0000    13.0000    6.0000     9.0000     15.0000
0.0000     0.0000     0.0000     0.0000     2.0000     2.0000
```

scheme (discriminant analysis). The algebraic details of these techniques are provided by Hope (1968), Rao (1952) and Cooley and Lohnes (1962); examples of their application are given by Rao and Slater (1949), Griffiths (1957), Casetti (1964), Mather (1969B) and Mather and Doornkamp (1970). It is worth remarking that the technique of 'multiple discriminant analysis' used by Casetti (1964) is mathematically dissimilar – though conceptually identical – to the technique of the

Table 11.2   *Standardized data matrix (Z-scores).* The 60 measurements on each variable are given in turn

-0.1455	-0.1455	-0.7692	-0.5821	-0.7536	-0.8939
2.1934	2.1934	2.0374	0.7901	0.7901	1.1019
-0.7380	-1.2369	-0.8004	-0.7692	-0.6132	-0.3014
-0.8627	0.6341	1.1643	1.5229	1.8067	1.4138
0.9460	0.7901	0.4782	-0.4573	-0.9251	-0.9251
1.1316	0.7671	0.6943	0.4027	-0.3262	-0.3991
2.2249	1.1316	2.2249	0.0383	-0.3262	-0.1075
-0.6906	-0.6906	-0.0346	0.0383	0.3298	0.7671
-0.6906	-0.3262	0.4027	1.5325	1.6418	0.3298
0.7671	0.1840	-0.2533	-0.9093	-1.7839	-2.1484
-0.7837	-0.5910	-1.3231	-0.9378	-0.7644	-0.8992
1.5280	2.1059	1.3354	0.9501	1.1427	1.4124
-0.5525	-1.1680	-0.9763	-0.9763	-0.9378	-0.7837
-0.7066	0.9501	1.2198	1.0657	1.3585	1.5666
0.7575	0.8730	0.7189	-0.0902	-0.2057	-0.3984
0.3579	-0.2928	-1.5942	-0.6182	-0.2928	-0.2928
-0.2928	1.6593	2.3099	1.6593	1.0086	0.0325
-0.9435	-0.2928	-0.2928	-1.5942	0.0325	0.3579
-1.2688	0.0325	0.6832	0.0325	0.6832	2.3099
0.0325	0.0325	0.0325	-0.6182	-0.9435	-1.5942
-0.9515	-0.9515	-1.0798	-0.9515	-0.9515	-0.4383
1.2295	1.3578	1.9992	1.2295	1.3578	0.8446
-0.5666	-0.9515	-1.0798	-0.9515	-0.9515	-0.8232
-0.4383	0.7163	1.2295	1.8710	1.8710	1.3578
0.8446	0.9729	0.2031	0.0748	0.8446	0.4597
-0.6871	-0.6871	-0.6871	-0.6871	-0.6871	-0.6871
-0.6871	0.2717	0.2717	1.2304	0.2717	0.2717
-0.6871	-0.6871	-0.6871	-0.6871	-0.6871	-0.6871
-0.6871	0.2717	0.2717	1.2304	0.2717	-0.6871
1.2304	-0.6871	0.2717	3.1480	3.1480	1.2304
-0.3508	-0.8975	-0.8975	-0.6241	-0.7608	-0.7608
0.7426	1.5626	1.4259	1.2892	1.6992	0.6059
-0.8975	-0.7608	-0.6241	-0.8975	-0.8975	-0.4875
-0.8975	-0.4875	0.1959	0.7426	2.5193	1.9726
0.6059	0.3326	0.0592	-0.3508	0.1959	0.3326
-0.2217	-0.2217	-0.2217	1.8072	0.2292	-0.2217
-0.2217	-0.2217	-0.2217	-0.2217	-0.2217	-0.2217
-0.2217	-0.2217	0.4546	-0.2217	-0.2217	-0.2217
-0.2217	-0.2217	-0.2217	-0.2217	-0.2217	-0.2217
-0.2217	-0.2217	-0.2217	-0.2217	-0.2217	-0.2217
-0.2471	0.4943	-0.2471	-0.2471	-0.2471	-0.2471
-0.2471	-0.2471	-0.2471	-0.2471	-0.2471	-0.2471
-0.2471	-0.2471	0.4943	-0.2471	-0.2471	-0.2471
-0.2471	-0.2471	-0.2471	-0.2471	-0.2471	-0.2471
-0.2471	-0.2471	0.4943	-0.2471	-0.2471	-0.2471
-0.0436	0.8802	0.3670	1.3935	2.6766	0.3670
2.4200	0.8802	2.9332	0.6236	-0.6595	-0.6595
-0.6595	2.4200	1.9067	1.1368	-0.4029	-0.5569
-0.2489	-0.6595	-0.0436	-0.6082	-0.6595	-0.6595
-0.6595	-0.4542	-0.5055	-0.6595	-0.6595	-0.4029

same name used by the remaining authors cited.

The use of these multivariate techniques is limited by their assumptions and requirements, which include multivariate-normality of the data, equality of group variance–covariance matrices, measurement of data on a continuous scale and independence of observations. Difficulties also arise when the variables are measured in different units. Kendall (1965) says: 'correlation of a pair of individuals over p variables of noncomparable units does not, in general, make sense.

Table 11.2 (*continued*)

-0.7380	-0.7380	-0.3014	-0.1455	0.0104	0.0416
1.4824	0.7901	0.7901	1.0583	0.9460	-0.7380
-0.5821	-0.8004	-0.9875	-1.0810	-1.1746	-0.8471
0.1664	-0.3014	0.9460	0.6341	0.7901	0.6341
-0.9251	-1.4240	-1.3617	-1.2369	-1.0810	-0.7692
0.4027	0.6943	0.6943	0.6943	0.9129	1.3502
1.4231	0.6943	0.6943	0.6943	0.6214	0.4027
0.0383	-0.3262	-0.4719	-0.6906	-1.0550	-1.1270
-0.3262	-0.8364	-0.8364	-2.5128	-0.6906	0.4027
-1.4924	-1.3466	-1.2737	-1.0550	-0.9822	-0.6177
-0.3599	-1.2845	-0.7451	-0.5525	-0.4754	-0.7066
1.0734	0.6033	0.6033	0.9347	0.8345	-1.1304
-0.7451	-0.8222	-0.9763	-0.9763	-0.8992	-0.4562
0.3722	0.0640	1.6051	2.1059	1.3354	0.5643
-0.3599	-1.0534	-1.0148	-0.9763	-0.8222	-0.6295
-0.9435	-0.9435	-0.6182	0.0325	1.0086	1.6693
1.0086	1.9846	0.0325	1.0086	-0.2928	-0.9435
-0.6182	0.3579	-0.9435	-0.6182	-0.9435	-0.2928
1.3339	0.6832	1.3339	0.3579	0.6832	0.3579
-1.5942	0.3579	-0.2928	-0.2928	-1.5942	-1.5942
-1.0798	-0.8232	-0.9515	-0.5666	-0.4383	-0.8232
1.3578	0.0748	0.0748	0.5880	0.3314	-0.9515
-0.9515	-0.9515	-1.0798	-0.9515	-0.9515	-0.0635
0.3314	0.3314	1.3578	2.1275	0.9729	0.4597
-0.3100	-0.5666	-1.0798	-1.0798	-0.9515	-0.8232
-0.6871	-0.6871	-0.6871	0.2717	-0.6871	-0.6871
1.2304	0.2717	-0.6871	1.2304	-0.6871	-0.6871
-0.6871	-0.6871	-0.6871	-0.6871	-0.6871	-0.6871
2.1892	1.2304	0.2717	2.1892	1.2304	0.2717
0.2717	-0.6871	-0.6871	-0.6871	-0.6871	-0.6871
-0.7608	-0.4875	-0.6241	-0.8975	0.4692	0.3326
1.9726	3.2026	0.0592	0.3326	-0.3508	-0.7608
-0.8975	-0.8975	-0.8975	-0.7608	-0.8975	-0.8975
0.6059	1.0159	0.8792	-0.0774	0.3326	1.1526
-0.8975	-0.8975	-0.8975	-0.8975	-0.6241	-0.6241
-0.2217	6.5412	3.1598	-0.2217	-0.2217	-0.2217
-0.2217	-0.2217	-0.2217	-0.2217	-0.2217	-0.2217
-0.2217	-0.2217	-0.2217	-0.2217	-0.2217	-0.2217
-0.2217	-0.2217	-0.2217	-0.2217	-0.2217	-0.2217
-0.2217	-0.2217	-0.2217	-0.2217	-0.2217	-0.2217
-0.2471	-0.2471	-0.2471	-0.2471	-0.2471	-0.2471
-0.2471	7.1670	-0.2471	0.4943	-0.2471	-0.2471
-0.2471	1.2357	-0.2471	1.2357	0.4943	-0.2471
-0.2471	-0.2471	-0.2471	-0.2471	-0.2471	-0.2471
-0.2471	0.4943	-0.2471	-0.2471	-0.2471	-0.2471
-0.6595	-0.6595	-0.6595	-0.6595	-0.6595	-0.2489
-0.6595	-0.6595	-0.6595	-0.6595	-0.6595	-0.6595
-0.4029	-0.5569	-0.6595	1.9067	2.1633	-0.1463
-0.6595	-0.6595	-0.1463	0.1103	-0.0436	-0.5569
-0.6082	-0.1976	0.6236	-0.6595	-0.6595	-0.1463

The difficulty is not overcome by standardization.' This author describes distribution-free (non-parametric) procedures for both classification and discrimination, illustrating the latter with Fisher's classic Iris problem.

Finally, attention may be turned to examination of within-group relationships in order to determine whether these vary from region to region. Again, the most widely used methods are those of multivariate analysis, with the limitations discussed in the previous paragraph.

Table 11.3 *Extract from distance coefficient (similarity) matrix*

```
0.00
0.48  0.00
0.72  0.59  0.00
0.89  0.75  0.81  0.00
1.03  0.75  0.93  0.76  0.00
0.63  0.56  0.59  0.70  0.79  0.00
0.5A  0.63  0.50  0.79  0.94  0.53  0.00
2.21  2.24  2.18  1.64  1.12  2.21  2.16  0.00
1.14  1.21  1.19  0.79  1.45  1.20  1.10  1.10  0.00
0.46  0.64  0.77  1.02  1.18  0.63  0.51  2.21  1.10  0.00
0.44  0.82  1.07  1.15  1.30  0.84  0.81  2.28  1.14  0.62  0.00
0.48  0.87  1.18  1.19  1.31  0.97  0.95  2.34  1.26  0.76  0.32  0.00
1.56  1.5?  1.82  1.70  1.67  1.78  1.82  2.83  1.35  1.69  0.76  1.57  0.00
1.62  1.72  2.11  1.95  1.86  2.11  1.91  2.92  2.03  1.66  1.45  1.72  1.60  0.00
1.89  1.93  2.34  2.10  2.00  2.13  2.25  3.14  2.11  2.02  1.40  1.77  1.45  1.72  0.00
1.35  1.46  1.86  1.66  1.62  1.45  1.61  2.70  2.43  1.30  1.22  1.12  1.40  1.21  0.83  0.00
1.34  1.51  1.76  1.71  1.77  1.41  1.61  2.63  1.87  1.27  1.02  1.02  0.97  1.22  0.75  0.89  0.00
1.16  1.26  1.49  1.52  1.60  1.22  1.41  2.51  1.78  1.02  0.97  1.47  1.35  1.47  0.95  0.97  0.58  0.00
1.46  1.66  1.92  1.90  2.02  1.60  1.15  2.74  1.57  1.40  1.18  1.32  1.33  1.32  1.63  1.32  0.78  0.51  0.00
1.26  1.16  3.07  3.02  2.91  1.73  1.70  3.61  1.92  2.80  2.58  1.30  1.30  1.33  0.74  0.74  0.70  0.67  0.81  0.00
2.75  2.71  3.07  3.02  2.90  2.91  1.70  3.61  3.05  2.80  2.58  2.61  2.88  2.82  2.51  2.82  2.51  2.51  2.60  2.47  0.00
0.69  0.85  1.16  1.21  1.36  0.92  0.75  2.32  1.28  0.67  0.57  0.80  1.19  1.29  1.70  1.70  1.07  1.07  0.87  1.06  2.64  0.00
```

Table 11.4 *Pairing sequence*

Retained sample	Deleted sample	Similarity coefficient	Retained sample	Deleted sample	Similarity coefficient	Retained sample	Deleted sample	Similarity coefficient
1	39	0.213	43	44	0.457	15	11	0.676
59	60	0.234	23	38	0.469	46	41	0.684
7	24	0.247	50	51	0.495	52	54	0.730
5	26	0.283	48	50	0.447	4	5	0.736
33	58	0.287	23	39	0.496	16	21	0.739
34	35	0.298	31	56	0.501	16	19	0.741
11	29	0.301	17	2	0.503	16	16	0.538
3	12	0.317	31	18	0.513	1	4	0.750
3	28	0.341	31	55	0.544	14	40	0.795
25	36	0.341	6	31	0.218	14	15	0.830
25	37	0.356	47	48	0.548	13	14	0.865
6	25	0.338	16	45	0.550	18	42	0.878
33	59	0.351	16	17	0.455	9	9	1.098
31	33	0.339	3	6	0.560	16	13	1.141
5	27	0.382	1	3	0.563	1	10	1.221
5	34	0.393	22	32	0.574	1	52	1.125
48	49	0.418	21	23	0.595	1	46	1.444
56	57	0.438	43	22	0.425	1	8	1.725
52	53	0.447	43	47	0.627	1	20	2.644
6	7	0.449	16	43	0.416			

Multiple regression, factor analysis and canonical correlation analysis are of most value where measurements are random and independent, while spectral and cross-spectral analysis can provide valuable insight into the structure of and relationships between time-series data. King (1969) is a useful source of reference on these techniques.

The establishment of a regional classification is thus seen to be a means to an end rather than an end in itself. The purpose of establishing such a classification is to allow the study of the relationships within and between a set of areal units, or regions, with the aim of identifying the causes of the observed relationships.

Table 11.5 *Comparison of western region (columns 55 and 5) and eastern region (columns 18 and 43) with squares 20, 46 and 53*

	Western		Eastern				
Variable	55	5	18	43	20	46	53
1	−0·93	−0·75	1·10	0·17	0·79	0·63	−0·92
2	−1·49	−0·32	−0·11	−0·33	0·69	−2·51	−1·78
3	−0·36	−0·76	1·41	0·37	0·60	2·11	−0·21
4	−1·59	−0·29	0·03	1·34	1·98	0·36	−0·94
5	−0·31	−0·95	0·84	0·33	0·07	2·13	0·84
6	0·27	−0·69	0·27	2·19	0·27	2·19	3·15
7	−0·90	−0·76	0·61	0·61	3·20	−0·08	0·20
8	−0·22	0·23	−0·22	−0·22	−0·22	−0·22	−0·22
9	−0·24	−0·25	−0·25	−0·25	7·17	−0·25	−0·25
10	−0·61	2·68	−0·66	−0·66	−0·65	0·11	−0·66

References

ANDERSON, H. W. (1957) Relating sediment yield to watershed variables; *Transactions of the American Geophysical Union* 38, 921–4.

BALL, G. H. and HALL, P. J. (1967) A clustering technique for summarizing multivariate data; *Behavioural Science* 2, 153–5.

BERRY, B. J. L. (1961) A method for deriving multifactor uniform regions; *Przeglad Geograficzne* 33, 263–82.

BERRY, B. J. L. (1966) Essays on commodity flows and the spatial structure of the Indian economy; *University of Chicago, Department of Geography, Research Paper* No. 111, 334p.

BRONOWSKI, J. and LONG, W. M. (1951) Statistical methods in anthropology; *Nature* 168, 794.

BROWN, E. H. (1960) *The Relief and Drainage of Wales* (University of Wales Press, Cardiff), 186p.

BUNGE, W. (1966) Locations are not unique; *Annals of the Association of American Geographers* 56, 375–6.

CASETTI, E. (1964) Multiple discriminant functions; *Office of Naval Research, Geography Branch, ONR Task No. 389–135, Contract Nonr 1228(26), Technical Report No. 11*, Northwestern University, Evanston, Illinois, 63p.

CHORLEY, R. J. (1965) A re-evaluation of the geomorphic system of W. M. Davis; In Chorley, R. J. and Haggett, P. (Eds.), *Frontiers in Geographical Teaching* (Methuen, London), 21–38.

CHORLEY, R. J. (1967) Models in geomorphology; In Chorley, R. J. and Haggett, P. (Eds.), *Models in Geography* (Methuen, London), 59–96.

CHORLEY, R. J. and HAGGETT, P. (Eds.) (1965) *Frontiers in Geographical Teaching* (Methuen, London), 379p.

CHORLEY, R. J. and HAGGETT, P. (Eds.) (1967) *Models in Geography* (Methuen, London), 816p.

COOLEY, W. W. and LOHNES, P. R. (1962) *Multivariate Procedures in the Behavioural Sciences* (Wiley, New York), 211p.

CROWSON, R. A. (1965) Classification, statistics and phyllogeny; *Systematic Zoology* 14, 144–8.

DUPRAW, A. J. (1965) Non-Linnean taxonomy and the systematics of honey bees; *Systematic Zoology* 14, 1–24.

FENNEMAN, N. M. (1938) *Physiography of Eastern United States* (McGraw-Hill, New York), 714p.

FRANK, R. E. and GREEN, P. E. (1968) Numerical taxonomy in marketing analysis; *Journal of Marketing Research* 5, 83–93.

GENGERELLI, J. A. (1965) A method for detecting subgroups in a population and specifying their membership; *Journal of Psychology* 55, 457–68.

GILBERT, E. W. (1957) Geography and regionalism; In Taylor, G. (Ed.), *Geography in the Twentieth Century* (Methuen, London), 345–71.

GILBERT, G. K. (1895) The inculcation of scientific method by example; *American Journal of Science* 31, 284–99.

GRIFFITHS, J. C. (1957) *Petrographic Investigation of the Salt Wash Sediments* (Final Report, U.S. Atomic Energy Commission, RME-3151), 37p.

GRIGG, D. (1967) Regions, models and classes; In Chorley, R. J. and Haggett, P. (Eds.), *Models in Geography* (Methuen, London), 461–509.

HARTSHORNE, R. (1939) The nature of geography; *Annals of the Association of American Geographers*, 29, 173–658.

HARTSHORNE, R. (1959) *Perspective on the Nature of Geography* (John Murray, London), 200p.

HOPE, K. (1968) *Methods of Multivariate Analysis* (University of London Press), 288p.

HORST, P. (1965) *Factor Analysis of Data Matrices* (Holt, Rinehart and Winston, New York), 730p.

IMBRIE, J. and PURDY, E. G. (1962) Classification of modern Bahamian carbonate sediments; *American Association of Petroleum Geologists*, Memoir 1, 253–72.

JOHNSON, R. J. (1968) Choice in classification; the subjectivity of objective methods; *Annals of the Association of American Geographers* 58, 575–89.

KENDALL, M. G. (1965) Discrimination and classification; In Krishnaiah, P. R. (Ed.), *Proceedings of the International Symposium on Multivariate Analysis*, Dayton, Ohio, 165–85.

KING, L. J. (1969) *Statistical Analysis in Geography* (Prentice-Hall, New York), 288p.

KRUMBEIN, W. C. and GRAYBILL, F. A. (1965) *An Introduction to Statistical Models in Geology* (McGraw-Hill, New York), 475p.

LANKFORD, P. M. (1969) Regionalization: theory and alternative algorithms; *Geographical Analysis* 1, 196–212.

MATHER, P. M. (1969A) Cluster Analysis; *Computer Applications in the Natural and Social Sciences* (University of Nottingham, Department of Geography).

MATHER, P. M. (1969B) *Analysis of Some Late Pleistocene Sediments from South Lancashire* (unpublished Ph.D. thesis, University of Nottingham).

MATHER, P. M. and DOORNKAMP, J. C. (1970) Multivariate analysis in geography; *Transactions of the Institute of British Geographers* 51, 163–87.

McCAMMON, R. B. (1968) The dendrograph – a new tool for correlation; *Bulletin of the Geological Society of America* 79, 1663–70.

MELLO, J. F. and BUZAS, M. A. (1968) An application of cluster analysis as a method of determining biofacies; *Journal of Paleontology* 42, 747–58.

PARKS, J. M. (1969) Multivariate facies maps; In Merriam, D. F. (Ed.), Symposium on Computer Applications in Petroleum Investigations, *Kansas Computer Contribution* No. 40, 6–12.

POCOCK, D. C. D. and WISHART, D. (1969) Methods of deriving multifactor uniform regions; *Transactions of the Institute of British Geographers* 47, 73–98.

RAO, C. R. (1952) *Introduction to Advanced Statistical Methods in Biometrics* (Wiley, New York), 390p.

RAO, C. R. and SLATER, P. (1949) Multivariate analysis applied to differences between neurotic groups; *British Journal of Psychology, Statistics Section* 2, 17–29.

REYNOLDS, R. B. (1956) Statistical methods and geographic research; *Geographical Review* 46, 129–32.

SAVIGEAR, R. A. G. (1965) A technique of morphological mapping; *Annals of the Association of American Geographers* 55, 514–38.

SCHAEFER, F. K. (1953) Exceptionalism in geography – a methodological examination; *Annals of the Association of American Geographers* 43, 226–49.

SISSONS, J. B. (1967) *The Evolution of Scotland's Scenery* (Oliver and Boyd, Edinburgh), 259p.

SNEATH, P. H. A. (1961) Recent developments in theoretical and quantitative taxonomy; *Systematic Zoology* 10, 118–39.

SNEATH, P. H. A. (1965) Application of numerical taxonomy to medical problems; In *Mathematics and Computer Science in Biology and Medicine*. Medical Research Council Conference, Oxford, July, 1964 (H.M.S.O., London), 81–91.

SNEATH, P. H. A. (1967) Some statistical problems in numerical taxonomy; *The Statistician* 17, 1–12.

SOKAL, R. R. (1966) Numerical taxonomy; *Scientific American* 215, 106–17.

SOKAL, R. R. and MICHENER, C. D. (1958) A statistical method for evaluating systematic relationships; *University of Kansas Science Bulletin* 38, 1409–38.

SOKAL, R. R. and SNEATH, P. H. A. (1963) *Principles of Numerical Taxonomy* (Freeman, San Francisco).

SPARKS, B. W. (1949) The denudation chronology of the dip-slope of the South Downs; *Proceedings of the Geologists' Association* 60, 165–215.

SPENCE, N. A. and TAYLOR, P. J. (1971) Quantitative methods in regional taxonomy; *Progress in Geography*, 2, 1–64.

TAYLOR, P. J. (1969) The location variable in taxonomy; *Geographical Analysis* 1, 181–95.

WILLIAMS, W. T. and DALE, M. B. (1965) Fundamental problems in numerical taxonomy; *Advances in Botanical Research* 2, 35–68.

WOOLDRIDGE, S. W. and LINTON, D. L. (1955) *Structure, Surface and Drainage in South-east England*, 2nd Edn (Philip, London), 176p.

WRIGLEY, E. A. (1965) Changes in the philosophy of geography; In Chorley, R. J. and Haggett, P. (Eds.), *Frontiers in Geographical Teaching* (Methuen, London), 3–20.

12 The average hexagon in spatial hierarchies

MICHAEL J. WOLDENBERG

Graduate School of Design, Harvard University

I. Introduction to hierarchical spatial systems

Geomorphologists and economic geographers have found that geometric progressions of modular areas organise geographic space. For example, for large river basins there is a geometric progression[1] of basin area with order (Horton 1945; Schumm 1956). To order basins, we follow Strahler's modification (1952) of Horton's method of stream ordering. Unbranched tributaries are of order one and flow within a first-order basin. Two streams of order one join to make a stream of order two (fig. 12.1). In general, two streams of equal order join to make a stream (and basin) of the next higher order; tributary streams of lower order are considered to be within the basin of higher order. The number of streams (areas) *declines* by geometric progression as order increases. Therefore, the size of basin area must *increase* geometrically with order.[2] The branching or bifurcation ratio (R_b) is approximately equal to the area ratio (R_A).

For systems of towns, organised into a hierarchy of central places, again, areas of market regions surrounding towns[3] increase geometrically with town order (Christaller 1933, 72; Baskin 1966, 67). Central place ordering is a more complex matter, not entirely free of subjectivity (see Palomäki 1964).

---

[1] $y = aR^x$ where x and y are variables and a and R are constants. R is the common ratio or base of the progression. Such progressions form straight lines on semilogarithmic paper.

[2] The ratios of these geometric progressions are not strictly inversely related because of inter-basin areas. See fig. 12.1.

[3] For a comparison of central place and fluvial systems, see Woldenberg (1971), Woldenberg and Berry (1967).

In addition to geometric progressions of area, there are geometric progressions of other properties such as the stream slope (an inverse series), cumulative length, stream relief (Fok 1971) and volume rate of flow[1] and the number of activities and number of establishments and populations of towns (Christaller 1933, 72; Baskin 1966, 67; Woldenberg and Berry 1967; Berry 1967).

Techniques of fluvial morphometry have been applied to the human

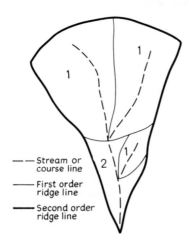

- - Stream or course line

—— First order ridge line

▬▬ Second order ridge line

Fig. 12.1 Ordering a basin. Each ordered stream segment defines a basin of the same order.

lung, and analogous geometric progressions with branch order have been found (Woldenberg *et al.* 1970).

How can we explain these persistent geometric progressions? If one can establish the necessity for geometric progressions of area with order, then other parameters which are related to area by power function[2] relationships must also progress geometrically with order. Thus a key to explaining Horton's (and Christaller's) laws is to explain the geometric progressions of modular areas.

II. The milieu of existing theories
Work related space filling
There have been two basic approaches used for explaining hierarchical spatial systems. One approach is based on concepts of work and two-dimensional space filling; the other is based on various stochastic

[1] For a review of fluvial morphometry, see Haggett and Chorley (1969), Strahler (1964). These geometric progressions are known as Horton's laws. If we plot the logarithm of stream number against arithmetic order, the absolute value of the antilogarithm of the slope of the least-squares regression line yields the bifurcation ratio (Maxwell 1955). Other ratios are calculated in similar fashion.

[2] $y = ax^b$ where x and y are variables and a and b are constants. These functions plot as straight lines on double logarithmic paper. b is the slope of the line.

models, which can operate more or less removed from considerations of
work and two-dimensional geometry though these factors can be built
into the models (e.g. Howard 1971).

Christaller (1933) in his theory of central places and their associated
market areas, assumed that the hexagon produced an optimum partitioning of space on a featureless plain. Lösch (1945) developed a proof,
showing that aggregate transport costs for a fixed area were minimised
for an hexagonal lattice with a given number of points. By assuming
that nth-order hexagons are contained within or shared by $n + 1$
order hexagons, Christaller reasoned that each large hexagon might
contain 3, 4 or 7 small hexagons (fig. 12.2). A pure hierarchy could
be developed on the basis of a ratio of 3, 4 or 7. Christaller (1933, 73)

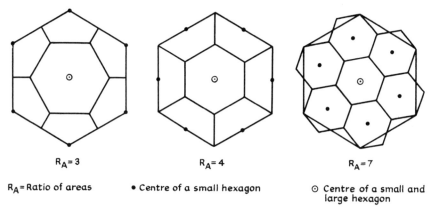

$R_A = 3$ $R_A = 4$ $R_A = 7$

R_A = Ratio of areas   • Centre of a small hexagon   ⊙ Centre of a small and
large hexagon

Fig. 12.2 Dividing an hexagonal area (after Christaller).

further deduced that if market area would follow a geometric progression with order, then the number of areas (an inverse relationship),
the populations, the functions, and the number of establishments should
similarly show geometric progressions. Hence the whole family of
geometric progressions was based ultimately on the optimal partitioning of space into hexagons. (In his empirical findings Christaller
concentrated on the number of areas or places.)

In the case of fluvial systems, Horton suggested a headward growth
model in which slopes were divided and subdivided into square basins
as long as overland flows were sufficient to erode and maintain a
channel (Horton 1945, 339–49). A channel was located on one diagonal
of each square. The repeated division and subdivision of square
drainage basin areas would tend to create a geometric series law of
stream[1] number, ($R_b = 4$), and stream length, ($R_L = 2$). Stream confluence angles would be 90°.

[1] Strahler orders are used here.

Horton does not hold rigidly to the division of space into hierarchies of square basin areas. He recognises that the variation of basin shape and slope and confluence angles all will have an effect on R_b and R_L. None the less, his spatial model serves as a hint towards explaining the geometric progressions which he found.

Random models

In recent years it has become customary to interpret complex systems by invoking some random process. Some investigators generate random walk graphs. In most models first-order streams *coalesce* to eventually form a main stream (Leopold and Langbein 1962; Schenk 1963; Smart, Surkan and Considine 1967). Howard (1971) simulates the branching of a stream by *headward erosion*, i.e. areal subdivision. These methods produce networks which, upon analysis, produce geometric progressions of number and length. Howard's method has the added advantage of producing symmetrical basins.

In other random models, the results are expressed in equations rather than graphically. Reasoning in a fashion analogous to Berry and Garrison (1959), who attempted to explain the rank–size relationship of cities[1] on the basis of a stochastic process, Woldenberg (1966) suggested that power function relationships (and hence their related geometric progressions) could be explained by Gibrat's law of proportional effect, that is, that an entity in growth is a random proportion of its previous size (Gibrat 1932). This reasoning was thought to apply to fluvial and urban systems (Woldenberg and Berry 1967).

Shreve (1966) suggested investigating *all* the topologically distinct networks[2] which could be created from n first-order streams. Then these networks were classified by number of orders. It was found that the most probable networks followed an inverse geometric progression of stream number with order. (This implies that basin area should follow a direct geometric progression.)

Shreve (1967, 184) suggests that R_b should approach 4 on the average, and R_L should be 2. Using a variant of Shreve's approach, Werner (1970) suggests R_b should be 3·618. For further developments see Shreve (1967; 1969). Other writers have developed the probability models and a comprehensive review of their work may be found in Haggett and Chorley (1969) and Howard (1971).

In effect, the random modellers believe real fluvial systems are so

---

[1] The rank[b] (size of population) = constant. Generally the constant $b = 1$. For a different probabilistic interpretation of the populations of a system of cities, see Curry (1964).

[2] 'Topologically distinct networks' are those whose schematic map projections cannot be continuously deformed and rotated in the plane of projection so as to become congruent (Shreve 1966, 27).

complex that our knowledge of individual events can never be complete enough to 'deduce the recession of a slope, the juncture of channels, or the velocity of the flow that caused them. Nevertheless, certain average relationships can be deduced from our incomplete knowledge of the individual processes, simply because the net effect of the many individual events is the same as if the individual events were to occur at random, although the events are, strictly speaking, entirely predetermined' (Scheidegger and Langbein 1966, C1).

This writer believes that a model based on the partitioning of space by hexagons can generate the numbers of areas per order that are actually found in nature (Woldenberg 1968; 1969; 1971). In these papers the reader was, in effect, asked to *assume* hexagons existed on the land surface. Deductions from this assumption led to the power to predict the number of areas per order.

The next section of the paper (*III*) is devoted to showing that the assumption of hexagonal areas is indeed reasonable. In Part IV the model of mixed hexagonal hierarchies will be reviewed.

III. The irregular hexagon as a spatial average

This portion of the paper will show that the polygonal areas met with in geography, for instance, river basins or market areas, must have on the average for large networks approximately six sides. In a word, on the average these regions tend to be hexagonal. The polygons will not be regular, though they will tend to be of approximately equal areas. The important point is that if there are hexagons present, then their corners and sides (in the case of river basins) and their centres (in the case of towns, each surrounded by market areas), must exist on the landscape.

By far the more extensive discussion is required to show that the river basins tend to be hexagonal. We must first define the parts of any completely differentiable surface (no cliffs or overhangs, no tears) and of course, these terms describe a network of river basins.

A description of a surface[1]

'A contour line may be regarded as the intersection of two surfaces, the conventional circumstance being that of the intersection of the variable surface under consideration and some specified constant-valued 'level' surface. Any two intersecting surfaces do so along a line. This line thus connects points of equal value on the variable surface.

[1] Much of this discussion is derived from the more extended treatment written by Warntz (1967; 1968). Warntz, in turn, based his discussion on the work of Reech (1858), Cayley (1859) and Maxwell (1870).

The summary of such values is conveniently displayed by means of a map on which selected contour lines are shown representing the intersections of level surfaces with the variable surface of the phenomenon and with each contour line distinguished by a numeral which shows the level surface to which it belongs.

Let us now relate the nature and significance of contour lines to certain absolute extremum points (local maximum or minimum) and mixed extremum points on a surface (fig. 12.3). The contour line at a peak becomes a point. A peak is a local maximum of elevation. . . . The contour line at a pit becomes a point. A pit is a local minimum of elevation. . . . Peaks and pits are to be regarded as singular points and constitute the category we shall call absolute extremum points.

The other kind of singular points, i.e., mixed extremum points, are passes and pales. A pass or saddle point exists, for example, at the self-crossing point of some contour line that forms two loops, one around each of two adjacent peaks (i.e., an outloop or figure '8'. . . .)

A pale exists between adjacent pits. The self-crossing contour line for adjacent pits may be either of the inloop or outloop type. The inloop type also consists of two closed curves, one of which, however, lies inside the other except for their shared point. . . .

On continuous surfaces the points of the pale and of the pass have at once the attributes of both a local maximum and a local minimum, each resulting in a self-crossing contour. . . . A profile, properly taken, shows the pass at the lowest value between two peaks; a profile through the same point taken at right angles to the first shows the value at the pass to be higher than any other along that cross section. The pale occurs at the high point between pits, but it is a low point along the line at right angles to the line joining the pits.' (Warntz 1968, 2, 3 and 5.)

Having described the most common occurrences of singular points on a surface, we will now discuss slope lines which pass through these points. Through every point on the surface one may consider that there is not only a contour line, but also a slope line. Slope lines indicate the direction of steepest gradient at a given point and, of necessity, intersect the contour lines at right angles. We therefore have two families of orthogonal curves on the surface, namely contour lines and slope lines. The property of orthogonality is preserved in a conformal projection to the plane of these two systems.

Suppose we are given an island, with only one peak. Let the sea rise and let sea-level contour lines be traced on a map. Slope lines must always be perpendicular to these contour lines. As the island gets smaller, the slope lines must converge to the peak of the island. Now let us bring two such islands together so that they just touch at a pass (fig. 12.3). Allow the water level to drop to considerable depth below the pass. Here we have a situation where slope lines must be perpendicular to the contours of one peak, or the other peak. There is a slope line of indifference which is traced across the pass. This particular slope line is called a course line (fig. 12.3). Since slope lines diverge

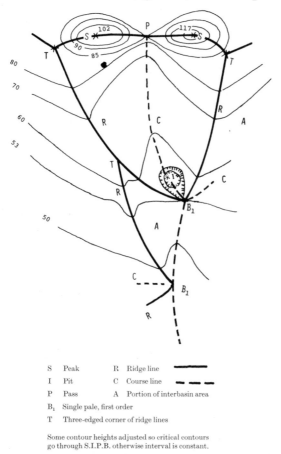

S	Peak	R	Ridge line	———
I	Pit	C	Course line	– – –
P	Pass	A	Portion of interbasin area	
B_1	Single pale, first order			
T	Three-edged corner of ridge lines			

Some contour heights adjusted so critical contours
go through S.I.P.B. otherwise interval is constant.

Fig. 12.3 Analysis of a first-order basin.

uphill from the course line to adjacent peaks, flows along the slope lines converge downhill towards the course line, although they meet the course line only at the pit. If water flows are great enough, the course line is excavated to produce a stream.

The course line will trace a path from a pass to a pit and then to a pale. In river systems, streams (course lines) flow into other streams at a pale (fig. 12.3). Course lines surround and delimit hills. Similarly, other slope lines may run from peaks to passes or pales. These are called

ridge lines. The ridge line is a divide, a line of indifference and divergence. Slope lines on one side of the divide proceed to one pit. Slope
lines on the other side proceed to another pit. Ridge lines delimit dales
(or in fluvial systems, river basins).

Warntz (1968) has shown that a course or ridge line must be traced
through an alternation of extremum and mixed points. Furthermore,
course lines and ridge lines must only intersect at passes and pales.
The intersections are at right angles.

A more detailed description of a fluvially eroded surface

Let us relate these ideas more specifically to fluvial systems. The first
qualification is that the whole land surface drains into some stream. The
stream network should communicate with the sea, or with some
large body of water which serves as a local base level. We purposely
exclude landscapes which are developed in very arid regions where
interior drainage prevails, and where wind erosion may be of great
importance. We also exclude regions of Karst topography.

Topographic maps show that when contours cross a stream or course
line, they are bent so that the contours are notched in the upstream
direction. On the other hand, when contours cross a ridge, the notch
points downhill. We may now review as a whole the simple first-order
basin in fig. 12.3. Note the two varieties of self-crossing contours; the
figure-eight typically surrounds the two peaks, and the inloop surrounds a pit in a stream near its mouth. Note that the ridge line
proceeds from peak to pass to peak to pale to peak. Note that the
course line would proceed from pass to pit to pale.

We now extrapolate beyond the discussion and nomenclature of
Warntz (1967; 1968). When a stream network is drawn we can identify
some of the essential singular points for our analysis of the polygons
which compose river basins. We continue to use the Strahler (1952)
system of stream ordering to help us classify the basins. An unbranched
(first-order) stream flows within a first-order basin area. The basin is
completely surrounded by a ridge line (fig. 12.3). The ridge line actually
crosses the stream at a pale located at the confluence with a stream of
equal order u, or higher order. The pale is denoted by a B with a numerical subscript which matches the order of the basin. Second-order
streams are formed by the confluence of two first-order streams. The
pales of each stream coincide at a point. Hence, there should be a
double symbol at this double pale. It follows that streams of order u
join at a double pale to form a stream of $u + 1$ order. It will be important to maintain the distinction to properly analyse the topology of
basins. The Strahler ordering system requires that if a stream of lower
order enters a higher-order stream, the order of the higher-order stream
is not changed. To some, this seems a very arbitrary rule. Clearly, the

flow of such a stream has been increased. (Rzhanitsyn 1964, 26; Scheidegger 1965, 187–8; Shreve 1967.)

It is crucial to show that the Strahler system of ordering is not arbitrary. For each stream of order u, there is a basin of order u. All lower-order basins are nested within higher-order basins. In addition, there are other small space-filling areas which do not flow into first-order streams, but flow directly into higher-order streams. These are called inter-basin areas (Schumm 1956). Inter-basin areas must be

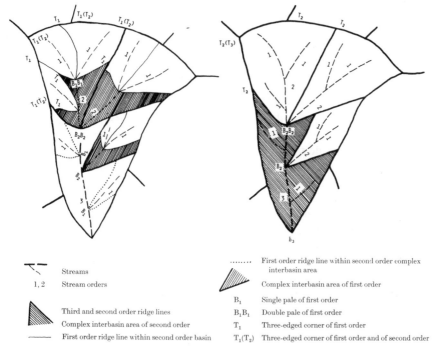

Fig. 12.4 Left. First-order polygons within second-order polygons.
Right. Basins and complex inter-basin areas of second order within a third-order basin.

found filling space within all basins greater than first order (Warntz 1968). It should be noted that inter-basin areas are also bounded by ridge lines and a course line (figs. 12.3 and 12.4).

We can see that while originally intended to order streams, the Strahler method actually orders basins. The seemingly arbitrary way of counting streams actually is an accurate and easy way to count basins.

Having counted basins, the next step is to delimit their areas in a rigorous way. We search for those corners and edges which will enable us to divide two dimensional space into polygonal areas. Two-edged corners cannot divide space into polygons by themselves. Corners with at least three edges are necessary. In addition, two-edged corners under deformation can form a straight line; the location of the corner is lost. Three-edged corners preserve a location for a point under all deformations as long as the surface is not torn. In our discussion of singular points, the question of how many lines meet at a point was not discussed.

Thus we must redefine our notion of what are the important points which will serve as corners between segments of ridge lines. A corner pale (equivalent to a double pale) is identified by the fact that three-line segments must meet at this pale. A single pale joins only two ridge-line segments and cannot be considered a corner in a network of ridge lines.

In a recent, as yet unpublished, paper, Christian Werner (1971) has suggested that polygons surround each link (a course line segment between two confluences). The boundaries of the polygons would be divides. The divides would be composed of ridge lines, and those slope lines which intersect a single pale. Such slope lines are not ridge lines or lines of divergent flow. They do not have topographic expression. However, using Werner's subdivision of the space, every point on the surface would drain into one and only one link. The polygons surrounding each interior link may be relatively small, compared to basins, and this is especially true when the analysis is shifted from the first to higher orders. These small areas surrounding interior links do not seem comparable to true basins in that they seem to have no minimum threshold size. Therefore for the purposes of this analysis, the single pales will have only two ridge lines, and no other divide lines.[1]

Werner's subdivision of the space is useful for his purposes, as he then shows that the number of areas equals the number of links, and he is able to apply relationships already developed for link analysis to the analysis of numbers of polygons.

Professor Cyril Smith (personal communication, June 1971) believes it an error not to count these single pales as corners. After all, the single pale corners are there, even though only two ridge-line edges meet at the corner. (Actually the corner is defined by three course-line edges or links.) But we must recognise that in the subdivision of a bounded surface into a network of polygons with ridge-line boundaries, at least one new interior three-edged corner is required for each new polygon created. Since no new polygon is created by the single pale (a two-edged corner), we choose to ignore it. It should be remembered that our 'average' of the number of sides refers to the number of sides enclosed only by three-edged corners. Should we count two-edged corners, we would have to add an additional two sides per single pale to the sum of

---

[1] The corners of Werner's polygons always have three edges.

the total number of sides for all the polygons in the net. This will, of course, increase the average for the number of sides per polygon.

The ridge-line segments which meet at a corner pale will be known as ridge-line edges, since they serve as the edges of polygonal areas. One ridge-line edge passes through a single pale, and three ridge-line edges meet at a double pale. All peaks, passes and pales are located on ridge lines. Some passes and pales join only two ridge-line and course-line segments. These points cannot serve as corners for a polygonal network of ridge lines. Yet there are many corners (besides double pales) where three ridge-line segments join at a point. Warntz (1968, 16a, 17) identifies this type of corner as a peak.

A careful analysis has shown that not all such tri-edged points are peaks. Since the technical naming of these points[1] other than double pales is not necessary for the task of simply locating corners for polygonal areas in the network, we shall call all places where three ridge-line edges come together a tri-corner; we may continue to use the term double pale. If more than three edges were to meet a point they would be called poly-corners. Then corner pales, tri-corners and just possibly poly-corners make up the total set of corners for the polygonal network of ridge lines. The edges of the network join the corners.

Suppose we want to define the sides or edges of a basin of order $u - 1$ (fig. 12.4). We trace a ridge line from a pale of order $u - 1$ to a tri-corner. A tri-corner is of order $u - 1$ if it is needed to delimit a basin of order u. A tri-corner may be simultaneously of more than one order. As this procedure is carried out on the several basins of order $u - 1$, we note that a considerable area lies outside the basins of order $u - 1$, but inside the basin of order u.

When analysing a second-order basin this outside space will be composed of simple inter-basin areas (Schumm, 1956) which drain directly into the second-order stream. Since we do not use course-line edges to delimit our polygons, we must identify the *complex inter-basin* area, that is, two or more simple inter-basin areas which form a continuous region bounded by ridge lines. This area crosses course lines. In this case the complex inter-basin area and the first-order basins lie within the second-order basin. Thus, we classify this complex inter-basin area as first order.

Similarly, when dealing with a third-order basin, the complex inter-basin area which lies outside the second-order basins is of second order. These complex inter-basin areas may contain isolated first-order basins as well. By analogy, when dealing with a basin of order u, there are basins of $u - 1$, and the complex inter-basin area is also of order $u - 1$. The latter may contain some isolated basins of order 1 to $u - 2$ as well as simple inter-basin areas.

[1] For a full discussion, see Woldenberg (1970, 9–11).

We now have defined the polygons of order $u - 1$ which lie within the basin of order u. Many of these polygons are basins of order $u - 1$. The polygon of order $u - 1$ is a complex inter-basin area. The fact that complex inter-basin areas have an order number probably implies geometrical progressions of their areas with order. This should be tested.

We have defined each tri-corner to be at the intersection of three ridge-line edges. In addition, each double pale is at the meeting point of three ridge-line edges. I will argue that three edges for each corner is the maximum for all but the most transitory of polygons.

The necessity of three-edged corners

Suppose, in fig. 12.5, corner F has four edges. Line AB passes through F. The ridge line is in a fairly stable position, although erosion on either side of the line may move the line to the right or left. Let CF meet AB at F. This line is likely to be fairly stable although erosion could move it to intersect AB at F_1 or F_2. The same things can be said for line DF. There seems to be no reason why CF and DF should meet AB at F. Although this *may* happen, differing conditions in each second-order basin would preclude the stability of a four-edged corner at F. Much more probable is a network with two three-edged corners (fig.

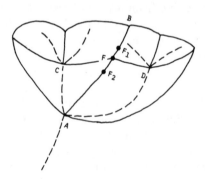

Fig. 12.5 Basin polycorners.

12.6). The reasoning here is perfectly general. Three-edged corners are clearly much more probable than four-edged corners, although it may be difficult to assign a value to the probability (Dacey 1963, 3).

Let us next consider whether more than three ridge-line edges may meet precisely at a pale. This implies three or more tributary streams join precisely at the pale. A stream may enter another stream in many alternative places. For it to join exactly where two other streams do so is highly improbable, unless there is some physical reason which makes this possibility much more probable than other possibilities. I can think of no compelling reason why this should ordinarily be the case. Therefore, in the absence of such a condition, we will assume that a third tributary joining at the existing double pale when there are so many other possible places to join the stream, creates a probability close to zero. The fact that the event of three stream branches joining to make one stream is so rare suggests that this reasoning may be correct.

Another network which is of interest is the pattern of market area boundaries which surrounds central places. Classical central place theory (Christaller 1933) makes clear that market areas of a given order are the same size, and they form a network of hexagons. The corners, of

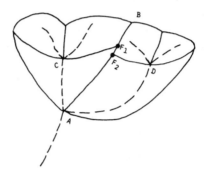

Fig. 12.6 Basin tri-corners.

course, are three-edged. This type of network ensures that the competition between the centres of the hexagons is minimised. For a given number of market centres, the total transport cost is minimised because the almost circular market areas are concentrated as close to their

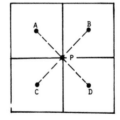

Fig. 12.7 Market area polycorners.

centres as possible. This implies that the money left over to buy goods after transport costs are subtracted must be at a maximum (Lösch 1954).

Let us see under what conditions the network of market areas might be square, rather than hexagonal. In fig. 12.7, in order for all points in each quadrant to be closer to the quadrant centres than to any other centre, there would have to be a rectangular or square lattice. If any three points (ABC, BCD, etc.) form a right triangle, the perpendicular bisectors of all three sides must intersect on the hypotenuse at point P. Point P is common to all four right triangles if there are horizontal and vertical axes of symmetry as illustrated in fig. 12.7.

In fig. 12.8 there is no axis of symmetry common to all four points. So as a result, point P must 'separate' into P_1 and P_2, each of which has three edges. Under what conditions could symmetry of the required sort persist over large areas for a group of market centres? Clearly a completely uniform surface is required even for first-order market areas. Because some of these points may serve as centres for high-order market areas, the first-order space will be distorted. The presence of roads will also distort the space. Thus the conditions of symmetry are immediately destroyed. Suppose that for some reason a four-edged corner appears. The dynamic nature of the economic system will cause the boundaries to move so that only in the most improbable circumstances will four-edged corners persist. The most probable state is the three-edged corner (Dacey 1963, 3).

The argument has not appealed to work (transport cost) minimisa-

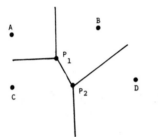

Fig. 12.8 Market area tri-corners.

tion. The hexagonal network is the optimum solution for this problem. This reinforces the tendency to produce the three-edged corners. It is interesting to observe that a uniform plain is also required for a perfect hexagonal net. However, should conditions not be perfectly uniform, the network will still have three edges per corner.

In networks of basin areas, or in networks of market areas, if conditions on the surface are reasonably uniform the threshold for first-order areas will be about the same for any given system. The same is approximately true for higher-order areas. Thus we are not confronted with a totally random pattern of corners and centres for our network. There would seem to be a bias against strong clustering of corners. There is, then, a fairly uniform scatter of three-edged corners, which approaches the condition necessary for a hexagonal lattice.

Having established the necessity for three-edged corners, I now will show that the average number of sides for all polygons in such a network tends to approach or equal six.

Why networks with three-edged corners tend to have six sides

This proof follows that of C. S. Smith (1952), who enlarged upon the original proof of Graustein (1932). Graustein's proof was developed for networks with corners of m edges and polygons of n sides, although he dwelt upon the case of $m = 3$, $n = 6$. Smith also developed some interesting corollaries in two dimensions, and similar theorems in three-dimensional space.

Euler's law for the number of polygons (P), edges (E) and corners (C) of a polyhedron is:

$$P - E + C = 2. \tag{1}$$

We divide the surface of the polyhedron into two regions. One region, which we will map on to a plane, will have:

$$P - E + C = 1. \tag{2}$$

the other region will simply be one polygon, whose edges and corners have already been counted.

We will assume that three edges meet at a corner for all corners. Interior corners are shared by three polygons and exterior corners belong to one or two polygons. Similarly, when counting the total numbers of sides for the polygons there are two sides per edge in the interior of the network, but on the periphery each boundary edge (Eb) bounds only one polygon.

Euler's law makes no distinction about interior or exterior edges or corners. Thus to count the numbers of corners and edges, if we assume the whole network behaves as the interior polygons do, we must make a correction for the peripheral polygons.

One way of correcting for the peripheral corners is simply to round all exterior edges and to preserve only those corners which have three edges, which bound, or are internal to, the network (fig. 12.9A). Thus two polygons meet at each exterior corner. Now, since three edges must meet at each corner and since each edge joins two corners:

$$C = \frac{2E}{3} \tag{3}$$

Also, since each polygon with n sides (Pn) contributes $n/2$ edges except for those edges at the boundary of the network (Eb), where the edge belongs to only one polygon, it follows that:

$$E = \frac{\Sigma \, n \, \mathrm{P}n}{2} + \frac{\mathrm{Eb}}{2} \tag{4}$$

Substituting (3) and (4) into equation 2 (where Σ Pn equals the number of polygons) gives:

$$\Sigma \, \mathrm{P}n - \frac{\Sigma \, n\mathrm{P}n + \mathrm{Eb}}{2} + \frac{\Sigma \, n\mathrm{P}n + \mathrm{Eb}}{3} = 1 \tag{5}$$

Simplifying:

$$\Sigma \, (6 - n)\mathrm{P}n - \mathrm{Eb} = 6. \tag{6}$$

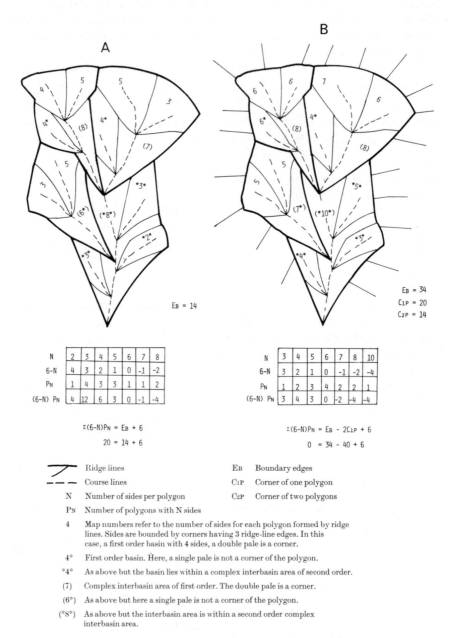

$$\Sigma(6-N)P_N = E_B + 6$$
$$20 = 14 + 6$$

$$\Sigma(6-N)P_N = E_B - 2C_{1P} + 6$$
$$0 = 34 - 40 + 6$$

		Ridge lines			E_B	Boundary edges

Ridge lines E_B Boundary edges

Course lines C_{1P} Corner of one polygon

N   Number of sides per polygon C_{2P} Corner of two polygons

P_N Number of polygons with N sides

4   Map numbers refer to the number of sides for each polygon formed by ridge lines. Sides are bounded by corners having 3 ridge-line edges. In this case, a first order basin with 4 sides, a double pale is a corner.

4*   First order basin. Here, a single pale is not a corner of the polygon.

4   As above but the basin lies within a complex interbasin area of second order.

(7)   Complex interbasin area of first order. The double pale is a corner.

(6*)   As above but here a single pale is not a corner of the polygon.

(*8*)   As above but the interbasin area is within a second order complex interbasin area.

Fig. 12.9 First-order polygons inside second-order polygons.

Equation 6 tells us that in a network which has *only* three-edged corners, if the total number of interior polygons is very large, Eb + 6 tends to be relatively small, and the average polygon approaches 6 sides. In all cases, we can deduce that the average number of sides cannot exceed 6.[1]

We pass to the case (fig. 12.9B) where some exterior corners belong to only one polygon (C_{1P}) and to 2 polygons (C_{2P}). In the expression for the total number of sides, $\Sigma\, n\, Pn$, we note that each side ends in two corners and that there are 6 sides (as distinguished from edges) which meet at every interior corner. Thus the number of interior corners, $C_{3P} = \Sigma\, n\, Pn/3$. At each C_{2P} only four sides meet, and at each C_{1P} only two sides meet. Therefore, to use the expression $\Sigma\, n\, Pn/3$ for the exterior as well as the interior polygons would undercount the number of exterior corners. A correction of $(6 - 4)/6$ or $\frac{1}{3}C_{2P}$ and $(6 - 2)/6$ or $\frac{2}{3}C_{1P}$ is necessary. Therefore,

$$C = \frac{\Sigma\, n\, Pn}{3} + \tfrac{1}{3}C_{2P} + \tfrac{2}{3}C_{1P} \qquad (7)$$

The number of peripheral edges is equal to $C_{2P} + C_{1P}$. These edges are unlike the interior edges in that they are not shared by two polygons. The total number of edges in the network is

$$E = \frac{\Sigma\, n\, Pn}{2} + \frac{C_{2P} + C_{1P}}{2} \qquad (8)$$

Combining equations (7) and (8) with (2) (ΣPn is the number of polygons)

$$\Sigma\, (6 - n)\, Pn = C_{2P} - C_{1P} + 6 \qquad (9)$$

Since $Eb = C_{1P} + C_{2P}$, we have

$$\Sigma\, (6 - n)\, Pn = Eb - 2C_{1P} + 6. \qquad (10)$$

This equation is used to facilitate counting.

Suppose we are given a perfect hexagonal network. We tear out an interior portion of the network, and we count the edges, corners and hexagons. This network will satisfy $C_{1P} - C_{2P} = 6$. Suppose we insert any polygonal network (where every corner has three edges) into the hole. The peripheral edges and corner angles of the network can be distorted to fit exactly and the average number of sides per polygon is exactly 6. This boundary condition is not difficult to satisfy. Slight changes in the number of C_{1P} can create the average values of $n < 6$, $n = 6$, $n > 6$.

Haggett (1965, 51; 1969, 49), Pedersen (in Haggett 1969, 50) and

---

[1] For concentrically arranged hexagons, 18 are required to produce an average of 5 sides per polygon. For less compact arrays of polygons, more than 18 are required. Networks of hexagons must be more than two deep in places to produce an average of 5 sides.

Boots (1970, 8–11), have found geographical networks which average between five and six sides per polygon. Boots recognised that three-edged corners are of critical importance but did not present a proof. Dacey (1963) assumes three-edged corners and derives an expected mean value 5·7888 based on a probability model. Dacey cites Meijering (1953), who used Smith's (1952) result which has been reproduced here.

We have seen that large compact polygonal networks with only three-edged corners must average between five and six sides. Of course, if $C_{1P} - C_{2P} = 6$, then the average number of sides will be six. Euler's law must take precedence over the probability model.

In suggesting why three-edged corners must be overwhelmingly predominant, I have not relied on the hexagonal model which is said to minimise work, though this idea has been exploited before by Lösch (1954) and Woldenberg (1968; 1969). We pass now to a short review of other phenomena whose three-edged corners may be attributed to a work minimisation constraint, or to the achievement of the most probable state for corners.

Some other hexagonal networks

D'Arcy Thompson in his magnificent book *On Growth and Form* (1942, 465–566) reviews the occurrence of hexagonal networks in nature. Starting with lessons learned from Plateaus's soap bubble experiments, Thompson suggests that cell membranes behave like soap bubbles in the process of achieving at equilibrium the utmost possible reduction of the surfaces in contact. This condition is arrived at when the surface tension energy per unit area is equalised for all surfaces. This leads to the following: (a) In every liquid system of thin films the sum of the areas of the films is a minimum. (b) The films meeting at any one edge are three in number. (c) The edges meeting in any one corner are four in number (Thompson 1942, 486, quoting La Marle). Smith (1952), following Thompson, makes the same argument for metal grains.

Suppose we pass a plane through the aggregate of soap bubbles. The films intersect the plane forming edges of polygons. As was stated, there are three polygons at each intersecting corner. The angle of intersection must be 120°. Four-edged corners occur as transitory forms, and are replaced by more stable three-edged forms.[1] This is true for soap bubbles, cell aggregates, metal grains and so forth. Such networks must have three-edged corners to minimise work.

We find three-edged corners in columnar basalt, ceramic cracks, frozen ground and mud cracks. The angles between edges are not always 120°. When a tension crack forms in ceramics, for example, a new crack

[1] For beautiful illustrations using soap bubbles, see Smith (1952).

will meet the old at right angles, but will not cross it. This is probably due to the slightly different conditions on either side of the first crack. In such cases, three-edged corners are overwhelmingly *probable.*

Sometimes polygonal networks with three-edged corners where angles do not equal 120° between edges may be modified so that the angles approach the ideal. In new tissues, new cell walls grow perpendicularly to old stable walls. Thompson shows that new cambium cells, looking exactly like a brick wall, eventually evolve to perfect hexagons (Thompson 1942, 480).

Summary and relevance to current research

In sum, the hypothesis of regular hexagonal areas deduced by Christaller and Lösch is modified by the argument presented in this paper. It was shown that it is extremely probable that no more than three edges meet at a corner in geographical networks of polygons. This fact leads us, through Graustein's (1932) application of Euler's law, to the fact that the *average* number of sides for polygons in such networks tends to approach six, as the network grows in size while remaining fairly compact.

Furthermore, since thresholds necessary to create first-order areas must be approximately the same in a given urban or fluvial system, the sizes of the polygons will remain above a minimum and below a maximum, and this quality should persist upward through the hierarchy of polygonal areas.

In addition, the reasoning of this paper about polygons on a surface can be applied to three-dimensional branching networks. Trees, or the pulmonary artery or airway system of the lung (Woldenberg *et al.* 1970), have discontinuous surfaces. For example, the leaves of a tree are discrete and not continuous with other leaves. Nevertheless, the areas exposed to the sun tend to fill with branches and leaves. When too many leaves cause too much shade, some leaves and branches die. In effect the leaves make a continuous surface served by twigs and branches; thus the tree mimics a fluvial system. One can make a similar kind of statement about alveolar or capillary surfaces in the lung. Apparently the cutting of these surfaces simply facilitates their packing into a small, portable volume. Certainly the branches may be bent and twisted so that a map of such a system looks not unlike a river system. The fact that the numbers of branches (basin areas) per order is in agreement with river systems suggests that a common geometric and topologic basis underlies rivers and these organic branching systems (Woldenberg, 1968; 1969; 1971; 1972; Woldenberg *et al.* 1970).

IV. The theory of mixed hexagonal hierarchies

The theory of mixed hexagonal hierarchies follows in the tradition of Christaller and Horton; that is, it is motivated by notions of space filling by a flow system which tends to accomplish the work it must do with a minimum expenditure of energy.

The data in table 12.1, to be explained by the theory, are the numbers of streams of branches of organic systems at each order, or the number of market areas at each order. Since each town of order u has one market area for all orders below order u, one simply cumulates the number of towns from highest to lowest order to obtain the number of market areas.

Table 12.1 A, B, C and D presents several sets of empirical data which may be compared to each other and to the convergent means (to be explained below) of the appropriate model systems, for each relative order. (Relative order is used here because only the top orders of the branches in the lung were measured.) I have observed that spatial hierarchies often resemble each other in terms of numbers of areas per order.

For branching systems (fluvial and organic in the tables to follow), the numbers of 'basins' are given by the number of Strahler ordered branches. The number of polygons (basins plus inter-basin areas) could be derived by applying the following rule: For every Strahler branch of order u there is a complex inter-basin area of order $u - 1$ (except when order $u = 1$). This can be seen in fig. 12.4. Thus cumulating the number of Strahler segments from highest order to lowest gives the number of polygonal areas per order.

The number of basins, rather than the total number of polygons, is compared with the models. This is done because there appears to be no minimum threshold sized area for a complex inter-basin area of any order, while there is a clear minimum threshold for a basin of order u. Each first-order stream must equal or exceed a threshold, and since at least two basins and a complex inter-basin area of order u are required to create a basin of order $u + 1$, there must be a minimum threshold size of basin of order $u + 1$. It is not difficult to imagine a very short stream of order $u + 1$, and hence a tiny inter-basin area of order u with no appreciable threshold size for even very high-order streams.

Because the ratios between orders for the numbers of 'basins', or the total number of polygons, is about the same, the fact that only 'basins' are counted will not affect the conclusion that the data could have been generated by the model.

Such consistency demands an explanation which can be summarised as follows: Assuming a continuous hexagonal area, one can generate a network of hexagons by dividing the original area into 3, 4 or 7 equal

Table 12.1A   *Model for pulmonary artery, left lung**

Order	Groups of powers of 3, 4, 7	Means		
		Geometric	Arithmetic	Convergent
u	1, 1, 1	1·00	1·00	1·00
$u - 1$	3, 4	3·46	3·50	3·48
$u - 2$	7, 9, 16	10·03	10·67	10·34
$u - 3$	27, 49	36·37	38·00	37·18
$u - 4$	64, 81, 243, 256	134·00	161·00	147·19
$u - 5$	343, 729	500·05	536·00	517·87

Empirical systems

Order	Branch from left lung of human pulmonary artery (*Singhal* in *Woldenberg* et al. *1970*)	Market areas of the central place system of *Vaasa, Finland* (*Palomäki 1963*)	Branch of paper birch
u	1	1	1
$u - 1$	3	3	3
$u - 2$	10	9	10
$u - 3$	38	36	36
$u - 4$	146	151	139
$u - 5$		513	

	Casselman River, Maryland (*Morisawa 1962*)		Piney Creek, Maryland (*Morisawa 1962*)		Mill Dam River, Maryland (*Schumm 1956*)	Seneca Lake Area, New York (*Horton 1945*)
u	1 × (·8) =	·8		1	1	1
$u - 1$	3 ,,	2·4	1 × (2) =	2	2	2
$u - 1$	11 ,,	8·8	4 ,,	8	8	10
$u - 3$	46 ,,	36·8	19 ,,	38	37	37
$u - 4$	190 ,,	152·0	77 ,,	154	150	141
$u - 5$	653 ,,	522·4	271 ,,	542		476

* Classified by Woldenberg (1971) as 729, 518, VI, 1.

Table 12.1B *Model for commonly found fluvial systems**

Order	Groups of powers of 3, 4, 7	Means		
		Geometric	Arithmetic	Convergent
u	1, 1, 1	1·00	1·00	1·00
$u-1$	3, 4, 7, 9	5·24	5·75	5·49
$u-2$	16, 17	20·78	21·50	21·14
$u-3$	49, 64, 81, 243	88·64	109·25	98·67
$u-4$	256, 343, 729	400·02	442·67	421·06

Empirical data

Order	Bell Canyon, Calif. (*Maxwell 1960*) (*Average*)		Wolfskill Basin, Calif. (*Maxwell 1960*)
u	$1 \times (1·1) =$	1·1	1
$u-1$	4 ,,	4·4	5
$u-2$	24 ,,	26·4	21
$u-3$	89 ,,	97·9	96
$u-4$	386 ,,	424·6	409

	Mill Creek, Ohio (*Morisawa 1962*)	Hiwasee River, Ga. $[\times \frac{1}{2}]$ (*Horton, 1945; Woldenberg 1969*)	Perth Amboy, N.J. $[\times \frac{1}{2}]$ (*Schumm 1956*)
u	1	1·0	1·0
$u-1$	5	5·5	4·0
$u-2$	22	21·5	22·5
$u-3$	104	94·5	101·5

* Classified by Woldenberg (1971) as 729, 421, V, 1.

Table 12.1C *Model for pulmonary artery, right lung**

Order	Groups of powers of 3, 4, 7	Means		
		Geometric	Arithmetic	Convergent
u	1, 1, 1	1·00	1·00	1·00
$u-1$	3, 4	3·46	3·50	3·48
$u-2$	7, 9, 16	10·03	10·67	10·34
$u-3$	27, 49	36·37	38·00	37·18
$u-4$	64, 81, 243	108·00	129·33	118·43
$u-5$	256, 343, 729	400·02	442·67	421·06

Empirical systems

Human pulmonary artery, right lung
(Singhal in Woldenberg et al. *1971*)

Order	Branch AB	Branch BC	Total of the two adjacent branches
u		1	1
$u-1$	1	3	4
$u-2$	4	8	12
$u-3$	13	25	38
$u-4$	46	75	121
$u-5$	158	261	419

	Upper Esopus Creek, N.Y. (Horton 1945)	Bovine liver, bile system (Williamson 1967, p. 43a)
u	1	1
$u-1$	2	3
$u-2$	11	12
$u-3$	36	38
$u-4$	126	117

* Classified by Woldenberg (1971) as 729, 421, VI, 1. This is a common system in biological branching structures.

Table 12.1D   *Model for airway systems, human lung**

Order	Groups of powers of 3, 4, 7	Means		
		Geometric	Arithmetic	Convergent
u	1, 1, 1	1·00	1·00	1·00
$u-1$	3, 4	3·46	3·50	3·48
$u-2$	7, 9	7·94	8·00	7·97
$u-3$	16, 27	20·78	21·50	21·14
$u-4$	49, 64	56·00	56·50	56·25
$u-5$	81, 243	140·30	162.00	150·95
$u-6$	256, 343, 729	400·02	442·67	421·06

Empirical systems
(Data collected by K. Horsfield, in Woldenberg *et al.* 1970).

Samples from left lung

Order	Branch BW	Branch BV	Total of adjacent branches
u			1
$u-1$	1	1	2
$u-2$	2	4	6
$u-3$	9	9	18
$u-4$	31	23	54
$u-5$	82	69	151
$u-6$	222	203	425

Samples from right lung

Order	Branch BK	Branch BI
u	1	1
$u-1$	2	3
$u-2$	6	7
$u-3$	18	19
$u-4$	57	55
$u-5$	142	143
$u-6$	424	421

* Classified by Woldenberg (1971) as 729, 421, VII, 1.

hexagons (fig. 12.2). Each area can then be redivided so that three geometric series of numbers are created: 1, 3, 9, 27, 81, 243, 729; or 1, 4, 16, 64, 256; or 1, 7, 49, 343. We mix these series and rank the numbers consecutively by size. Each number represents the number of hexagonal areas into which the large region may be divided. Choosing the first of the models in fig. 12.10, which corresponds to the data in table 12.1A, note that *consecutive* numbers are now formed into groups so that the group geometric or arithmetic means seem to approximate the empirical values of numbers of areas per order. A still better fit is

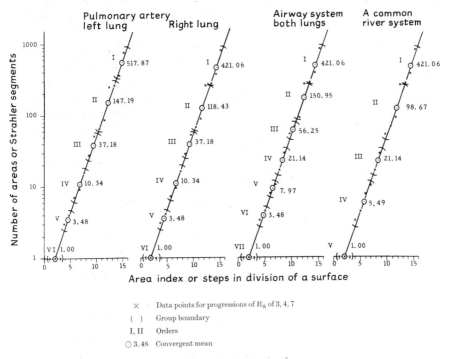

Fig. 12.10 Model hierarchies.

given by the convergent mean – a compromise between the group geometric and arithmetic means achieved by taking successive arithmetic and geometric means until the values converge. The compromise ensures the balance between many small areas, suggested by the arithmetic mean, which maximises accessibility to the region served by a branch (or town), and fewer larger areas, suggested by the geometric mean, which increases the possibility for increasing efficiency through size.

Only two groups are arbitrarily chosen to fit the data, the first and the highest order. The other groups are selected to approach as closely as possible the values of a perfect geometric progression which extends from order one to n. Alternative hierarchies and hence alternative groups are tested as follows: Each group convergent mean is divided by the corresponding value for the perfect geometric progression. The sum of the absolute deviations from unity is calculated, and the hierarchy is chosen which minimises this sum. This rule, hopefully the most general one,[1] supersedes similar but less general rules (Woldenberg 1968; 1969; 1971).[2]

This ensures that the values of powers of 3, 4 and 7 are as evenly represented as possible. The convergent mean values approach the sloping line in fig. 12.10 and the intervals between them are fairly constant.

Thus the mixed hexagonal hierarchy model can duplicate the numbers of areas per order. The question remains, why should some systems have high values for R_b and others have low values for R_b? Woldenberg (1971) has shown that for organic systems where the surface has been cut (as in the lung) R_b must exceed 2. When the surface is uncut (as in fluvial and central place systems) R_b must be greater than 3 and be less than 7 (see also Woldenberg 1969).

Table 12.2 shows the ratio of diameters[3] (R_d) in the airway and arterial systems of the lung compared to R_b.

Table 12.2　R_d/R_b *for the pulmonary artery and airways**

	Left lung	*Right lung*
Pulmonary artery	$\dfrac{1\cdot8121}{3\cdot5902} = 0\cdot50$	$\dfrac{1\cdot6252}{3\cdot2384} = 0\cdot50$
Airways	$\dfrac{1\cdot3998}{2\cdot7592} = 0\cdot51$	$\dfrac{1\cdot4006}{2\cdot8188} = 0\cdot50$

* Woldenberg, Cumming, Harding, Horsfield, Prowse and Singhal 1970, 51.

In the lung, R_d seems to be $\frac{1}{2}$ of R_b. Since resistance to flow is inversely proportional to the fourth power of the diameter (for laminar flow), we might suspect that diameters in the lung are related to the work the system must do.[4] Note that the airway system carries a gas

[1] For any later developments, see Woldenberg (in preparation).

[2] For a table of convergent means, see Woldenberg (1971, 171–3).

[3] Diameter of resin casts are measured at the point of confluence with a branch of equal order.

[4] Flow is turbulent in the arteries and airways. Equations for these flows are complex and still subject to disagreement. None the less diameters are critical to the resistance.

which has a low viscosity compared to the blood carried in the arterial system. Thus diameters can increase slowly with order in the airway system, but must increase more rapidly with the arterial system in order to gain scale economies.

We note that R_b is always twice R_d. The reason for this is unclear. However, it seems reasonable to suggest, for the lung at least, that R_b is dependent upon R_d. Could not measurements of cross-sectional area, taken at the confluence of streams of equal order, be made? The average square root of this value for each order might yield an analogue to R_d in the lung. Perhaps R_b will be seen as a function of R_d in fluvial systems as well. In the smaller left lung Rb for the pulmonary artery has a value 3·59; in the right lung it is 3·24 (taken over 6 orders). The higher ratio may be associated with a smaller space; the diameters of the system must increase from capillary size to the main pulmonary artery; this causes the system to branch more rapidly to 'finish' before the space is exhausted. This is not true of the airway system. The airways branch in similar fashion in both lungs. Perhaps the differences in space could be resolved sufficiently by simply having a smaller lung of similar structure.

I suggest, then, that the branching of a fluvial system is a function of work[1] and space requirements. The branching is accomplished (in the absence of structural controls) according to the rules of mixed hexagonal hierarchies. All other dimensionless ratios which are tied to order are inevitably related to R_b (for instance, the length ratio, R_L, is less than R_b, though it may well be greater than one half).

Acknowledgements

The author is grateful to the Geography Programs, Office of Naval Research, which funded this research under task order number NR 389–147. Miss Evelyn Pruitt, of the O.N.R., whose name is rarely mentioned in these acknowledgements, deserves my special thanks. In addition, Mr David Barer of the Queen Elizabeth Hospital, Birmingham, England, has generously given his time for programming and refining the hierarchy model. I am especially thankful to William Warntz, who provided me the opportunity and stimulation to conduct studies of spatial hierarchies. All errors in these papers are the author's responsibility.

References

BASKIN, C. W. (1966) Translation of *'Central Places in Southern Germany'* by Walter Christaller (Prentice-Hall, New Jersey).

[1] In the absence of a more refined model a random component can be added to predictions made on the basis of the hexagonal least-work model.

BERRY, B. J. L. (1967) *Geography of Market Centers and Retail Distribution* (Prentice-Hall, New Jersey), 146p.

BERRY, B. J. L. and GARRISON, W. L. (1958) Alternate explanations of urban rank–size relationships; *Annals of the Association of American Geographers* 48, 83–91.

BOOTS, B. N. (1970) An approach to the study of patterns of cellular nets; *Rutgers University, Department of Geography, Discussion Paper* No. 1, 24p.

CAYLEY, A. (1859) On contour and slope lines; *The London, Edinburgh and Dublin Philosophical Magazine and Journal of Science* 18, 264–8.

CHRISTALLER, W. (1933) *Die zentralen orte in Süddeutschland* (Gustav Fischer Verlag, Jena).

CURRY, L. (1964) The random spatial economy: An exploration in settlement theory; *Annals of the Association of American Geographers* 54, 138–46.

DACEY, M. F. (1963) Certain properties of edges on a polygon in a two-dimensional aggregate of polygons having randomly distributed nuclei; *University of Pennsylvania, Wharton School*, Mimeo, 20p.

FOK, YU-SI (1971) Law of stream relief in Horton's stream morphological system; *Water Resources Research* 7, 201–3.

GIBRAT, R. (1932) *Les inégalitiés economiques* (Sirey, Paris).

GRAUSTEIN, W. C. (1932) On the average number of sides of polygons of a net; *Annals of Mathematics* 32, 149–53.

HAGGETT, P. (1965) *Locational Analysis in Human Geography* (Arnold, London). 339p.

HAGGETT, P. and CHORLEY, R. J. (1969) *Network Analysis in Geography* (Arnold, London), 348p.

HORTON, R. E. (1945) Erosional development of streams and their drainage basins: Hydro-physical approach to quantitative morphology; *Bulletin of the Geological Society of America* 56, 275–370.

HOWARD, A. D. (1971) Simulation of stream networks by headward growth and branching; *Geographical Analysis* 3, 29–50.

LEOPOLD, K. B. and LANGBEIN, W. B. (1962) The concept of entropy in landscape evolution; *United States Geological Survey Professional Paper* 500-A, 1–37.

LÖSCH, A. (1954) *The Economics of Location* (trans. of 2nd edn by W. Woglom and W. Stolper) (Yale University Press, New Haven), 520p.

MAXWELL, J. CLERK (1870) On hills and dales; *The London, Edinburgh, and Dublin Philosophical Magazine and Journal of Science* 40 (4th Series), 421–7.

MAXWELL, J. C. (1955) The bifurcation ratio in Horton's law of stream numbers (Abst.); *Transactions of the American Geophysical Union* 36, 250.

MAXWELL, J. C. (1960) Quantitative geomorphology of the San Dimas experimental forest, California; *Office of Naval Research, Geography Branch, Task No. 389–042, Technical Report 19.* Republished, 1967, as 'Quantitative geomorphology of some mountain chaparral watersheds of Southern California', In Garrison, W. L., and Marble, D. F. (Eds.), *Quantitative Geography, Part II: Physical and Cartographic Topics* (Department of Geography, Northwestern University), 108–226.

MEIJERING, J. L. (1953) Interface area, edge length, and number of vertices in crystal aggregates with random nucleation; *Philips Research Reports* 8, 270–90.

MORISAWA, M. E. (1962) Quantitative geomorphology of some watersheds in the Appalachian Plateau; *Bulletin of the Geological Society of America* 73, 1025–46.

PALOMÄKI, M. (1964) The functional centers and areas of South Bothnia, Finland; *Fennia* 88, 1–235.

REECH, M. (1858) Demonstration d'une propriété général des surfaces fermées; *Journal de l'École Polytechnique* 37, 169–78.

RZHANITSYN, N. A. (1964) *Morphological and Hydrological Regularities of the River Net* (trans. from the Russian by D. R. Krimgold), U.S. Government Printing Office, Washington, D.C. (for U.S. Department of Agriculture).

SCHEIDEGGER, A. E. (1965) The algebra of stream-order numbers; *United States Geological Survey Professional Paper* 525-B, p. B187–B189.

SCHEIDEGGER, A. E. and LANGBEIN, W. L. (1966) Probability concepts in geomorphology; *United States Geological Survey Professional Paper* 500-C.

SCHENK, H. S., JR (1963) Simulation of the evolution of drainage-basin networks with a digital computer; *Journal of Geophysical Research* 68, 5739–45.

SCHUMM, S. (1956) Evolution of drainage systems and slopes in badlands at Perth Amboy, New Jersey; *Bulletin of the Geological Society of America* 67, 597–646.

SHREVE, R. L. (1966) Statistical law of stream numbers; *Journal of Geology* 74, 17–38.

SHREVE, R. L. (1967) Infinite topologically random channel networks; *Journal of Geology* 75, 178–86.

SHREVE, R. L. (1969) Stream lengths and basin areas in topologically random channel networks; *Journal of Geology* 77, 397–414.

SMART, J. S. (1971) Channel networks; Report issued by I.B.M. Yorktown Heights, New York, for the Office of Naval Research. (Forthcoming in *Advances in Hydroscience*.)

SMART, J. S., SURKAN, A. J. and CONSIDINE, J. P. (1967) Digital simulation of channel networks; In *Symposium of River Morphology; International Association of Scientific Hydrology, Bern*, 87–98.

SMITH, C. S. (1952) Grain shapes and other metallurgical applications of topology; In *Metal Interfaces* (American Society of Metals, Cleveland), 65–113.

STRAHLER, A. N. (1952) Hypsometric (area-altitude) analysis of erosional topography; *Bulletin of the Geological Society of America* 63, 1117–42.

STRAHLER, A. N. (1964) Quantitative geomorphology of drainage basins and channel networks: In Chow, V. T. (Ed.), *Handbook of Applied Hydrology* (McGraw-Hill, New York), 439–76.

THOMPSON, D'ARCY, W. (1942) *On Growth and Form*, 2nd Edn (Cambridge University Press).

WARNTZ, W. (1968) A note on stream ordering and contour mapping; *Harvard Papers in Theoretical Geography, No. 18, Office of Naval Research Technical Report, Project NR389–147*, also issued by the Laboratory for Computer Graphics and Spatial Analysis, Harvard University.

WARNTZ, W. and WOLDENBERG, M. (1967) Concepts and applications – spatial order; *Harvard Papers in Theoretical Geography, No. 1, Office of Naval Research Technical Report, Project NR389–147, also issued by the Laboratory for Computer Graphics and Spatial Analysis, Harvard University.*

WERNER, C. (1970) Horton's law of stream numbers for topologically random channel networks; *Canadian Geographer* 14, 57–66.

WERNER, C. (1971) *Patterns of drainage areas with random topology;* Unpublished Report, Department of Social Sciences, University of California, Irvine, California.

WILLIAMSON, M. E. (1967) *The Venous and Biliary Systems in the Bovine Liver* (unpublished Master's Thesis, Cornell University).

WOLDENBERG, M. J. (1966) Horton's laws justified in terms of allometric growth and steady state in open systems; *Bulletin of the Geological Society of America* 77, 431–4.

WOLDENBERG, M. J. (1968) Energy flow and spatial order – mixed hexagonal hierarchies of central places; *Geographical Review* 58, 552–74.

WOLDENBERG, M. J. (1969) Spatial order in fluvial systems: Horton's laws derived from mixed hexagonal hierarchies of drainage basin areas; *Bulletin of the Geological Society of America* 80, 97–112.

WOLDENBERG, M. J. (1970) The hexagon as a spatial average; *Harvard Papers in Theoretical Geography,* No. 42, Office of Naval Research Technical Report, Project NR 389–147, also issued by the Laboratory for Computer Graphics and Spatial Analysis, Harvard University.

WOLDENBERG, M. J. (1971) A structural taxonomy of spatial hierarchies; In Chisholm, M., Frey, A. and Haggett, P. (Eds.), *Regional Forecasting,* Colston Papers, Vol. XXII (Butterworths, London), 147–75.

WOLDENBERG, M. J. (in preparation) *Spatial Hierarchies in Physical, Biological and Socio-Economic Systems* (Columbia University Press, New York).

WOLDENBERG, M. J. and BERRY, B. J. L. (1967) Rivers and central places; analogous systems?; *Journal of Regional Science* 7, 129–39.

WOLDENBERG, M. J., CUMMING, G., HARDING, K., HORSFIELD, K., PROWSE, K. and SINGHAL, S. (1970) Law and order in the human lung; *Harvard Papers in Theoretical Geography, No. 41, Issued by the Laboratory for Computer Graphics and Spatial Analysis, Harvard University.*

PART VI
Simulation

13 Some spatial aspects of the analysis of coastal spits

C. A. M. KING

Department of Geography, University of Nottingham

Introduction

Some geomorphological features are more amenable to spatial analysis than others, in that they occur in large numbers and can be represented as point patterns. Others, however, occur as individuals and each must be studied on its own merits. A large drumlin field exemplifies the first situation, while coastal spits fall into the second category. Each spit must be considered individually, although the operation of processes common to them all can often be recognised in the form of spits in general. They are all formed by waves and usually respond in a similar way to similar processes. Spits have another characteristic that makes their analysis rewarding. They develop very rapidly and changes in form can be recorded over short time intervals. This rapid speed of development means that both time and space can be introduced into the analysis of spit forms.

This contribution aims to describe methods of studying the form of spits over both time and space. In the first section trends in the development of the spit at Gibraltar Point in Lincolnshire are considered, as an example of simple time-series analysis. In the second section the form of the spit at one point in time is investigated by curve fitting methods. In the final section both time and form are considered together in a simulation model of Hurst Castle Spit in Hampshire. The simulation model allows prediction by extrapolation, and thus enables changes in form to be considered over time both in the past and future.

Some general points concerning spit development should be made before the examples are cited. The relative abundance and complexity of spits around the coasts of the world at present, is the result of the immaturity of all coastlines. It is only at most about 3000 years since sea level attained close to its present height along the stable coasts of the world, following the rapid Flandrian transgression, during which sea level rose by nearly 100 m during the previous 15,000 years. Thus modern spits have had at most a few thousand years to form.

Another result of the very large and recent fluctuations of sea level has been the creation of an intricate coastline in many areas. River incision and glacial deposition at times of low sea level, provided an uneven surface across which the sea has risen. Spits are most likely to form where the coastline is irregular, and their form will also be more complex as a result. Where the coastline is irregular longshore movement of material will be unlikely to be uniform, and spits will form in those areas where more material is reaching a stretch of coast than is leaving it. Such areas will be determined by the wave pattern in association with the coastal outline. In some areas the tidal streams also play an important part. Of the two areas exemplified, the Lincolnshire coast is influenced both by the tide and the waves, while Hurst Castle Spit is probably influenced more by the waves than the tide, although the latter does play a part. In both areas the position of the spit is influenced by the form of the coast, as each spit prolongs the direction of the mainland coast where this turns abruptly inland. The features are, however, of very different ages, although both are fairly small.

The spit at Gibraltar Point can only have been forming since 1922, when it was initiated by a storm, which truncated the old marsh and dune deposits, driving the latter over the former. The spit developed from the angle of the coast at Gibraltar Point (fig. 13.1). Its development has been studied in repeated surveys at roughly annual intervals since 1951. It is dependent for its growth on material supplied from the north along the shore by wave action. Changes that are taking place on the beach to the north are relevant, therefore, in analysing the growth of the spit.

Where the movement is predominantly in one direction along a coast, as it is on the Lincolnshire coast, the updrift beaches supply the downdrift ones. The situation here is such that it can lead to cyclic development of coastal features. In the Gibraltar Point area, accretion is taking place to the north of the spit and material that should reach the spit is being increasingly absorbed further north, or diverted along the beach at too low a level, to build the spit up further. The spit may therefore decline in time and a new one will form further offshore, as has happened in the past. The present spit lies further seaward than one which formerly prolonged the earlier line of dunes to the west of the mature marsh (fig. 13.1). A reverse cyclic procedure, associated with coastal erosion and inland transference of successive spits, has

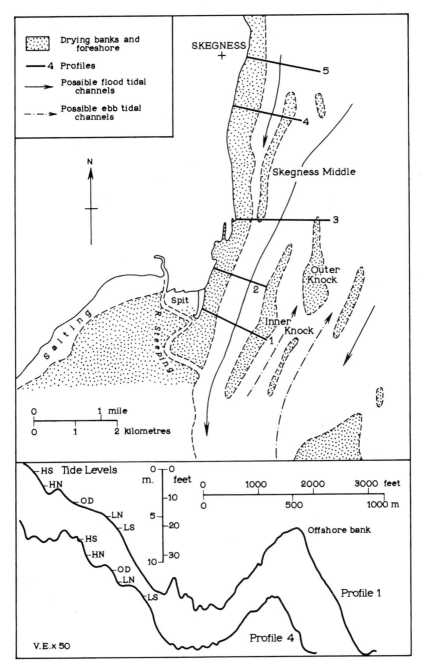

Fig. 13.1 Location of profiles and the spit at Gibraltar Point,
Lincolnshire.

been established by G. de Boer (1964) for Spurn Point at the mouth of the Humber to the north.

Spit changes over time

The spit is a small feature, only just over 350 m long when it was first surveyed in 1951. It consists of sand with some shingle, and at times has a recurved narrow distal end, which is very mobile, while its proximal end is lower and flatter. At high tide the distal end often forms an island as the tide washes across the low root area of the spit. Detailed measurements reveal that the spit has in general been increasing in length, area and volume. The increases in length have been rather variable although the growth in volume has been more regular. The trend equations indicate the rate of change of the different dimensions. The increase in area has been slow above 10 ft, indicating that the spit has gained material mainly by an increase in height and length. The increase in height has enabled vegetation to renew its foothold on the spit, a process that further aids the upgrowth of the spit, by trapping sand blown on to the spit. The maximum difference between the trend line and the actual volume of accretion is $22\cdot8 \times 1000$ ft^3, a value less than the mean annual increment of $24\cdot4 \times 1000$ ft^3 over the period from 1951 to 1969.

The supply of material is regular. Sediment is carried by waves southwards along the beach from the area of accretion to the north. Much of this material is brought into the area by tidal streams, which have formed a complex offshore relief of tidal banks and channels. The offshore relief has a twofold influence on beach accretion. It directs the tidal streams along the channels, allowing the flood stream residual, in the channel nearest the shore, to carry material up on to the lower foreshore a short distance to the north of the spit. The banks also shelter the beach from destructive waves, and ensure that some of the sand moving into the area along the coast from the north is trapped in this zone. The material is forming a growing ness of accumulation just south of Skegness a few miles north of the spit.

The relationship between the measured growth on the spit, and the accretion on the beach at the north of the spit, provides a useful means of studying the process of spit growth. The trend for the volume of the spit above 10 ft O.D. is given by $Y = 310 + 24\cdot4X$ for the period 1951–69. The value 310 gives the volume in the middle of the period of observation, 1960, in 1000s ft^3, while $24\cdot4$ is the mean annual addition of material in the same units. Y is the volume and X is the year. Over the period 1952–68 the spit increased in length at the 10-ft level by a mean amount of 30 ft/year, as the trend equation is $Y = 686 + 30X$, where 686 ft represents the mean length from a fixed position. The trends for the other one-foot contours are respectively:

11 ft − $Y = 525 + 40X$, 12 ft − $Y = 383 + 47 \cdot 5X$, for 1952–68.

The values of the coefficients indicate that the higher contours have grown in length more rapidly, thus steepening the distal end. The trend equations for the areas are:

Above 10 ft − $Y = 259 + 1 \cdot 56X$ Above 11 ft − $Y = 139 + 8 \cdot 07X$
Above 12 ft − $Y = 62 \cdot 3 + 9 \cdot 9X$, for the period 1951–69.

These equations show that the maximum increase in area took place at the higher levels, with little increase at the 10-ft level. Only 1560 ft² were added each year at 10 ft, compared with 8070 ft² at 11 ft and 9900 ft² at 12 ft.

The material that increased the volume and height to the spit has come from the north and has been added to the uppermost part of the foreshore, as the mean high spring tide level is 10·9 ft O.D. Liverpool. The process by which sediment moves to the spit is related to wave drift along the foreshore, assisted by tidal streams, which have a due southerly direction at high·tide. The upper foreshore is, therefore, only influenced by southerly flowing tidal streams. The importance of material derived from the north in the build-up of the spit can be assessed quantitatively by relating the accretion on the spit, in terms of its volume, to the accretion on the foreshore in the area both to the north and immediately in front of it. Three profiles have been surveyed over a long period of time across the foreshore in this area. Profile 1 lies immediately in front of the spit. Profile 4 lies 5 km further north, just south of Skegness, and profile 1 lies about 800 m north of profile 1 and the spit (fig. 13.1). The annual mean change of volume on these profiles has been measured by recording the net loss and gain between surveys over a unit width of beach, and constructing trend equations for the results. The trend equations for the three profiles are as follows:

Profile 1 $Y = -1 + 43 \cdot 42X$, Profile 2 $Y = 826 + 114 \cdot 68X$,
Profile 4 $Y = 1504 + 190 \cdot 71X$ for the period 1953–69.
Values are in 50s ft²/ft.

The results show that profile 1, immediately in front of the spit, has gained the least material, and in fact over the whole period the constant of the equation is slightly negative. The coefficient is positive, because the slow and steady loss of sand between 1953 and 1965 was changed to a net gain over the period 1965–9. Over the period 1952–65 the trend coefficient was negative. The changes on profiles 2 and 4 are very different and show a rapid rate of accretion, which is considerably higher at profile 4 than at profile 2, amounting respectively to an annual gain of 190·71 and 114·68 in the same units.

There is, therefore, material accumulating and available to the north to build up the spit, but it is not so easy to show conclusively that the material moves south to the spit. The pattern of ridge movement, however, indicates a regular southerly transference of material. The ridges are closely associated with the accretion, as shown by the correlation of ridge height with accretion (King 1970). As the ridges increase in height, so the volume of accretion increases. The accretion explains 80% of the variation in ridge size, when the profiles are controlled by analysis of covariance. The cause of the higher ridges on profile 4 is

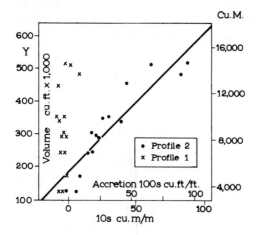

Fig. 13.2 Correlation of accretion on profiles 1 and 2 with changes in the volume of the spit.

the coarser material on this part of the beach. The sequence of surveys shows that the ridges move steadily landwards on each profile. This movement results from the bodily transfer of the ridges southwards by the pushing of sand over their crests by the waves. The ridges move landwards on each profile because they diverge slightly from the shore southwards, in response to the direction of approach of the long refracted swells from north and north-east, the direction of maximum fetch. The level of the foreshore at which the maximum amount of sand moves, therefore, is related to the position of the ridges on the beach at any one time.

The relationship between accretion on profile 1 and profile 2 and the increase in the volume of the spit above 10 ft O.D. is shown in fig. 13.2. The graph shows that there is a close correlation between the accretion on profile 2 and the spit volume but there is no relation between the changes on the foreshore in front of the spit and the increase in spit volume. The correlation between accretion on profile 2 and the spit volume is 0·90. Because both variables may be related to similar changes over time in this instance, the correlation may be spurious and it does not necessarily indicate a causal relationship between the variables. A test for autocorrelation of time series shows that in fact both sets of values are highly significantly autocorrelated. The K value for accretion on profile 2 is 0·0192 and for spit volume is 0·00145. Both values are far below the 99% significance level of 0·84 for $n = 10$. The time series are thus positively autocorrelated, and the individual values cannot be considered independent.

In this area, however, where southerly sediment transfer can be demonstrated by ridge movement, it is likely that the relationship is causal. The lack of relationship between accretion on profile 1 and the spit volume indicates that the sand moving south moves only along the upper foreshore to build up the spit. This may be due to lack of wave action on the foreshore in front of the spit because an offshore bank protects it from the waves when the tide falls only a short distance down the beach. At high water, when sand movement must take place, the water depths offshore will be greatest and waves, therefore, less refracted, so that longshore drift can take place most effectively. The waves will also be higher, The tidal streams in addition give a net movement south, especially near the high water level.

Spit form

The outline of the spit has changed during the period of annual surveys from 1951 to 1970. Sometimes the spit has been relatively straight, while at others it has been hooked. A method of assessing the form of the beach outline has been developed by Yasso (1964). It has been applied to the outline of the crest of the Gibraltar Point spit. A geometrical form that fits the equilibrium beach outline, in situations where one dominant wave type can form a fairly constant refraction pattern, is the logarithmic spiral. The beach must be free to align itself parallel to the direction from which the refracted waves reach the foreshore. This situation applies to spits that are wave-formed features and hence reflect the pattern of wave approach. The spit at Gibraltar Point is built by the long, constructive refracted swells that come from a northerly quarter. This significance in terms of ridge alignment and spit nourishment has already been considered. They also are important in determining the shape of the spit outline.

The logarithmic spiral has an increasing radius of curvature away from its initial point, which represents the distal point of the spit. This is usually recurved shorewards as the waves become refracted around the distal end. The equation for a logarithmic spiral is $r = e^{\theta \cot a}$ where r is the length of the radius vector and θ is the angle of the vector from some convenient datum direction. The angle α is the spiral angle.

It is the angle between the radius vector and the tangent to the spiral curve at that point. For any one logarithmic spiral it is a constant, and for curves fitted to beach outlines it normally varies between 30 and 90 degrees.

One problem in fitting spirals to beach outline is to establish the optimum centre from which the spiral best fits the beach outline. The exponential form of the spiral equation, $r = e^{\theta \cot a}$, can be written in logarithmic form, $\log_{10} r = \theta \cot \alpha \log_e$, or $\log_e r = \theta \cot \alpha$. Cot α is a constant for any one spiral curve. The linear relationship between $\log r$ and θ must be optimised. The relationship between $\log r$ and θ can be expressed as $\log_e r = a\theta + b$, where a is the slope of the regression line and b is the regression constant. The value of r and θ vary with the origin of the spiral. The highest value of r, the correlation coefficient, can be found from repeated trials centres, and the mean squared difference can be reduced to a minimum in the same way.

The method was used to fit the best spiral form to the 1969 outline of the spit, when it had a well-developed recurved form. Repeated trials revealed a point at which the value of r was 0·9980, and the mean squared error was 0·031. Another point had a correlation of 0·9970 and a mean squared error of 0·029. The spiral curves drawn from these two points are entered in fig. 13.5, which shows the outline of the spit for comparison. The spiral angle was 48·4 and 38·5 for the two curves. The angle increases as the spiral starting point moves away from the distal tip of the spit, because the tangent becomes more nearly normal to the vector as the centre moves away from the distal tip. The fitting of the spiral from a number of centres is facilitated if a computer program is used, and one was written by P. M. Mather to perform the necessary calculations in this instance.

The spit in 1969 showed a close approximation to the logarithmic spiral form, as it had a well-developed recurved end at this time. The high value of r and low value of the mean squared error shows that the form was well developed. The close approximation suggests that prior to the survey, long waves from the north had been acting to drive material along the spit. This is shown by the increase in length between 1968 and 1969, an increase that followed three years during which the length had been reduced. The reduction was partly due to the cutting of the River Steeping across the beach just south of the spit. The elongation of the spit was achieved by the addition of the recurve to the narrow and steep distal tip of the spit. As the spit has grown, it has been moved by the waves increasingly inland over the marsh deposits accumulating in its lee. This movement means that refraction has pushed the distal tip further to the west. As a result the distal curvature has increased, because refraction has become greater as the spit has moved west. The horizontal movement of the spit crest has

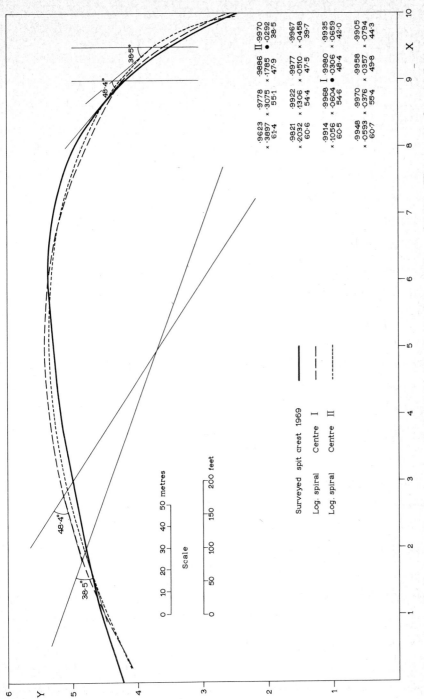

Fig. 13.3 Logarithmic spirals compared with the Gibraltar Point Spit outline of 1969. The upper figure of each set gives the correlation coefficient between r and θ, the middle figure gives the mean squared difference and the lower figure gives the spiral angle.

amounted to more than 200 ft (60 m) at a point 1000 ft (300 m) from the longest distal tip in 1964, while the tip itself has moved even further.

The outline of the spit has not always shown a good spiral fit. Curves have been fitted for the years 1960, 1962, 1967, 1968 and 1970. In 1960 the spit was straight with only a minor and tight recurve, which was not of log spiral form, in view of the relatively low values of the correlation coefficient obtained, the best being 0·970. By 1962 a large recurve had formed but the degree of curvature was again excessive. The distal tip lay parallel to the outer part of the spit. The values for the correlation coefficient were also low in 1966, when the spit was almost straight, with no recurve at all. By 1967 a small tight recurve had formed, very similar to that of 1960. A larger recurve had formed by 1968, but again the spiral fit was not very close, because the spiral did not bend round enough near the tip. A correlation of 0·9823 was obtained, however, showing a fair relationship between the curve and the spiral. The 1970 curve had straightened out somewhat from the best fit of 1969, but the log spiral form was reasonably well maintained, with a maximum correlation of 0·9951 and a mean squared error of 0·029, values only marginally less good than those for 1969.

The results suggest tentatively that the spit is now reaching a stage where it is almost in equilibrium with the long waves that form it, and has become adjusted to their refracted form. This has been achieved by the movement of the spit westwards over the marsh so that the alignment now reflects the pattern of the refracted swell more closely than it did in the early years of the last decade. Since 1965 it has had a much steeper distal end, which would allow more effective waves to reach it at high tide, and hence to form a spiral of rather more regular form. The very tight recurve characteristic of the spit outlines analysed in the earlier years were formed when only very short waves could reach the distal part of the spit. These short waves would come from the east or south-east. They would not be related to the long ones that formed the main line of the spit. The lack of good fit along the whole of the spit outline would have been related to the operations of two different wave types. A good fit is only possible when the whole spit reflects one wave throughout its length. This can occur when the water round the tip is steep enough to allow the long waves coming from the north to be refracted round it to form one complete logarithmic spiral.

Changes of spit form over time – simulation model

Some spits have a form that is determined by the action of several distinct wave types. Hurst Castle Spit in Hampshire provides an example of a complex recurved spit of this type. A computer simulation model has been developed to study the processes that are responsible for the spit form and to predict its possible future development. The details of the program and of its application to the study of the spit form have been published elsewhere (McCullagh and King 1970; King and McCullagh 1971). Only a brief summary of the most essential elements of the study are given to exemplify the possibilities of this

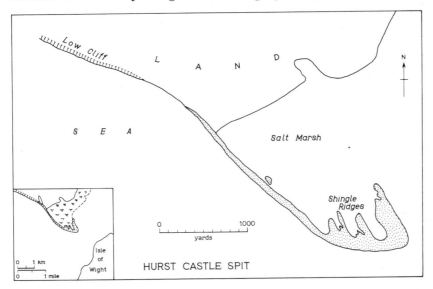

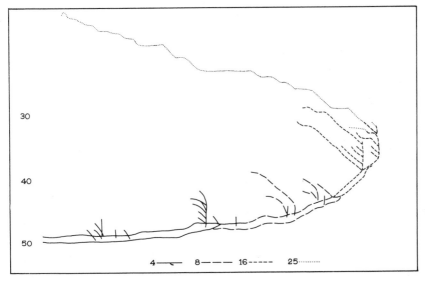

Fig. 13.4 The best-fit computed Hurst Castle Spit (below) compared with the real spit. The spit outline at four stages is indicated, stage 16 being the stage of best fit and stage 25 the extrapolated form.

type of model. The shingle spit is $2\frac{1}{2}$ km long and it consists of one main ridge, which has a large curve leading up to the distal point. There are a series of recurved ridges that join the main ridge at an acute angle, and which become more numerous and longer towards its distal end, as shown in fig. 13.4. The spit is largely supplied by shingle from the west, brought into the area by waves with a westerly component by beach drifting. The main spit crest is built up by storm waves, which approach from the direction of maximum fetch to the south-west. The waves from south or south-east are short, owing to the shelter of the Isle of Wight, and these move shingle around the distal tip on to the recurves. The recurve ridges are built up by waves from the north-east coming down the Solent. Apart from the different wave types that influence the form of the spit, it is controlled also by the increasing depth offshore, which has the effect of slowing down its easterly growth. This factor increases the number and length of recurves towards the distal end of the spit. The other process of importance is the refraction of the long waves, which determine the alignment of the main ridge. This factor determines the curvature of the spit. All these variables have been incorporated in the computer simulation model.

The simulated spit is built up element by element on a 50 by 60 matrix. Random numbers determine the order in which the subroutines that represent the different wave types are called. In addition the depth and refraction factors are added to the subroutine that simulates the operation of westerly waves. The depth factor operates by fixing a column at which the depth function changes from linear to exponential, and another column where depth becomes infinite, thus fixing a limit to spit growth. It operates through selection of random numbers that make the elongation of the spit by waves from the west less likely. The refraction factor operates by allowing the spit to move up the matrix if the random number selected lies within a specified range, which can be modified in the data input. It is thus possible to test various proportions of different wave types, variations in the depth pattern and in the operation of refraction. Different initial random numbers from which the others are derived allow the effect of varying the order of operation of the subroutines to be tested.

The results of a number of tests indicate that the initial random number does not affect the form of the spit very much, thus leading to the conclusion that the order in which the different types of wave operate is not important in determining the form of the spit. Tests to determine the effect of different proportions of the various subroutines showed that, when there was a superabundance of westerly waves, the spit became too long and insufficient recurves formed. When too many south-east or north-east waves operated too many recurves were created. The optimum proportions agreed closely with recorded wind frequencies, if allowance was made for wasting many westerly wave random numbers by the depth and refraction factors. The optimum values for the depth factors were found by testing a variety of possible values. The best-fit spit was formed when the increase of depth was linear to column 35 and exponential to column 55. By comparison with the chart of the area, these values can be related to the true spit by means of a scale of 10 columns of the matrix to 0·5 km in reality. This would place the depth changes at 1·7 km and 2·7 km from the root of the spit. In fact the depth increases linearly to a point 2·1 km from the root and then more rapidly to 2·6 km at the tip of the spit. At this point the depth increases very quickly to 57 m only 350 m from the distal tip. The depth is 12 m at the first point and 20 m at the second. The values simulated are, therefore, realistic. The correct simulation of the depth pattern is very important in producing a good approximation of the real spit in the computer model. If the depth is too shallow not enough recurves form, while if it is too great too many occur too close to the spit root in the model.

The curvature of the main line of the spit is influenced by the refraction of the storm waves, coming from the south-west, which build up the main part of the spit. The refraction in turn is influenced by the offshore relief. The presence of the Shingles Shoal is important in this respect. At present the shoal is situated off the proximal two-thirds of the spit. It retards the waves approaching this part of the spit, but they can advance faster at the distal end of the spit where the offshore depths are greater, hence producing the bend of the spit. The waves coming from the south-west thus are refracted to approach more from the south in the distal part of the spit. The refraction factor simulates changes in the position of the shoal water area. When the refraction factor is low, the shoal and deep water areas are represented as being further west. When it is high the shoal area is represented as being continuous, thus preventing waves from swinging round in deep water to approach from further south. The intermediate value that gave the best curvature represents the shoal in its correct position relative to the real spit. The effect of changes in the position of the shoal could thus be simulated by varying the refraction factor.

The standard simulated spit compares very well with the outline of the real spit (fig. 13.4), indicating that the optimum values for the input variables have been adequately established. These values can also be shown to be realistic when they are compared with the controlling variables that determine the form of the real spit. The future growth of the spit may be suggested by continuing the development beyond the stage of best fit. More recurves were built at the distal end of the spit. The great length of some of these is not, however, realistic. This is due to the omission of some essential controls from the model. No allowance has been made for the growth of marsh behind the spit, for instance. The value of such models depends, therefore, on the correct simulation of reality. In view of the great simplification of the situation in this model, which in reality is highly complex, the results are remarkably consistent. Their application must, however, be treated with caution. Nevertheless such simulation models do provide a useful means of rapidly testing a large number of different conditions and controls, once the computer program has been made operational.

Conclusions

The analysis of form in geomorphology can have several purposes and can be carried out in many different ways. The examples that have been considered briefly are concerned with aspects of coastal spit formation. They are intended to illustrate how form can be related to process, and to the development of the feature through time. The first method of form analysis was directly related to time, via trend analysis of the changing dimensions of the spit over a period of nearly two decades. The trends can be used to relate the changes on the foreshore to the accretion on the spit, thus suggesting processes by which the material reaches the spit. Problems of autocorrelation in time series cause some complications, when processes are related to time and care must be used not to draw false conclusions from correlations of this type. In the example cited independent evidence reduces this danger.

The second example was aimed to express more accurately the form of a spit, which is one of its important spatial characteristics, by a curve fitting technique. The logarithmic spiral curve fits the spit shape closely at some periods, and increasingly so through time. The best fit obtained was for the 1969 and 1970 outlines of the spit crest. This suggests that now the spit has deeper water off its distal tip one set of long waves can influence its form so that its whole outline is closely approximated by a single spiral. The degree of curvature is probably a function of the length of waves to which it has become adjusted. The more open curvature of 1970 may be the effect of less refracted waves than those that formed the outline in 1969, when the distal curvature was quite strong. The very tight curvature of former years was probably due to small waves from another quarter, when the water at the end of the spit was very shallow.

The third example was concerned with simulation of spit formation. A computer program was based on a number of subroutines that simulated four individual wave types as well as other factors. These variables are thought to be responsible for the distinctive form of Hurst Castle complex recurved spit. The computer model produced a spit very similar in outline to the real spit, when the optimum range of random numbers were allocated to the different subroutines, and when the depth and refraction conditions were simulated optimally. These last two variables were found to be particularly important in the optimum simulation of the spit.

References

BOER, G. DE (1964) Spurn Head: Its history and evolution; *Transactions of the Institute of British Geographers* 34, 71–89.

KING, C. A. M. (1970) Changes in the spit at Gibraltar Point, Lincolnshire, 1951–1969; *East Midland Geographer* 5 (1 and 2), 33 and 44, 19–30.

KING, C. A. M. and MCCULLAGH, M. J. (1971) A simulation model of a complex recurved spit; *Journal of Geology* 79, 22–37.

MCCULLAGH, M. J. and KING, C. A. M. (1970) Spitsym, a Fortran IV Computer program for spit simulation; *Kansas Computer Contribution* No. 50, 20p.

YASSO, W. E. (1964) Geometry and development of spit-bar shorelines at Horseshoe Cove, Sandy Hook, New Jersey; *Office of Naval Research, Geographical Branch, Task No. 388–657, Contract Nonr 266(68) Technical Report No. 5, Department of Geology, Columbia University*, 166p.

14 Digital simulation of drainage basin development

B. SPRUNT

Department of Geography, Portsmouth Polytechnic

Introduction

In recent years, quantitative geomorphologists have carried out a great deal of research into the nature of fluvial processes and their effects on the natural landscape. Drainage basins have been studied as basic geomorphic units within the landscape (Chorley 1969B), and many of their properties have been determined through the techniques of morphometric analysis (Strahler 1964). Despite these efforts, little information has come to light which provides quantitative evidence of the way in which drainage basins have evolved through time.

Several empirical and theoretical laws have been derived which relate geometric and hydrological variables for drainage basins as we find them at the present time. Detailed studies of erosion processes on hillslopes and in drainage channels, have established empirical laws which enable erosion rates to be expressed, for example, in terms of slope magnitude and contributing areas (Kirkby 1969 and 1971).

Digital simulation studies in this field are attempts to investigate the logical consequences of accepting these empirical laws and applying them over a period of time to real or synthetic drainage basins. For precisely defined initial and boundary conditions, landforms will be predicted by the laws, and numerical representations of these landforms may be computed. If the predicted landforms for a particular law are adequate approximations to real landforms, then the computations may be performed again to observe the outcomes when the form of the law is varied in a controlled and systematic manner.

The nature of the simulation process provides a sequence of developing basin forms which, at any stage, may be subjected to morphometric analysis and compared with observations from real basins. It is hoped that the observed behaviour of the simulated basin will lead to insights into the evolutionary processes governing real drainage basins.

The potential value of an adequate simulation study is extremely high for the geomorphologist because it enables him to simulate in minutes with the aid of a digital computer, sequences of events covering thousands of years in real time. Several attempts have been made to simulate drainage basin development, or to simulate certain structures within a basin, but to date the results have been of little value other than to stimulate further investigation. This chapter reviews some of the work that has taken place and describes a working model for future research.

The nature of simulation studies

The term simulation has been applied with a variety of shades of meaning in geomorphological research, as in other fields of research. A very general view of the nature of simulation regards simulation as 'essentially a working analogy' (Chorafas 1965). This is too general for many applications so it is convenient here to restrict attention to what is meant by the nature of simulation as it arises when digital computers are used in the study of systems.

Parts of a system which interact and behave collectively like the system are called components of the system. Components may sometimes be systems themselves, i.e. subsystems. At a particular instant in time, the system is in a specific state which is determined by the instantaneous states of its components. A record of the state of a system is a state description and applies to an instant in time so that a succession of state descriptions in chronological order may be regarded as a state history of the system.

Essential to simulation studies is the concept of a model of a system. Models in geomorphology have been discussed at length elsewhere (Chorley 1967), so that only a few comments will be made here. There may be several possible models of a system but no single model can represent all aspects of a real system. Even when a suitable model of a system has been constructed, that model does not represent the activity or behaviour of the system. It is a state history which provides the necessary representation of behaviour of the system.

Using the terminology introduced above, Evans, Wallace and Sutherland (1967) have produced a definition of simulation which states: 'Given a system and a model of that system, simulation is the use of the model to produce chronologically a state history of the model, which is regarded as a state history of the modeled system.'

From the state history produced by the simulation study it is possible to make an assessment of the adequacy of the model used. If the state history represents the behaviour of the real system in a way that is sufficiently accurate for the purposes of the study then it may be assumed that the model is an adequate one. Without a statement of the purpose of the study and a subsequent testing of the adequacy of the model, a simulation study has little value except as an educational exercise.

Investigation of the correspondence between the components of the model and the components of the system is very important when interpreting the results of a simulation study. It is possible to regard the model as a black box which produces an adequate state history, but the behaviour of the components of the model may not correspond to the behaviour of the components of the real system. In terms of drainage basin simulation, the study may produce a succession of landform surfaces which look real and are an adequate representation of reality, but the network and flow characteristics which produced them may not be of the same form as real stream networks.

Simulation by digital computer has the great advantage that it is completely repeatable and free from the physical limitations of the system being studied. Quantitative data for collection and processing are readily available at all stages of the simulation study making it possible to monitor the behaviour of the system in graphical form via on-line displays.

Inherent in the nature of simulation is the fact that a state history represents a sequence of discrete changes of state within the system being simulated and that two consecutive state descriptions are separated by a finite time step. The consequences of these two facts are discussed in detail by Evans *et al.* (1967, 118–25), and are summarised in the following paragraph.

The choice of the size of the time step in relation to the separation time of events which change the state of the system may be critical in determining the form of the state history produced. Ideally, a 'next event' formulation is required in which the time step is chosen so that events are processed as they occur. Unfortunately this does not appear to be practicable in drainage basin simulation so that a 'fixed time step' formulation has to be used. An important consequence of the fixed time step formulation is that events occurring within the time step are batch processed so that the choice of the time step may radically alter the interrelationships of event occurrences. One further consequence of using the fixed time step formulation is that almost any time step chosen will enable the model to go through the process of simulating.

These comments emphasise the need for a precise statement of the assumptions made in a simulation study, especially if it is the purpose of the study to test a particular hypothesis. Without this information, it is impossible to make a critical assessment of the adequacy of another person's simulation study.

The drainage basin as an open system

A general systems approach to geomorphological problems (Chorley 1962) provides a suitable conceptual framework for building models of drainage basins. The basic concept of the drainage basin as a system is central to all morphometric studies, both real and simulated. Strahler (1964, 4–40) writes:

> Of fundamental importance is the concept of the drainage basin as an open system tending to achieve a steady state of operation. . . . An open system imports and exports matter and energy through system boundaries and must transform energy uniformly to maintain operation. In a drainage basin the land surface within the limits of the basin perimeter constitutes a system boundary through which precipitation is imported. Mineral matter supplied from within the system and excess precipitation leave the system through the basin mouth. In a graded drainage basin the steady state manifests itself in the development of certain topographic characteristics which achieve a time-independent state.

These are the ideas which a successful simulation study must try to illuminate. Early attempts will inevitably make simulation runs simply to see what happens, but ultimately it is hoped that they will be more concerned with energy distributions as well as with the dynamic carving of surfaces.

Chorley (1967, 78) highlights another important feature of geomorphic systems:

> Geomorphic systems can all be considered part of 'supersystems' (e.g. whole landform assemblages) and as being composed of 'subsystems' (e.g. slope or channel segments). . . . In geomorphology subsystems are commonly combined by 'cascading' the output of one subsystem into another to form its input.

This cascading effect is a central feature of the writer's simulation model described in a later section of this chapter.

The distinctive feature of a drainage basin system which makes it a particularly *geographical* system is the characteristic way in which the elements of its three-dimensional surface are spatially interrelated by an implied transportation network resting on the surface. One of the most difficult problems of drainage basin simulation is to simulate the delicate interdependence between the surface and the network. Any adequate simulation of drainage basin development must solve this problem but no published work has yet achieved a satisfactory solution.

Development of simulation models

During the 1960's, simulation studies relevant to drainage basin development followed two main lines of investigation. On the one hand, network models were developed in order to simulate the spatial variation of network characteristics in the landscape, and were, in the majority of cases, unrelated to any underlying surface form. On the other hand, models of slope development in two dimensions provided a more fundamental approach by concentrating on the simulation of erosion processes from which a characteristic form might be produced. Network models have tended to be stochastic in nature, while process models have been predominantly deterministic.

An account of simulation studies which use network models has already been published (Haggett and Chorley 1969). Some aspects of these studies are repeated here because they serve to illuminate the nature and purpose of simulation when applied to three-dimensional models of drainage basins.

Simulations of slope development are of an entirely different nature, using numerical methods of solution to differential equations in order to produce sequences of slope profiles. The profiles may be regarded as state histories of points on a stream or hillslope which indicate stages of development with time.

Network simulations

During the course of drainage basin simulation studies, network patterns are frequently generated within the limits of the study area. It is useful to know the properties of the population of possible networks to which a generated network belongs in order to assess the adequacy of the simulation. An adequate simulation would be expected to generate sample networks with properties similar to those of real drainage networks. Several studies have been made to find the properties of network populations by using simulation techniques to draw samples from them.

Leopold and Langbein (1962, A14), made the first study of this kind and generated sample networks which exhibited some of the properties of real stream networks. Constrained random walks were used to simulate

(a) the way in which initially parallel rills might unite on a uniform slope, and

(b) the way in which areas are occupied by stream networks.

The results were found to be consistent with their theory that network systems develop in a way which tends to maximise entropy in the system.

By applying different kinds of constraints to the random walks, a variety of network forms can be obtained so that, if it is desirable, particular network patterns can be imitated. Scheidegger (1967) and Armstrong (1971), have used random walk techniques in this way.

The methods introduced by Leopold and Langbein were soon automated by Schenck (1963) and were later developed to allow biased, constrained random walks by Smart, Surkan and Considine (1967). An important innovation in the latter publication was the use of the computer to tabulate and analyse the data generated by the simulation process.

Hack (1965), tested the Leopold–Langbein theory of stream network development against a real stream network developed on the Ontonagon Plain, Michigan. This piece of work is of particular interest in that it provides the first attempt to simulate a specific drainage system, in order to test a hypothesis about that system.

The Ontonagon Plain has a steep slope towards Lake Superior and drainage is constrained by parallel grooves tending to produce long narrow drainage basins.

Hack (1965, 16) made the hypothesis that the merging of adjacent channels could have been caused by random fluctuations in the level of Lake Duluth as it retreated. Each stand of the lake could have produced erosion at the lake shore of part of the ridge between grooves, giving adjacent streams the opportunity of uniting. This system was simulated by assuming that a set of parallel grooves one-tenth of a mile apart collects all surface water, and that at the end of each one-mile stretch, the water has equal probabilities of being diverted to the left, to the right, or continuing straight on.

Streams generated by this simulation study were examined at various points in the model and the length of the principal stream at these points was compared with its corresponding drainage area. For these sample points, the law $L = 4A^{0.67}$ was obtained, compared with $L = 3.6A^{0.6}$ for a typical grooved area on the plain. Part of the simulated network and part of the real network are reproduced together in fig. 14.1 for comparison.

The first attempt to simulate a developing network pattern which is related to an underlying surface form appears to have been made by Howard (reported in Haggett and Chorley 1969, 292). This simulation takes as its initial state a set of parallel drainage lines emptying into a single transverse drainage line which together drain a concave surface down to an outlet point. Development of the system proceeds by a sequence of captures determined from randomly selected points of possible capture. Streams in squares adjacent to selected points were assumed to be able to capture the chosen stream segment with probabilities proportional to the areas drained by them. If a capture was allowed, the upstream drainage area was regraded according to an

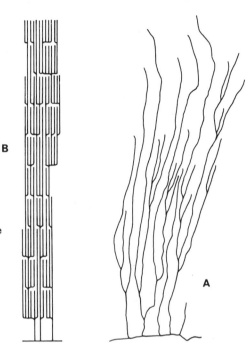

Fig. 14.1 Drainage systems on the Ontonagan Plain, Michigan, U.S.A. (after Hack 1965).

A. Part of the network between Mineral and Cranberry Rivers.
B. Part of the simulated network.

empirical law relating slope to area drained. After a given number of repetitions of this procedure the simulation run ceases.

This particular study illustrates well the problem of locating dependent events in a realistic way on a simulated time scale. By regrading an upstream area in an instant of time, a large number of points have their probabilities of capture altered immediately. Some readers may not be prepared to accept this as a reasonable approximation to reality.

Final network forms generated in this way were reported to have properties which agreed quite well with those of real networks.

The most promising simulation study to date has been started by Moultrie (1970), in which future development of Poplar Gap drainage basin has been investigated. One hundred and six control points were obtained by sampling elevations over a rectangular grid from a topographic map of the basin. The model is a series of precipitation and hydraulic flow equations which allow appropriate amounts of removal of material to be calculated on the basis of slope, depth of overland flow, and shear force. At each cycle of the program which implements this model, elevations are reduced the appropriate amount and progress is monitored periodically.

Early tests of this model have simulated a 6000-year period in which it was found that slopes were maintained, and, for sample points, slopes did not vary more than 0·25 degrees from their initial value.

No statement is given of the way in which control points are spatially related, so that it is difficult to assess the adequacy of this simulation study. However, it is the first three-dimensional drainage basin simulation study which has attempted to indicate the possible development of a real basin, and promises to yield useful information on this method of geomorphic research.

Models of slope development

Several attempts have been made to express erosion processes in the form of differential equations which may be solved by analytical or numerical methods to yield sequences of slope profiles. The equations are important to simulation studies because they provide a means of generating data during a simulation run.

In order to keep a standard notation in this account, space co-ordinates are taken to be measured with respect to three mutually perpendicular axes. The vertical, z-axis, is taken to represent altitude, longitudinal profiles of channels and hillslopes are assumed parallel to the x-axis, and transverse profiles are assumed parallel to the y-axis. Time is measured by the variable t, and the rate of transport of material from a point on a profile is the variable q.

A general three-dimensional landform changing through time will be of the form $z = f(x, y, t)$, $\partial z/\partial t$ is the rate of lowering of a point with time, $\partial z/\partial x$ is the rate of change of altitude along a profile slope, i.e. the profile slope at a point, and $\partial z/\partial y$ is the slope of a transverse profile. The general approach to the formation of differential equations of slope development is to assume that the rate of lowering of a point is the result of the differences between tectonic movements and removal of material. Also this change in elevation must be reflected by the change in discharge, $\partial q/\partial x$, of material downslope at that point.

Culling (1960 and 1963) has given a detailed mathematical treatment of slope development based on the diffusion equation

$$\frac{\partial z}{\partial t} = k\frac{\partial^2 z}{\partial x^2}$$

which states that the rate of lowering is proportional to the rate of change of slope along the slope profile. Solutions were found for the two-dimensional case with certain restricted boundary conditions and the three-dimensional form of the equation was applied to landforms having radial symmetry. In the former case, steady-state forms were linear indicating parallel retreat, while in the latter case concave logarithmic profiles were characteristic. Culling (1963, 158) observed that: 'The type of problem likely to arise in predicting the probable development of an actual landform with complicated boundaries can only be tackled by numerical methods.' Computational details and examples of solutions to the diffusion equation in one and two dimensions have been given by Harbaugh and Bonham-Carter (1970).

Numerical solutions to differential equations of landform development had already been obtained by Scheidegger (1961). He considered three simple cases (ibid., 99):

1. Denudation independent of the slope.
2. Denudation proportional to the height of the point above a certain base level.
3. Denudation proportional to the steepness of the slope.

Using the cases as particular forms of a function $f(z)$, allowing for tectonic movements by a function $F(x)$ and assuming weathering effects normal to the slope, equations of the form

$$\frac{\partial z}{\partial t} = \sqrt{\left\{1 + \left(\frac{\partial z}{\partial x}\right)^2\right\}} \cdot f(z) + F(x)$$

were solved. It was necessary to express all but the simplest forms as difference equations and then solve the difference equations by digital computer.

Profiles predicted by deterministic theory do not explain the wide variety of profile forms found in the field. One explanation of this discrepancy is that deterministic laws simply explain what would happen if there were no chance variations in the factors determining erosion processes. The presence of chance variations produces a wide variety of forms which may be averaged to find a most probable state. Leopold and Langbein (1962) have used the technique of random walk generation to simulate possible stream profiles on the assumption that the downstream rate of production of entropy per unit mass is constant.

The profiles produced by random walks may be interpreted as being solutions to the differential equation

$$\frac{\partial z}{\partial x} = f(z),$$

where f is a stochastic, not a deterministic function. In effect, the finite difference form of $\partial z/\partial x$, $(z_{n+1} - z_n)$, becomes a stochastic variable which is simulated by the random walk. With $f(z)$ set so that the probability of a unit change in z was proportional to z, the equivalent of the well-known exponential solution, $z = ae^{-bx}$, was found as the most probable solution.

Following the approach of Scheidegger, Ahnert (1966) used a digital computer to generate slope profiles. The nature of this particular study seems to support Thornes (1971, 50), who suggests that queueing theory is likely to have a wide and fruitful application in geomorphology, particularly in slope and fluvial studies. Ahnert's model could be interpreted as a sequence of queueing models in which the queues correspond to debris thickness, C, at a point on a slope profile. Arrivals at the queue correspond to material produced by weathering and departures correspond to material transported from the surface.

Two rates of weathering are suggested, one a linear function of C, and the other an exponential function of C. The former represents chemical weathering and the latter mechanical weathering. Removal of

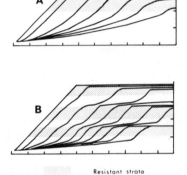

Fig. 14.2 Simulated profile development in strata with differing resistances to erosion (after Ahnert, 1966).

A. Shaded strata twice as resistant as unshaded strata.
B. Shaded strata ten times as resistant as unshaded strata.

Resistant strata

debris downslope is expressed as a function of the product of C with the sine of the slope angle at the point where C is measured. Details of the implementation of this model give useful insights into the problems of computer simulation (Ahnert 1966, 127–8). Three constraints were built into the program:

1. To ensure numerical stability, no point was allowed to be lowered below the elevation of the next point downslope.
2. No slope was allowed to develop an angle greater than forty-five degrees.
3. When the linear weathering rate was used, it was taken as zero if C exceeded a chosen value.

In this study it was possible to simulate the effect of horizontal strata of different resistances. Figure 14.2A shows the effect of one rock overlying another of half its resistance, while fig. 14.2B shows the effect of strata ten times as resistant as the ones with which they alternate.

The problem of placing events realistically on a time scale is recognised when it is stated that although weathering and removal occur at the same time in reality, during simulation, these processes must be dealt with alternately (Ahnert 1966, 128).

Ahnert does not use differential equations, but uses finite difference representations of the process being simulated. The change from C to C' at each iteration is expressed in the form

$$C'_j = C_j - \frac{1}{\sin \alpha_{max}}(C_j \sin \alpha_j - C_{j+1} \sin \alpha_{j+1}).$$

where α is the slope angle.

In an interesting theoretical paper, Devdariani (1967, 183) uses a law of the form

$$q = -k\frac{\partial z}{\partial x} + k_0$$

for stream erosion where q is the discharge of sediment and k is a function depending on the discharge of water and on the properties of the rocks of the river bed. Below a critical value of $\partial z/\partial x$ erosion is assumed to cease so that ultimately an equilibrium form will result. If z_r indicates the relative elevation above the limiting profile, the equation can be written

$$q = -k\frac{\partial z_r}{\partial x}.$$

This equation may be further generalised to

$$q = -k(x)\frac{\partial z_r}{\partial x},$$

so that k is a function of the distance along the profile. A function v is defined to represent rate of tectonic movement and together with a form of continuity equation

$$\frac{\partial q}{\partial x} = -\frac{\partial z}{\partial t} + v(x, t),$$

an equation

$$\frac{\partial z_r}{\partial t} = \frac{\partial}{\partial x}\left[k(x)\frac{\partial z_r}{\partial x}\right] + v(x, t)$$

is obtained which is similar to the equation used by Pollack (1969, in Harbaugh and Bonham-Carter 1970), and reduces to the diffusion equation as a special case. Solutions are found of the form

$$z(x, t) = z_r(x) + z_0(x, t)$$

which is similar to the solutions found by Kirkby (1971, 21) in that $z_r(x, t) \rightarrow 0$ as $t \rightarrow \infty$, leaving the 'characteristic form' $z_0(x)$. Devdariani's computed solutions show a rapid change from an initial convex profile (inequilibrium stage) to a pronounced concave profile (dynamic equilibrium). The concavity decreases and the profile lowers at an exponential rate of decay down to the final characteristic form of the profile (static equilibrium).

Pollack (1969, in Harbaugh and Bonham-Carter 1970, 534) used a modified form of the diffusion equation to produce a sequence of transverse profiles simulating development of the Grand Canyon of Arizona. The form of the equation used,

$$\frac{\partial z}{\partial t} = \frac{\partial}{\partial y}\left[K(z)\frac{\partial z}{\partial y}\right] + A(y, z),$$

uses two functions, K and A, to allow adjustment to particular local environments. Resistance factors for different rock types are incorporated in the function K so that slope wasting reflects the stratigraphic sequence of the simulated cross-section. Downcutting rates for streams are determined by the function A which takes the value zero unless the value of y corresponds to the position of a stream on the profile. A series of realistic profiles for the Grand Canyon was produced by this simulation study.

A detailed analytical treatment of fluvial processes has been given by Kirkby (1971) in which he developed process-response models for hillslopes. The models are based on the continuity equation and use systems of differential equations to relate production of debris on slopes to the rate of removal of debris from the slope. As a consequence of the theory developed, characteristic slope forms were derived corresponding to different empirical process laws. In some applications it was possible to show that final forms are independent of the initial form of the slope. To illustrate the rate at which a characteristic form is approached, a table is given showing how the divide elevation approaches that of the predicted final form for an initially straight slope. By the time that the initial elevation had been halved, the form is within 0.2% of the predicted final form (Kirkby 1971, 22).

This analysis contains two particularly important contributions to the subject of model building for drainage basin simulation studies. Firstly, it was recognised that hillslope processes can be represented in terms of a general measure of the transportation capacity, C, which is applicable to *all* lines of greatest slope within a geomorphological unit. The transportation capacity $C = f(a).[(-\partial z/\partial x)^n]$ can take one of several particular empirical forms to suit local conditions and can be modified to operate over restricted ranges of slope angle (Kirkby 1971, 20). Secondly, the process laws described are functions of area drained, $f(a)$ in the above equation, so that they apply to three-dimensional models (ibid., 19).

A computer model therefore needs to be able to determine contributing areas and to simulate transportation along all lines of greatest slope on the surface. The model described in the final section of this chapter attempts to incorporate these features within a general simulation framework. Given a suitable framework for simulation, it should be possible to determine the predictions of each of the symbolic models described in this section when extended to three dimensions.

A framework for drainage basin simulation studies

Published work on drainage basin simulation studies suggests that no adequate solution has yet been found to the problem of representing spatial interrelationships within a developing basin system. Studies of hillslope processes, studies of network development and the development of digital representations of terrain have been carried out largely as separate activities. In an attempt to integrate these activities into a single framework, the writer is currently developing general-purpose computer programs for drainage basin simulation studies.

It is desirable that a general-purpose program for drainage basin studies should meet the following requirements:

1. The drainage basin system should be represented within the program as an assemblage of subsystems.
2. Basic information for each subsystem, e.g. spatial co-ordinates, runoff, resistance to erosion and process laws operating within this system, must be accessible in order to determine the state of each subsystem at a given time.
3. It should be possible at any time to determine a graph structure showing the interrelationships between all subsystems, so that the order of processing events can be determined.
4. The program should contain a procedure for processing events simulated in the study in accordance with the ordering specified by the graph.
5. Provision should be made for adequate monitoring during a simulation run to provide data for comparison with real drainage basin systems.
6. To avoid excessively long computer runs, the program should have the facility for dumping information during a run so that it can be restarted at the appropriate point on a subsequent occasion.
7. It should be possible to change program functions and parameters from run to run without having to re-structure the program.

The above list is not intended to be comprehensive, nor is it a program specification, but it is hoped that it might serve as a useful point of departure for those developing similar programs.

Most of the above requirements are met by the writer's FORTRAN IV program called 'DBSIM', written for an ICL 4130 computer, and currently undergoing testing and modification.

Within this program, a landform surface is defined on a two-dimensional array of 32 × 32 elements, each element representing the altitude of a point on the surface. The array dimensions allow for a 30 × 30 central study area with a border of elements in which to set up boundary conditions. It is convenient to let the positions of array elements correspond to the x, y, co-ordinates of points on the surface being studied in order to conserve storage space. Each point on the surface is in turn associated with a unit square area of the plan view of this surface, and in this way, the whole system and local system environment being studied is represented by an assemblage of subsystems (sub-areas of the surface). Information about each subsystem is then conveniently stored as a stack of arrays of identical dimensions.

At any given time, the spatial interrelationships between subsystems are determined by the altitudes associated with the subsystems at that time, and interpreted as a graph structure. The graph is formed by assuming that arcs join any two adjacent subsystems of differing altitudes, with the orientation of the arc indicating 'downhill'. Later, this structure is modified to suit local conditions within the study area. If, for example, the graph were simplified by including only those arcs indicating maximum difference in altitude for a subsystem, fig. 14.3B would result for the basin indicated in fig. 14.3A.

Flow through the system, in the form of runoff and transported debris, is routed in accordance with the transportation network so defined. The advantage of this form of network generation is that it is completely objective, unambiguous, and can be achieved without human intervention. It clearly provides only an approximation to reality but it avoids having to identify flowlines from contours as described by Onstad and Brakensiek (1968) in their simulation of a small watershed.

For given inputs of runoff in each subsystem, a simple cascade algorithm computes the cumulative effect in the main system, by starting with those subsystems most remote from the basin outlet and working its way systematically to the outlet. Inputs could be adjusted to produce any desired flow characteristics although the program has not been used in this way at the time of writing.

As a result of the flow pattern produced, material is assumed to be transported and ultimately removed from the system, either by crossing the boundaries of the study area or by passing through the outlet. The program calculates the appropriate amount by which altitudes are to be lowered by referring to the particular law or laws being used in the simulation.

A difficult problem during simulation is the determination of the particular law for rate of removal that is appropriate for a given sub-system. One solution is to group subsystems according to the amounts of cumulative runoff received and to supply a list giving the law for each group. In 'DBSIM', cumulative runoff is used in this way to divide subsystems into two distinct groups according to whether the runoff is greater or less than a selected critical value. For the purposes of the test study, the arbitrary division was used to indicate the appearance of channelled flow. Figure 14.3C illustrates the effect of reducing the network of fig. 14.3B to only those parts which indicate channelled

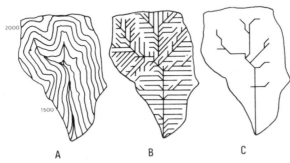

A   B   C

Fig. 14.3 Automatically derived networks for a basin digitised at points on a 250-ft grid.

  A. Source basin defined by 100-ft contours.
  B. Computer-drawn network showing relationships between grid points.
  C. Computer-drawn network for all grid points preceded by at least six other grid points on the network.

flow. A further assumption used in test runs was that channelled flow indicated $\partial z/\partial x$ as the appropriate measure of slope, but that for other parts, $\sqrt{\{(\partial z/\partial x)^2 + (\partial z/\partial y)^2\}}$ seemed more representative.

Figure 14.3 is an illustration of the network technique used in the early stages of development of 'DBSIM'. It suffers from the disadvantage that by restricting outputs from a subsystem to just one approximation to the steepest slope, gross misrepresentations occur on long slopes with parallel contours oblique to principal grid lines and diagonals. To overcome this, in the current version all downslope routes are included in the network for areas of non-channelled flow. Total discharge from a subsystem is allocated as outputs to all neighbouring 'lower' subsystems with values in proportion to the sine of the slope in the direction of the output. In this way, the effect of contour curvature is incorporated into the model in that convex contours cause dispersion of flow through the network tubes, while concave contours concentrate it to define channels.

The procedures for computing changes in state which are implemented in 'DBSIM' may be summarised as follows.

Initially, the user may request an inclined plane to be generated as the study area, or he may request a previously written surface to be read in from backing store. Then the following sequence is carried out:

1. The current states of the subsystems are held in a stack of arrays.
2. A distribution representing runoff inputs for each subsystem is generated as a stochastic variable.
3. The network is defined showing the relationships between subsystems.
4. Starting with a list of all subsystems with input requirements fulfilled, new states are computed according to the process laws, and outputs are carried forward into adjacent subsystems. When a subsystem has received all its expected inputs, it is added to the list to be processed. New states are held in separate arrays until all subsystems have been processed.
5. The values giving the new states of the subsystems become the current states for the next iteration.

In practice, this sequence is repeated for a selected number of iterations and the state descriptions are recorded on magnetic tape to provide state histories for subsequent analysis.

Adjustments could be made to the altitude matrix at the end of each iteration in order to simulate the effect of base level fluctuations.

Results of a production run

As an exploratory test run for the program 'DBSIM', a simple law of removal was applied which was jointly proportional to the area drained and a function of slope. A slope function $f(S) = \sin \alpha/(\tan^{0.3} \alpha)$, was used (Horton 1945, 321), where α is the slope angle. The intention was to examine the resulting relationship between slope and contributing area if recognisable landforms were produced. Erosion was allowed to take place on an inclined plane with base level 1200 ft and summit level 2100 ft to see if any basins comparable to fig. 14.3A were formed. After 100 iterations the run was terminated. At the beginning of each iteration, 900 units of runoff were distributed randomly to the 900 subsystems, as inputs. The critical value which determined the beginning of channelled flow was arbitrarily set to $6.0 + 40.0/$(iteration number).

In the first few iterations, a system of parallel streams developed on the inclined plane. Approximate calculations of volumes indicated that the amount of material removed per iteration reached a maximum after only four iterations and then rapidly declined to a very small rate of decrease which was maintained from the tenth to the hundredth iteration.

The spatial variation of runoff inputs caused local channel deepening, which in turn caused capturing to take place whenever an adjacent stream was provided with a steeper path than its current one. At the end of 100 iterations, twenty first-order basins, six second-order basins and one third-order basin had been formed. Development of the only third-order basin is illustrated in fig. 14.4. Analysis of this basin using

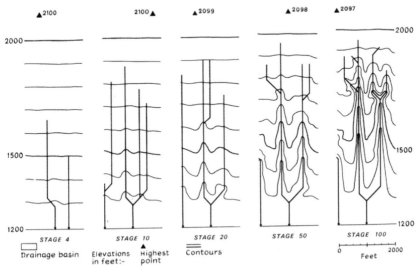

Fig. 14.4 Simulated development of a third-order drainage basin on an initially uniform plane.

sample points within the drainage network indicated that the tangent of channel slope angle was proportional to the inverse square of the area. This unrealistic result was probably a consequence of the unrealistic assumption of infinite transportation capacity and a static base level. In subsequent development of this project it is intended to allow for more realistic adjustments of the system than can be expressed by a simple deterministic law of erosion.

It is possible to infer from the appearance of the results that the simulation does reflect the interrelationships between the surface form and a developing network on it. Within the framework described, it seems likely, therefore, that more realistic and more specific simulations are possible.

Conclusion

Digital simulation is a powerful research tool which can give the geomorphologist greater experimental control and flexibility than can be achieved with laboratory hardware models. Its application to the study of drainage basin evolution presents several problems which may account for its slow rate of development. Firstly, it seems unlikely that any long-term predictions made as the result of a simulation study in this field could ever be verified by observation. Secondly, considerable amounts of computer time and space are essential when systems of the complexity of drainage basins are to be simulated, so that this form of research tends to be costly compared with other forms of geomorphic research. And thirdly, information about a particular, real basin tends to be sparse, making the production of an adequate model a very difficult problem.

Progress has been made, however, and the work presented in this chapter indicates that most of the problems of transcribing drainage basin systems into symbolic and digital form, have been solved. What is lacking at the present time is a simulation study which appears to show more than was put into it, and which is of benefit to geomorphologists generally and not just to the individual who carried out the research. It is to be hoped that there will be many studies in the near future which have these characteristics and that their results will both complement and stimulate work on real drainage basins.

References

AHNERT, F. (1966) Zur Rolle der elektronischen Rechenmaschine und des mathematischen Modells in der Geomorphologie; *Geographische Zeitschrift* 54, 118–33.

ARMSTRONG, A. C. (1971) Simulation of drainage patterns on homogeneous non-isotropic surfaces (Abst.); *Area* 3, 18.

CHORAFAS, D. N. (1965) *Systems and Simulation* (Academic Press, New York).

CHORLEY, R. J. (1962) Geomorphology and general systems theory; *United States Geological Survey Professional Paper* 500-B, 10p.

CHORLEY, R. J. (1967) Models in geomorphology; In Chorley, R. J. and Haggett, P (Eds.), *Models in Geography* (Methuen, London), 59–96.

CHORLEY, R. J. (Ed.) (1969A). *Water, Earth and Man* (Methuen, London), 588p.

CHORLEY, R. J. (1969B) The drainage basin as the fundamental geomorphic unit; In Chorley, R. J. (Ed.), *Water, Earth and Man* (Methuen, London), 77–99.

CHORLEY, R. J. and HAGGETT, P. (Eds.) (1967) *Models in Geography* (Methuen, London), 816p.

CHOW, V. T. (Ed.) (1964) *Handbook of Applied Hydrology* (McGraw-Hill, New York).

CULLING, W. E. H. (1960) Analytical theory of erosion; *Journal of Geology* 68, 336–44.

CULLING, W. E. H. (1963) Soil creep and the development of hillside slopes; *Journal of Geology* 71, 127–61.

DEVDARIANI, A. S. (1967) A plane mathematical model of the growth and erosion of an uplift; *Soviet Geography* 8, 183–98.

EVANS, G. W., WALLACE, G. F. and SUTHERLAND, G. L. (1967) *Simulation Using Digital Computers* (Prentice-Hall, New Jersey).

HACK, J. T. (1965) Postglacial drainage evolution and stream geometry in the Ontonagon Area, Michigan; *United States Geological Survey Professional Paper* 504-B.

HAGGETT, P. and CHORLEY, R. J. (1969) *Network Analysis in Geography* (Arnold, London), 348p.

HARBAUGH, J. W. and BONHAM-CARTER, G. (1970) *Computer Simulation in Geology* (Wiley–Interscience, New York), 575p.

HORTON, R. E. (1945) Erosional development of streams and their drainage basins: Hydrophysical approach to quantitative morphology; *Bulletin of the Geological Society of America* 56, 275–370.

KIRKBY, M. J. (1969) Erosion by water on hillslopes: In Chorley, R. J. (Ed.), *Water, Earth and Man* (Methuen, London), 229–38.

KIRKBY, M. J. (1971) Hillslope process-response models based on the continuity equation; In *Slopes: form and process, Institute of British Geographers, Special Publication Number 3.*

LEOPOLD, L. B. and LANGBEIN, W. B. (1962) The concept of entropy in landscape evolution; *United States Geological Survey, Professional Paper* 500-A, 20p.

MORISAWA, M. E. (1968) *Streams: Their dynamics and morphology* (McGraw-Hill, New York)

MOULTRIE, W. (1970) Systems, computer simulation, and drainage basins; *Bulletin of the Illinois Geographical Society* 12(2), 29–35.

ONSTAD, C. A. and BRAKENSIEK, D. L. (1968) Watershed simulation by stream path analogy; *Water Resources Research* 4, 965–71.

SCHEIDEGGER, A. E. (1961) Mathematical development of slope development; *Bulletin of the Geological Society of America* 72, 37–49.

SCHEIDEGGER, A. E. (1967) A stochastic model for drainage patterns into an intramontaine trench; *Bulletin of the International Association of Scientific Hydrology* Year 12(1), 15–20.

SCHENCK, H. (1963) Simulation of the evolution of drainage basin networks with a digital computer; *Journal of Geophysical Research* 68, 5739–45.

SMART, J. S., SURKAN, A. J. and CONSIDINE, J. P. (1967) Digital simulation of channel networks; *International Association of Scientific Hydrology, General Assembly of Berne, Sept.–Oct. 1967, Symposium on River Morphology*, 87–98.

STRAHLER, A. N. (1964) Quantitative geomorphology of drainage basins and channel networks; In Chow, V. T. (Ed.), *Handbook of Applied Hydrology* (McGraw-Hill, New York).

THORNES, J. B. (1971) State, environment and attribute in scree-slope studies; In *Slopes: form and process, Institute of British Geographers, Special Publication Number 3.*

British Geomorphological Research Group

The British Geomorphological Research Group was founded in 1961 to encourage research in geomorphology, to undertake large-scale projects of research or compilation in which the co-operation of many geomorphologists is involved, and to hold field meetings and symposia.

Publications of the Group are listed below. Details of membership of the Group may be obtained from:

> Dr E. Derbyshire, Hon. Secretary, B.G.R.G.
> Department of Geography, University of Keele,
>   Keele, Staffordshire ST5 5BG.

Publications

The following publications are available postage free if payment sent with order from:

> Geo Abstracts, University of East Anglia, Norwich NOR 88C, England.

Technical Bulletins: (£0.40 each)

1. Field methods of water hardness determination
   IAN DOUGLAS, 1969
2. Techniques for the tracing of subterranean drainage
   DAVID P. DREW & DAVID I. SMITH, 1969
3. The determination of the infiltration capacity of field soils using the cylinder infiltrometer
   RODNEY C. HILLS, 1970
4. The use of the Woodhead sea bed drifter
   ADA PHILLIPS, 1970
5. A method for the direct measurement of erosion on rock surfaces
   C. HIGH & F. K. HANNA, 1970
6. Techniques of till fabric analysis
   J. T. ANDREWS, 1970
7. Field method for hillslope description
   LUNA B. LEOPOLD & THOMAS DUNNE
8. The measurement of soil frost-heave in the field
   PETER A. JAMES

Current Register of research 1970–71 (£0.60)

Index